The Digital Shopfloor: Industrial Automation in the Industry 4.0 Era

Performance Analysis and Applications

RIVER PUBLISHERS SERIES IN AUTOMATION, CONTROL AND ROBOTICS

Indexing: All books published in this series are submitted to the Web of Science Book Citation Index (BkCI), to SCOPUS, to CrossRef and to Google Scholar for evaluation and indexing.

The "River Publishers Series in Automation, Control and Robotics" is a series of comprehensive academic and professional books which focus on the theory and applications of automation, control and robotics. The series focuses on topics ranging from the theory and use of control systems, automation engineering, robotics and intelligent machines.

Books published in the series include research monographs, edited volumes, handbooks and textbooks. The books provide professionals, researchers, educators, and advanced students in the field with an invaluable insight into the latest research and developments.

Topics covered in the series include, but are by no means restricted to the following:

- Robots and Intelligent Machines
- Robotics
- Control Systems
- Control Theory
- Automation Engineering

For a list of other books in this series, visit www.riverpublishers.com

The Digital Shopfloor: Industrial Automation in the Industry 4.0 Era
Performance Analysis and Applications

Editors

John Soldatos

Athens Information Technology
Greece

Oscar Lazaro

Innovalia Association
Spain

Franco Cavadini

Synesis-Consortium
Italy

Published 2018 by River Publishers
River Publishers
Alsbjergvej 10, 9260 Gistrup, Denmark
www.riverpublishers.com

Distributed exclusively by Routledge
4 Park Square, Milton Park, Abingdon, Oxon OX14 4RN 605
Third Avenue, New York, NY 10017, USA

The Digital Shopfloor: Industrial Automation in the Industry 4.0 Era Performance Analysis and Applications / by John Soldatos, Oscar Lazaro, Franco Cavadini.

Routledge is an imprint of the Taylor & Francis Group, an informa business

ISBN 978-87-7022-041-5 (print)

Contents

PART II

PART III

Foreword

As the Technical Director of the European Factories of the Future Research Association (EFFRA), it is with great pleasure and satisfaction that I witness the completion of this book on digital automation, cyber physical production systems and the vision of a fully digital shopfloor. EFFRA is an industry-driven association promoting the development of new and innovative production technologies. It is the official representative of the private side in the 'Factories of the Future' Public-Private Partnership (PPP) under the Horizon 2020 program of the European Commission. As such it has been also supporting the three research projects (FAR-EDGE, AUTOWARE, DAEDALUS) that produced the book, which have formed the Digital Shopfloor Alliance (DSA).

The book provides insights on a variety of digital automation platforms and solutions, based on advanced technologies ICT technologies like cloud/edge computing, distributed ledger technologies and cognitive computing, which will play a key role in supporting automation in the factories of the future. Moreover, solutions based on the promising IEC 61499 standards are described. Overall, the presented results are fully aligned with some of the research priorities that EFFRA has been setting and detailing during the last couple of years. In particular, two years ago, EFFRA launched the ConnectedFactories Coordination Action, with a view to providing more insight in priorities and steps towards the digital transformation of production systems and facilities. ConnectedFactories has generated a first set of generic pathways to digital manufacturing.

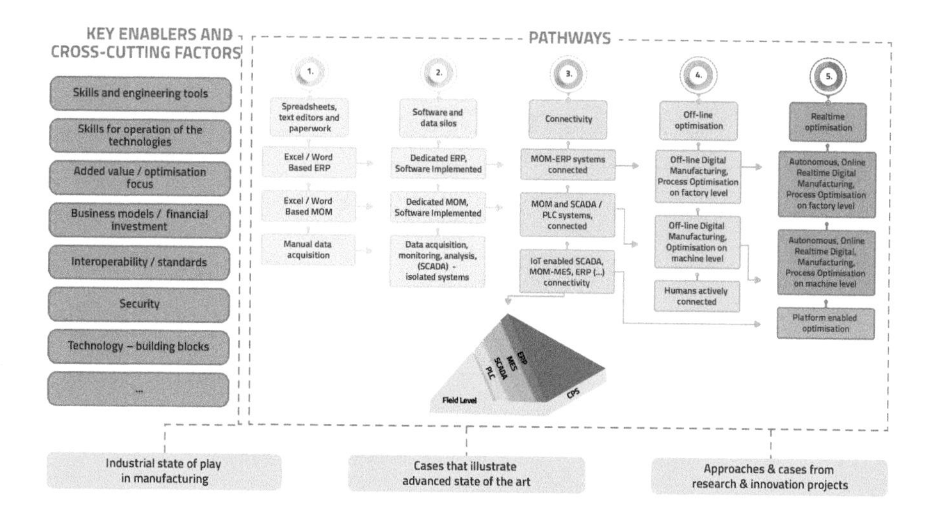

These pathways reflect our main directions for transforming factories in the Industry 4.0 era, and include:

- **The Autonomous Smart Factories pathway**, which focuses on optimised and sustainable manufacturing including advanced human-in-the-loop workspaces.
- **The Hyperconnected Factories pathway**, which boosts the networking enterprises towards formulating complex, dynamic supply chains and value networks.
- **The Collaborative Product-Service Factories pathway**, which emphasizes data-driven product-service engineering in knowledge intensive factories.

As part of the ConnectedFactories initiative, we have also illustrated a solid initial set of key cross-cutting factors and enablers that should be addressed in order to progress on the pathways. Likewise, we have also described a rich set of relevant industrial and research cases.

The work reflected in the book is perfectly aligned to our "Autonomous Smart Factories" pathway, as the presented technologies and use cases of the DSA are boosting significant improvements in production time, quality, sustainability and cost-efficiency at the same time. The co-editors have done a good job in presenting the added-value of the solutions developed by the three projects. At EFFRA we appreciate seeing results aligned to our research and development roadmaps. In the case of the results presented in this book, we are also happy to see the development of complementary services and community building initiatives, which could provide value to our members. We are happy to support the three projects in their dissemination and community building initiatives.

It's also very positive that this book is offered based on an Open Access publication modality, which could help it reach a wider readership and will boost its impact.

EFFRA is a growing network of actors that play key roles on national, regional, European and even global initiatives, as a contribution to knowledge exchange and experience sharing. I believe that many of these actors will find the book a very interesting read.

Chris Decubber
Technical Director
European Factories of the Future Research Association
Brussels
April 4th, 2019

Preface

In today's competitive global environment, manufacturers are offered with unprecedented opportunities to build hyper-efficient and highly flexible plants, towards meeting variable market demand, while at the same time supporting new production models such as make-to-order (MTO), configure-to-order (CTO) and engineer-to-order (ETO). In particular, the on-going digitization of industry enables manufacturers to develop, deploy and use scalable and advanced manufacturing systems (e.g., highly configurable production lines), which are suitable to support the above-listed production models and enable mass customization at shorter times and lower costs, without compromising manufacturing quality.

During the last few years, the digital transformation of industrial processes is propelled by the emergence and rise of the fourth industrial revolution (Industry 4.0). The latter is based on the extensive deployment of Cyber-Physical Production Systems (CPPS) and Industrial Internet of Things (IIoT) technologies in the manufacturing shopfloor, as well as on the seamless and timely exchange of digital information across supply chain participants. CPPS and IIoT technologies enable the virtualization of manufacturing operations, as well as their implementation based on IT (information technology) services rather than based on conventional OT (operational technology).

The benefits of Industry 4.0 have been already proven in the scope of pilot and production deployments in a number of different use cases including flexibility in automation, predictive maintenance, zero-defect manufacturing and so on. Recently, the digital manufacturing community has produced a wide array of standards for building Industry 4.0 systems, including standards-based Reference Architectures (RA), (such as RAMI 4.0 (Reference Architecture Model Industry 4.0) and the RA of the Industrial Internet Consortium (IIRA).

Despite early implementations and proof of concepts based on these RAs, CPPS/IIoT deployments are still in their infancy for a number of reasons, including:

- **Manufacturers' poor awareness about digital manufacturing solutions and their business value potential**, as well as the lack of relevant internal CPPS/IIoT knowledge.
- **The high costs associated with the deployment, maintenance and operation of CPPS systems in the manufacturing shopfloors**, which are particularly challenging in the case of SME (small and medium-sized enterprises) manufacturers that lack the equity capital needed to invest in Industry 4.0.
- **The time needed to implement CPPS/IIoT and the lack of a smooth and proven migration path** from existing OT solutions.
- **The uncertainty over the business benefits and impacts of IIoT and CPPS technologies**, including the lack of proven methods for the techno-economic evaluation of Industry 4.0 systems.
- **Manufacturers' increased reliance on external integrators, consultants and vendors**.
- **The absence of a well-developed value chain needed to sustain the acceptance of these new technologies for digital automation**.

In order to alleviate these challenges, three EC co-funded projects (namely H2020 FAR-EDGE (http://www.far-edge.eu/), H2020 DAEDALUS (http://daedalus.iec61499.eu) and H2020 AUTOWARE (http://www.autoware-eu.org/)) have recently joined forces towards a "Digital Shopfloor Alliance". The Alliance aims at providing leading edge and standards-based digital automation solutions, along with guidelines and blueprints for their effective deployment, validation and evaluation.

The present book provides a comprehensive description of some of the most representative solutions offered by these three projects, along with the ways these solutions can be combined in order to achieve multiplier effects and maximize the benefits of their use. The presented solutions include standards-based digital automation solutions, following different deployment paradigms, such as cloud and edge computing systems. Moreover, they also comprise a rich set of digital simulation solutions, which are have been explored in conjunction with the H2020 MAYA project (http://www.maya-euproject.com/). The latter facilitate the testing and evaluation of what-if scenarios at low risk and cost, without disrupting shopfloor operations. As already outlined, beyond leading edge scientific and technological development solutions, the book comprises a rich set of complementary assets that are indispensable to the successful adoption of IIoT/CPPS in the shopfloor. These assets include methods for techno-economic analysis, techniques for migrating for traditional technologies to IIoT/CPPS system, as well as ecosystems providing training and technical support to prospective deployers.

The book is structured in the following three parts, which deal with three distinct topics and elements of the next generation of digital automation in Industry 4.0:

- **The first part of the book is devoted to digital automation platforms.** Following an introduction to Industry 4.0 in general and digital automation platforms in particular, this part presents the digital automation platforms of the FAR-EDGE, AUTOWARE and DAEDALUS projects. As part of these platforms, various automation functionalities are presented, including data analytics functionalities. Moreover, the concept of a fully digital shopfloor is introduced.
- **The second part of the book focuses on the presentation of digital simulation and digital twins' functionalities.** These include information about the models that underpin digital twins, as well as the simulators that enable experimentation with these processes over these digital models.
- **The third part of the book provides information about complementary assets and supporting services that boost the adoption of digital automation functionalities in the Industry 4.0 era.** Training services, migration services and ecosystem building services are discussed based on the results of the three projects of the Digital Shopfloor Alliance.

The various topics in all three chapters are presented in a tutorial manner, in order to facilitate readers without deep technical backgrounds to follow them. Nevertheless, a basic understanding of cloud computing, Internet, sensors and data science concepts facilitates the reading and understanding of the core technical concepts that are presented in the book.

The target audience of the book includes:

- **Researchers in the areas of digital manufacturing and more specifically in the areas of digital automation and simulation**, who wish to be updated about latest Industry 4.0 developments in these areas.
- **Manufacturers, with an interest in the next generation of digital automation solutions** based on cyber-physical systems.

- **Practitioners and providers of Industrial IoT solutions**, who are interested in the implementation of use cases in automation, simulation and supply chain management.
- **Managers wishing to understand technologies and solutions that underpin Industry 4.0**, along with representative applications in the shopfloor and across the supply chain.

In general, the book provides insights into automation and simulation platforms towards a digital shopfloor. Moreover, it discusses the elements of a fully digital shopfloor, which is the vision of the DSA for the years to come. We hope that you will find it useful as a tutorial introduction to several digital automation topics and technologies, including cloud computing, edge computing, blockchains, software technologies and the IEC 61499 standard, along with their role in the future of digital automation. The book will be published as an open-access publication, which could make it broadly and freely available to the Industry 4.0 and Industrial Internet of Things communities. We would like to thank River Publishers for the opportunity and their collaboration in making this happen.

Finally, we take the chance to thank all members of our project for their valuable inputs and contributions in developing the presented systems and platforms, as well as in documenting them as part of the book. Likewise, we would also like to acknowledge funding and support from the European Commission as part of the H2020 AUTOWARE, DAEDALUS, MAYA and FAR-EDGE contracts.

September 2018,
John Soldatos
Oscar Lazaro
Franco Cavadini

List of Contributors

Aikaterini Roukounaki, *Kifisias 44 Ave., Marousi, GR15125, Greece; E-mail: arou@ait.gr*

Aitor Gonzalez, *Asociacion de Empresas Tecnologicas Innovalia, Rodriguez Arias, 6, 605, 48008-Bilbao, Spain; E-mail: aitgonzalez@innovalia.org*

Alessandro Brusaferri, *Consiglio Nazionale delle Ricerche (CNR), Institute of Industrial Technologies and Automation (STIIMA), Research Institute, Via Alfonso Corti 12, 20133 Milano, Italy; E-mail: alessandro.brusaferri@itia.cnr.it*

Ambra Calà, *Siemens AG Corporate Technology, Erlangen, Germany; E-mail: ambra.cala@siemens.com*

Andrea Barni, *Scuola Universitaria Professionale della Svizzera Italiana (SUPSI), The Institute of Systems and Technologies for Sustainable Production (ISTEPS), Galleria 2, Via Cantonale 2C, CH-6928 Manno, Switzerland; E-mail: andrea.barni@supsi.ch*

Andrea Passarella, *Institute of Informatics and Telematics, National Research Council (CNR), Pisa, Italy; E-mail: andrea.passarella@iit.cnr.it*

Anton Ružić, *Jožef Stefan Institute, Department of Automatics, Biocybernetics, and Robotics, Jamova 39, 1000 Ljubljana, Slovenia; E-mail: ales.ude@ijs.si*

Arndt Lüder, *Otto-von-Guericke University Magdeburg, Magdeburg, Germany; E-mail: arndt.lueder@ovgu.de*

Batzi Uribarri, *Software Quality Systems, Avenida Zugazarte 8 1-6, 48930-Getxo, Spain; E-mail: buribarri@sqs.es*

Begoña Laibarra, *Software Quality Systems, Avenida Zugazarte 8 1-6, 48930-Getxo, Spain; E-mail: blaibarraz@sqs.es*

Bojan Nemec, *Jožef Stefan Institute, Department of Automatics, Biocybernetics, and Robotics, Jamova 39, 1000 Ljubljana, Slovenia; E-mail: bojan.nemec@ijs.si*

Dario Piga, *Scuola Universitaria Professionale della Svizzera Italiana (SUPSI) Dalle Molle Institute for Artificial Intelligence (IDSIA) Galleria 2, Via Cantonale 2C, CH-6928 Manno, Switzerland; E-mail: dario.piga@supsi.ch*

Diego Rovere, *TTS srl, Italy; E-mail: rovere@ttsnetwork.com*

Elias Montini, *Scuola Universitaria Professionale della Svizzera Italiana (SUPSI), The Institute of Systems and Technologies for Sustainable Production (ISTEPS), Galleria 2, Via Cantonale 2C, CH-6928 Manno, Switzerland; E-mail: elias.montini@supsi.ch*

Elisa Negri, *Politecnico di Milano, Milan, Italy; E-mail: elisa.negri@polimi.it*

Filippo Boschi, *Politecnico di Milano, Milan, Italy; E-mail: filippo.boschi@polimi.it*

Franco A. Cavadini, *Synesis, SCARL, Via Cavour 2, 22074 Lomazzo, Italy; E-mail: franco.cavadini@synesis-consortium.eu*

Gernot Kollegger, *nxtControl, GmbH, Aumühlweg 3/B14, A-2544 Leobersdorf, Austria; E-mail: gernot.kollegger@nxtcontrol.com*

Giacomo Pallucca, *Consiglio Nazionale delle Ricerche (CNR), Institute of Industrial Technologies and Automation (STIIMA), Research Institute, Via Alfonso Corti 12, 20133 Milano, Italy; E-mail: giacomo.pallucca@itia.cnr.it*

Giovanni dal Maso, *TTS srl, Italy; E-mail: dalmaso@ttsnetwork.com*

Giuseppe Landolfi, *Scuola Universitaria Professionale della Svizzera Italiana (SUPSI), The Institute of Systems and Technologies for Sustainable Production (ISTEPS), Galleria 2, Via Cantonale 2C, CH-6928 Manno, Switzerland; E-mail: giuseppe.landolfi@supsi.ch*

Giuseppe Montalbano, *Synesis, SCARL, Via Cavour 2, 22074 Lomazzo, Italy; E-mail: giuseppe.montalbano@synesis-consortium.eu*

Horst Mayer, *nxtControl, GmbH, Aumühlweg 3/B14, A-2544 Leobersdorf, Austria; E-mail: horst.mayer@nxtcontrol.com*

Jan Wehrstedt, *SIEMENS, Germany; E-mail: janchristoph.wehrstedt@siemens.com*

Javier Gozalvez, *UWICORE Laboratory, Universidad Miguel Hernández de Elche (UMH), Elche, Spain; E-mail: j.gozalvez@umh.es*

John Kaldis, *Athens Information Technology, Greece; E-mail: jkaldis@ait.gr*

John Soldatos, *Kifisias 44 Ave., Marousi, GR15125, Greece; E-mail: jsol@ait.gr*

Jürgen Elger, *Siemens AG Corporate Technology, Erlangen, Germany; E-mail: juergen.elger@siemens.com*

Lara González, *Asociacion de Empresas Tecnologicas Innovalia, Rodriguez Arias, 6, 605, 48008-Bilbao, Spain; E-mail: lgonzalez@innovalia.org*

Marco Conti, *Institute of Informatics and Telematics, National Research Council (CNR), Pisa, Italy; E-mail: marco.conti@iit.cnr.it*

Marco Macchi, *Politecnico di Milano, Milan, Italy; E-mail: marco.macchi@polimi.it*

Marco Taisch, *Politecnico di Milano, Milan, Italy; E-mail: marco.taisch@polimi.it*

Maria del Carmen Lucas-Estañ, *UWICORE Laboratory, Universidad Miguel Hernández de Elche (UMH), Elche, Spain; E-mail: m.lucas@umh.es*

Marino Alge, *Scuola Universitaria Professionale della Svizzera Italiana (SUPSI), The Institute of Systems and Technologies for Sustainable Production (ISTEPS), Galleria 2, Via Cantonale 2C, CH-6928 Manno, Switzerland; E-mail: marino.alge@supsi.ch*

Martijn Rooker, *TTTech Computertechnik AG, Schoenbrunner Strasse 7, A-1040 Vienna, Austria; E-mail: martijn.rooker@tttech.com*

Marzio Sorlini, *Scuola Universitaria Professionale della Svizzera Italiana (SUPSI), The Institute of Systems and Technologies for Sustainable Production (ISTEPS), Galleria 2, Via Cantonale 2C, CH-6928 Manno, Switzerland; E-mail: marzio.sorlini@supsi.ch*

Mauro Isaja, *Engineering Ingegneria Informatica SpA, Italy; E-mail: mauro.isaja@eng.it*

Michele Ciavotta, *Università degli Studi di Milano-Bicocca, Italy; E-mail: michele.ciavotta@unimib.it*

Miguel Sepulcre, *UWICORE Laboratory, Universidad Miguel Hernández de Elche (UMH), Elche, Spain; E-mail: msepulcre@umh.es*

Nikos Kefalakis, *Kifisias 44 Ave., Marousi, GR15125, Greece; E-mail: nkef@ait.gr*

Oscar Lazaro, *Asociacion de Empresas Tecnologicas Innovalia, Rodriguez Arias, 6, 605, 48008-Bilbao, Spain; E-mail: olazaro@innovalia.org*

Paola Maria Fantini, *Politecnico di Milano, Milan, Italy; E-mail: paola.fantini@polimi.it*

Paolo Pedrazzoli, *Scuola Universitaria Professionale della Svizzera Italiana (SUPSI), The Institute of Systems and Technologies for Sustainable Production (ISTEPS), Galleria 2, Via Cantonale 2C, CH-6928 Manno, Switzerland; E-mail: paolo.pedrazzoli@supsi.ch; pedrazzoli@ttsnetwork.com*

Pedro Malo, *Unparallel Innovation Lda, Portugal; E-mail: pedro.malo@unparallel.pt*

Silvia Menato, *Scuola Universitaria Professionale della Svizzera Italiana (SUPSI), The Institute of Systems and Technologies for Sustainable Production (ISTEPS), Galleria 2, Via Cantonale 2C, CH-6928 Manno, Switzerland; E-mail: silvia.menato@supsi.ch*

Theofanis P. Raptis, *Institute of Informatics and Telematics, National Research Council (CNR), Pisa, Italy; E-mail: theofanis.raptis@iit.cnr.it*

Tiago Teixeira, *Unparallel Innovation Lda, Portugal; E-mail: tiago.teixeira@unparallel.pt*

Torben Meyer, *VOLKSWAGEN, Germany; E-mail: torben.meyer@volkswagen.de*

Valeriy Vytakin, *Department of Computer Science, Electrical and Space Engineering, Luleå tekniska universitet, A3314 Luleå, Sweden; E-mail: Valeriy.Vyatkin@ltu.se*

Veronika Brandstetter, *SIEMENS, Germany; E-mail: veronika.brandstetter@siemens.com*

Volkan Gezer, *German Research Center for Artificial Intelligence (DFKI), Germany; E-mail: Volkan.Gezer@dfki.de*

List of Figures

List of Tables

List of Abbreviations

ARX	Auto Regressive Exogenous
API	Application Programming Interface
BoM	Bill of Materials
BoP	Bill of Processes
CAD	Computer-Aided Design
CMM	Capability Maturity Model
CMMI	Capability Maturity Model Integration
CPS	Cyber-Physical System
CV	Controlled Variable
DREAMY	Digital REadiness Assessment MaturitY
DV	Disturbance Variable
EC	Edge Computing
ELC	Extended Linear Complementary
ERP	Enterprise Resource Planning
FAR-EDGE	Factory Automation Edge Computing Operating System Reference Implementation
FB	Functional Block
HA	Hybrid Automata
HMI	Human–Machine Interface
HMPC	Hybrid Model Predictive Control
ICT	Information and communication technologies
IMC-AESOP	ArchitecturE for Service-Oriented Process - Monitoring and Control
IoT	Internet of Things
IT	Information technologies
LAN	Local Area Network
LC	Linear Complementary
LP	Linear Programming
MASHUP	MigrAtion to Service Harmonization compUting Platform
MES	Manufacturing Execution System
MILP	Mixed-Integer Linear Programming

MIMO	Multi-Input Multi-Output
MIP	Mixed-Integer Programming
MIQP	Mixed-Integer Quadratic Programming
MLD	Mixed Logical Dynamical
MMPS	Max-Min-Plus-Scaling
MOMOCS	MOdel driven MOdernisation of Complex Systems
MPC	Model Predictive Control
MV	Manipulated Variable
OCM	On-line Control Modeller
OCS	On-line Control Solver
OIS	On-line Identification System
OT	Operational Technology
OV	Output Variable
PERFoRM	Production harmonizEd Reconfiguration of Flexible Robots and Machinery
PLC	Programmable Logic Controller
PLM	Product Lifecycle Management
PWA	Piece Wise Affine
QP	Quadratic Programming
RHC	Receding Horizon Control
SCADA	Supervisory Control And Data Acquisition
SDK	Software Development Kit
SISO	Single-Input Single-Output
SMART	Service-Oriented Migration and Reuse Technique
SOA	Service-oriented Architecture
SOAMIG	Migration of legacy software into service-oriented architectures
WAN	Wide Area Network
XIRUP	eXtreme end-User dRiven Process

1

Introduction to Industry 4.0 and the Digital Shopfloor Vision

John Soldatos

Kifisias 44 Ave., Marousi, GR15125, Greece
E-mail: jsol@ait.gr

This chapter is an introduction to the fourth industrial revolution (Industry 4.0) in general and digital automation platforms in particular. It illustrates the main drivers behind Industry 4.0 and presents some of the most prominent use cases. Accordingly, it introduces the scope and functionalities of digital automation platforms, along with digital technologies that enable them. The chapter ends by introducing the vision of a fully digital shopfloor, which sets the scene for understanding the platforms and technologies that are presented in subsequent chapters.

1.1 Introduction

In the era of globalization, industrial organizations are under continuous pressure to innovate, improve their competitiveness and perform better than their competitors in the global market. Digital technologies are one of their most powerful allies in these efforts, as they can help them increase automation, eliminate error prone processes, enhance their proactivity, streamline their business operations, make their processes knowledge intensive, reduce costs, increase their smartness and overall do more with less. Moreover, the technology acceleration trends provide them with a host of opportunities for innovating in their processes and transforming their operations in a way that results not only in marginal productivity improvements, but rather in a disruptive paradigm shift in their operations. This is the reason why many industrial organizations are heavily investing in the digitization of their processes as part of a wider and strategic digital transformation agenda.

In this landscape, the term Industry 4.0 has been recently introduced. This introduction has signalled the "official" start of the fourth industrial revolution, which is based on the deployment and use of Cyber-Physical Systems (CPS) in industrial plants, as means of fostering the digitization, automation and intelligence of industrial processes [1]. CPS systems facilitate the connection between the physical world of machines, industrial automation devices and Operational Technology (OT), with the world of computers, cloud data centres and Information Technology (IT). In simple terms, Industry 4.0 advocates the seamless connection of machines and physical devices with the IT infrastructure, as means of completely digitizing industrial processes.

In recent years, Industy 4.0 is used more widely, beyond CPS systems and physical processes, as a means of signifying the disruptive power of digital transformation in virtual all industries and application domains. For example, terms like Healthcare 4.0 or Finance 4.0 are commonly used as derivatives of Industry 4.0. Nevertheless, the origins of the term lie in the digitization of industrial organizations and their processes, notably in the digitization of factories and industrial plants. Note also that in most countries Industry 4.0 is used to signify the wider ecosystem of business actors, processes and services that underpin the digital transformation of industrial organizations, which makes it also a marketing concept rather than strictly a technological concept.

The present book refers to Industry 4.0 based on its original definition i.e. as the fourth industrial revolution in manufacturing and production, aiming to present some tangible digital solutions for manufacturing, but also to develop a vision for the future where plant operations will be fully digitized. However, it also provides insights on the complementary assets that should accompany technological developments towards successful adoption. For example, the book presents concrete examples of such assets, including migration services, training services and ecosystem building efforts. This chapter serves as a preamble to the entire book and has the following objectives:

- To introduce the business motivation and main drivers behind Industry 4.0 in manufacturing. Most of the systems and technologies that are presented in this book are destined to help manufacturers confront such business pressures and to excel in the era of globalization and technology acceleration.

- To present some of the main Industry 4.0 use cases in areas such as industrial automation, enterprise maintenance and worker safety. These use cases set the scene for understanding the functionalities and use of the platforms that are presented in this book, including use cases that are not explicitly presented as part of the subsequent chapters.
- To illustrate the main digital technologies that enable the platforms and technologies presented in the book. Note that the book is about the digitization of industrial processes and digital automation platforms, rather than about IT technologies. However, in this first chapter, we provide readers with insights about which digital technologies are enabling Industry 4.0 in manufacturing and how.
- To review the state of the art in digital automation platforms, including information about legacy efforts for digitizing the shopfloor based on technologies like Service Oriented Architectures (SOA) and intelligent agents. It's important to understand how we got to today's digital automation platforms and what is nowadays different from what has been done in the past.
- To introduce the vision of a fully digital shopfloor that is driving the collaboration of research projects that are contributing to this book. The vision involves interconnection of all machines and complete digitization of all processes in order to deliver the highest possible automation with excellent quality at the same time, as part of a cognitive and fully autonomous factory. It may take several years before this vision is realized, but the main building blocks are already set in place and presented as various chapters of the book.

In-line with the above-listed objectives, the chapter is structured as follows:

- Section 2 presents the main business drivers behind Industry 4.0 and illustrates some of the most prominent use cases, notably the ones with proven business value;
- Section 3 discusses the digital technologies that underpin the fourth industrial revolution and outlines their relation to the systems that are presented in the latter chapter;
- Section 4 reviews the past and the present of digital automation platforms, while also introducing the vision of a fully digital shopfloor;
- Section 5 is the final and concluding section of the chapter.

1.2 Drivers and Main Use Cases

The future of manufacturing is driven by the following trends, which stem from competitive pressures of the globalized environment:

- **New production models and mass customization**: Manufacturers are increasingly seeking ways of producing more customized products that are tailored to customer needs. As a result, there is a shift from mass production to mass customization. Likewise, conventional Made-to-Stock production models are giving their place to more customized ones such as Made-to-Order, Configure-to-Order and Engineering-to-Order.
- **Production Reshoring**: Globalization has led to the off-shoring of production operations for the places of innovation to low-labour countries. This was typically the case with several Western countries (including the USA (United States of America) and many EU (European Union) countries), which opted to keep the innovative design processes at home, while outsourcing manufacturing and production operations to Eastern countries (e.g., China, India). In recent years, several organizations are working towards reversing this trend through moving production processes back to the place of innovation, which is commonly called reshoring as opposed to off-shoring. Increased automation is a key enabler of reshoring strategies as it reduces the significance of the labour cost in the overall production process.
- **Proximity Sourcing**: Manufacturers are also employing proximity sourcing strategies as an element of their competitiveness. These strategies strive to ensure that sourcing is performed in close proximity to the plant that will use the source materials. This requires intelligent management of information about supply chain and logistics operations, which is also a main driver of the Industry 4.0.
- **Human-centred manufacturing**: Workers remain the major asset of the production process, yet a shift from laborious tasks to more knowledge intensive tasks is required. In addition to supporting other trends (such as mass customization and reshoring) this can be a key to improving workers' engagement, safety and quality of life. The digitalization of industrial processes obviates the need for laborious error-prone tasks and provides opportunities for improving workers' knowledge about the production processes. Hence, it's a key for placing the worker at the centre of the knowledge-intensive shopfloor and for transitioning to human centred processes.

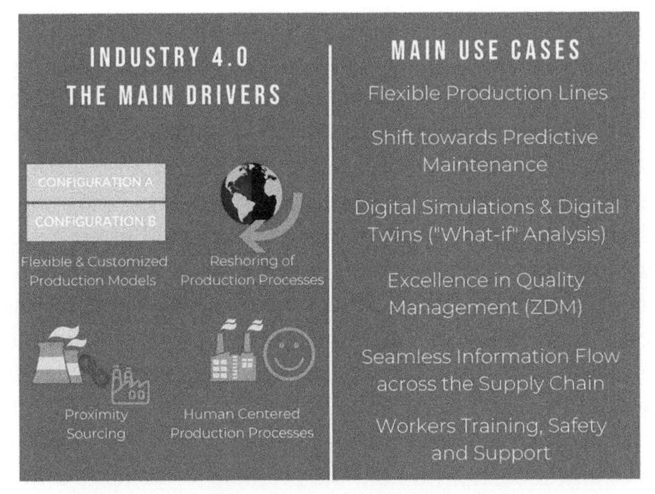

Figure 1.1 Main drivers and use cases of Industry 4.0.

The deployment of CPS systems in the shopfloor enables the seamless collection of digital data about all production processes, which increases the agility of automation operations, enabling the acquisition of knowledge about processes and facilitates optimal decision making. At the same time, CPS systems are able to initiate and execute digitally driven operations in the shopfloor. Coupled with digital technologies that are described in the next section, CPS systems can deliver endless possibilities for automation, optimizations and complete restructuring of industrial processes.

The fourth industrial revolution has an horizon of several decades, where it will deliver its full potential based on the interconnection all machines and OT systems, but also based on the employment of the ever evolving digital technologies such as Artificial Intelligence (AI), Big Data and the Industrial Internet of Things (IIoT). Nevertheless, during the first years of the Industry 4.0 movement, manufacturers have successfully deployed and validated the first set of use cases, which can directly deliver quick wins and business value. These use cases span the following areas:

- **Flexibility in Automation Architectures and Configuration:** Agility and flexibility in automation are key prerequisites for the transition to the range of future production models that enable mass customization. These models ask for flexibility in the way each individual product is produced, effectively reducing production lot to size one. In this context, digital technologies can be used to change the configuration of production lines at the digital/IT rather than at the physical/OT layer of production systems, yielding the configuration of a production line much faster and much more flexible. Hence, digitally transformed productions lines are able to produce products with different (rather than fixed) configurations.

- **Shift towards Predictive Maintenance:** Nowadays, most industrial organizations are employing preventive maintenance in order to avoid the catastrophic consequences of unplanned downtime and unscheduled maintenance. Hence, they replace tools and parts, at regular intervals before their estimated End of Life. Even though preventive maintenance techniques are much more effective than reactive maintenance, they are still far from delivering optimal Overall Equipment Efficiency (OEE), as they tend to perform maintenance earlier than actually required. Digital technologies and Industry 4.0 hold the promise to facilitate a transition to predictive maintenance that will enable the accurate prediction of parameters such as the End-of-Life (EoL) and the Remaining Useful Life (RUL) of machines and their parts, as a means of optimizing OEE, minimizing unscheduled downtimes and scheduling maintenance and repairs at the best point in time. Predictive maintenance is usually based on the collection and analysis of large digital datasets about the condition of the equipment, such as data from vibration, acoustic and ultrasonic sensors, data from thermal images, power consumption data, oil analysis data, data from thermal images, as well as quality data from enterprise systems. As such predictive maintenance is a classical Big Data and Artificial Intelligence problem in the industry, which is relevant not only in manufacturing, but also in other industrial sectors such as energy, mining, oil & gas and more.
- **Quality Management Excellence and Zero Defect Manufacturing (ZDM):** The advent of CPS systems and Industry 4.0 will enable manufacturers to collect large datasets about their processes, including data about the physical aspects of these processes. Equipment maintenance data is one example of such datasets. Other examples include datasets about the quality of the operations and of the resulting products, supply chain indicators, data about the quality of the source materials, data about the accuracy and consistency of assembly processes and more. By consolidating and analysing these datasets, manufacturers will be in a position to optimize their quality management processes and to meet stringent goals set from their quality management standards such

as SixSigma and Total Quality Management (TQM). Early quality management and predictive maintenance deployments that take advantage of CPS systems provide such evidence. Moreover, the expanded digitalization of the shopfloor will in the future enable the proactive identification of defect causes, as well as the activation of related remedial actions on the fly, as means of achieving the vision of Zero Defect Manufacturing (ZDM). Likewise, digital technologies will facilitate the implementation of continuous improvement disciplines, through continuous collection of data and the employment of self-learning systems that continually improve themselves based on past data and evidence. Overall, in the Industry 4.0 era, manufacturers will become able to implement more efficient and cost-effective ZDM processes, while lowering the barriers of transition from current approach to quality management excellence.

- **Digital Simulations and Digital Twins:** Industrial processes are generally inflexible given that it is practically impossible to cancel or undo an action once the latter has taken place in the shopfloor. Therefore, it's extremely difficult to test and validate alternative deployment configurations without disrupting production. Digital simulations provide the means of circumventing field testing, through using digital data for what-if analysis at the digital world and without a need of testing all scenarios in the field. Industry 4.0 technologies empower much more reliable and faster digital simulations, based on the use of advanced technologies for the collection, consolidation and analytics of very large datasets. Moreover, the Industry 4.0 era will be characterized by the wider use of a new disruptive concept i.e. the concept of a "digital twin". A digital twin is a faithful digital representation of a physical entity, which is built based on the development of a proper digital model for the physical item and the subsequent collection of a host of digital data about the item, in-line with the specified model. The design of a digital twin can be very challenging as a result of the need to consolidate the physical properties of an item, its behaviour, aspects of the processes where it is used and business aspects regarding its use in a single model. Digital twins provide plant operators and automation solution providers with the means of running credible simulations in the digital world, prior to deploying new automation ideas and algorithms in the physical world. In several cases, digital twins' instances can be connected and fully synchronized with their physical item counterparts as a means of configuring systems and processes at the IT rather than the OT layer of the Inudstry4.0 systems. As already outlined, this can greatly facilitate automation flexibility, as well.

- **Seamless and accurate information flows across the supply chain:** For over two decades, enterprises are heavily investing in the optimization of their supply chain operations, as a core element of their competitiveness. Supply chain management has always been a matter of properly acquiring, exchanging and managing information across the manufacturing chain, based on information sources and touch points of all supply chain stakeholders. Industry 4.0 comes to disrupt this information management, through adding an important element that was typically missing in traditional supply chain management: The information about the status of the physical world, such as the status of machines, equipment, processes and devices. Indeed, the advent of CPS systems and Industrial Internet of Things technologies enable the integration of this information across the supply chain. Furthermore, CPS systems and Industry 4.0 provide the means of influencing the status of the physical processes across the supply chain, in addition to changing the status of business information systems [e.g., production schedules in an Enterprise Resource Planning (ERP) system or materials information in a Warehouse Management System (WMS)]. This gives rise to disruptive supply chain innovations, which result in increased automation, less errors, increased efficiency and reduced supply chain costs.
- **Worker Training, Safety and Well Being:** Industry 4.0 emphasizes the importance of keeping employees engaged and at the centre of industrial processes, while alleviating them from the burden of laborious, tedious and time-consuming tasks. In this direction, several Industry 4.0 use cases entail the deployment of advanced visualization technologies such as ergonomic dashboards, Virtual Reality (VR) and Augmented Reality (AR) in order to ease the workers' interaction with the digital shopfloor and its devices. Note that AR and VR are extensively used in order to train employees under safe conditions i.e. through interaction with cyber representations of the physical equipment and/or with remote guidance from experienced colleagues or other experts. Likewise, wearables and other pervasive devices are extensively deployed in order to facilitate the tracking of the employee in the shopfloor towards ensuring that he/she works under safe conditions that do not jeopardise his/her well-being.

While the presented list of use cases is not exhaustive, it is certainly indicative of the purpose and scope of most digital manufacturing deployments in recent years. Later chapters in this book present practical examples of Industry 4.0 deployments that concern one or more of the above use cases. However, we expect that these use cases will gradually expand in sophistication as part of the digital shopfloor vision, which is illustrated in a following section of this chapter. Moreover, we will see the interconnection and interaction of these use cases as part of a more cognitive, autonomous and automated factory, where automation configuration, supply chain flexibility, predictive maintenance, worker training and safety, as well as digital twins co-exist and complement each other.

1.3 The Digital Technologies Behind Industry 4.0

Industry 4.0 is largely about the introduction of CPS systems in the shopfloor, in order to digitally interconnect the machines and the OT technology with IT systems such as Enterprise Resource Planning (ERP), Computerized Maintenance Management (CMM), Manufacturing Execution Systems (MES) and Customer Relationship Management (CRM), Supply Chain Management (SCM) systems. Based on CPS systems, the entire factory or plant can become a large scale CPS system that employs Industrial Internet of Things (IIoT) protocols and technologies for data collection, processing and actuation. In practice, an Industry 4.0 deployment takes advantage of multiple digital technologies in order to endow the digital automation systems with intelligence, accuracy and cost-effectiveness. Hence, Industry 4.0 is largely propelled by the rapid evolution of various digital technologies, which enable most of the use cases listed above. For example, predictive maintenance is greatly boosted by Big Data technologies that provide the means for analysing maintenance related data from a host of batch and streaming data sources. As another example, Industry 4.0 quality management and supply chain management use cases ask for fast exchange of data from and to the shopfloor, including interactions with numerous devices. The latter are propelled by advanced connectivity technologies such as 5G and LPWAN (Low Power Wide Area Networks).

In following paragraphs, we provide a list of the main digital technologies that empower the Industry 4.0 vision and highlight their importance for the factories of the future.

- **CPS and Industrial Internet of Things:** As already outlined, CPS systems are considered as the main building blocks of Industry 4.0 systems. In the medium term, most machines will be CPS systems that will provide the means for collecting digital data from the physical worlds, but also interfaces for actuating and control over them. CPS systems are conceptually Industrial Internet of Things (IIoT) systems, which enable interaction and data exchange with physical devices. Note however that IIoT systems provide also the means for interconnecting legacy machines with IT systems and ultimately treating them as CPS systems. This is mainly achieved through the augmentation of physical devices with middleware that implements popular IoT protocols, such as MQTT, OPC-UA, WebSocket and more. Overall, CPS and IIoT systems will be at the very core of all Industry 4.0 deployments in the years to come.
- **5G Communications:** Industrial plants are characteristic examples of device saturated environments, since there are likely to comprise thousands of sensors, edge gateways, machines and automation devices. Early Industry 4.0 involves only a small subset of these devices and hence can dispose with state of the art connectivity technologies such as Wi-Fi and 4G/LTE (Long Term Evolution) technologies. Nevertheless, in the medium and long term, a much larger number of machines and devices should be supported, as they will gradually connect to Industry 4.0 deployments. Likewise, much larger volumes of data and mobility of smart objects (e.g., drones and autonomous guided vehicles) should be handled, in several cases through high performance and lower latency. For these reasons, future deployments will require the capabilities advocated by 5G technologies which are currently being tested by several telecom operators worldwide. In particular, 5G technologies will enable low-latency data acquisition from thousands of devices at plant scale, which offering spectrum efficiency and ease of deployment.
- **Low Power Wide Area Networks:** In recent years, low power wide area network technologies (such as LoraWAN, NB-IoT and SigFox) have emerged, in order to support IoT devices connectivity at scale, notably the connectivity of low power devices. These technologies offer flexible and cost effective deployment, while at the same time supporting novel applications in both indoor and outdoor environments, including the accurate localization of items in indoor environments. We envisaged that such technologies will be also used in order to provide "location-as-a-service" capabilities in industrial plants. Their deployment will

come to enhance rather than replace the connectivity capabilities that are currently provided by 4G and WiFi technologies, notably in the direction of accurate item localization that existing technologies cannot deliver.

- **Cloud Computing:** CPS manufacturing systems and applications are very commonly deployed in the cloud, in order to take advantage of the capacity, scalability and quality of service of cloud computing. Moreover, manufacturers tend to deploy their enterprise systems in the cloud. Likewise, state of the art automation platforms (including some of the platforms that are presented in this book) are cloud based. In the medium term, we will see most manufacturing applications in the cloud, yielding cloud computing infrastructure an indispensable element of Industry 4.0.
- **Edge Computing:** During the last couple of years, CPS and IIoT deployments in factories implement the edge computing paradigm. The latter complements the cloud with capabilities for fast (nearly real time) processing, which is performed close to the field rather than in the cloud [2]. In an edge computing deployment, edge nodes are deployed close to the field in order to support data filtering, local data processing, as well as fast (real time) actuation and control tasks. The edge computing paradigm is promoted by the major reference architecture for IIoT and Industry 4.0 such as the Industrial Internet Consortium Reference Architecture (IIRA) and the Reference Architecture of the OpenFog consortium.
- **Big Data:** The vast majority of Industry 4.0 use cases are data intensive, as they involve many data flows from multiple heterogeneous data sources, including streaming data sources. In other words, several Industry 4.0 use cases are based on datasets that feature the 4Vs (Volume,Variety, Velocity, Veracity) of Big Data. As mentioned in earlier sections, predictive maintenance is a classic example of a Big Data use case, as it combines multi-sensor data with data from enterprise systems in a single processing pipeline. Therefore, the evolution of Big Data technologies and tools is a key enabler of the fourth industrial revolution. Industry 4.0 is typically empowered by Big Data technologies for data collection, consolidation and storage, given that industrial use cases need to bring together multiple fragmented datasets and to store them in a reliable and cost-effective fashion. However, the business value of these data lies in their analysis, which is indicative of the importance of Big Data analytics techniques, including machine learning techniques.

- **Artificial Intelligence:** Even though there is a lot of hype around the use of AI in the industry, most manufactures and plant operators are familiar with this technology. Indeed, AI has been deployed in industrial plants for over two decades, in different forms such as fuzzy logic and expert systems. In the Industry 4.0 the term is revised and extended in order to include the use of deep learning and deep neural networks for advanced data mining. The use of these techniques is directly enabled from the Big Data technologies that have been outlined in the previous paragraph. Hence, they have a very close affiliation with Big Data, as deep learning can be used in conjunction with Big Data technologies. AI data analytics is more efficient than conventional machine learning in identifying complex patterns such as operation degradation patterns for machines, patterns of product defect causes, complex failure modes and more. In industrial environments, AI can be embedded in digital automation systems, but also in physical devices such as robots and edge gateways.
- **Augmented Reality:** AR is another technology that has been used in plants since several decades. It is also revisited as a result of the emergence of more accurate tracking technologies and of new cost-effective devices. It can be used in many different ways in order to disrupt industrial processes. As a prominent example, AR can be used for remote support of maintenance workers in their service tasks. In particular, with AR the worker needs no longer to consult paper manuals or phone supports. He/she can rather view on-line the repair or service instructions provided by an expert (e.g., the machine vendor) from a remote location. As another example, AR can be used for training workers on complex tasks (e.g., picking or assembly tasks), through displaying them cyber-presentations of the ways these tasks are performed by experts or more experienced workers.
- **Blockchain Technologies:** Blockchain technologies are in their infancy as far as their deployment in industrial settings is concerned. Despite the hype around blockchains, their sole large scale, enterprise application remains their use in cryptocurrencies such as Bitcoin and Ethereum. Nevertheless, some of the projects that are presented in this book are already experimenting with blockchains in industry, while also benchmarking their performance. In particular, the FAR-EDGE project is using blockchain technology for the decentralization and

synchronization of industrial processes, notably processes that span multiple stations in the factory. However, other uses of the blockchain are also possible, such as its use for securing datasets based on encryption, as well as its use for traceability in the supply chain. It's therefore likely that the blockchain will play role in future stages of Industry 4.0, yet it has not so far been validated at scale. Note also that in the scope of Industry 4.0 applications, permissioned blockchains can be used (like in FAR-EDGE), instead of public blockchains. Permissioned blockchains provide increased privacy, authentication and authorization of users, as well as better performance than public ones, which makes them more suitable for industrial deployment and use.

- **Cyber Security:** Industry 4.0 applications introduce several security challenges, given that they are on the verge of IT and OT, which pose conflicting requirements from the security viewpoint. Any Industry 4.0 solutions should come with strong security features towards protecting datasets, ensuring the trustworthiness of new devices and protecting the deployment for vulnerabilities of IT assets.
- **3D Printing and Additive Manufacturing:** Along with the above-listed IT technologies, CPS manufacturing processes benefit from 3D printing, as an element of the digital automation platforms and processes. 3D printing processes can be driven by the digital data of an Industry 4.0 deployment, such as a digital twin of a piece of equipment or part that can be printed. Additive manufacturing processes can be integrated in a digital manufacturing deployment in support of the above-listed use cases. For example, 3D printing can be used to accelerate the maintenance and repair process, through printing parts or tools, rather than having to order them or to keep significant inventory. Likewise, printing processes can be integrated in order to flexible customize the configuration of a production line and subsequently of the products produced. This can greatly boost mass customization.

None of the chapters of the book is devoted to the presentation of digital technologies, as the emphasis is on digital automation systems and their functionalities. However, all the presented systems comprise one or more of the above digital building blocks. Moreover, some of the chapters are devoted to automation solutions that are built round the above listed technologies such as edge computing, cloud computing and blockchain technology.

1.4 Digital Automation Platforms and the Vision of the Digital Shopfloor

1.4.1 Overview of Digital Automation Platforms

The vision of using digital technologies towards enhancing the flexibility and configurability of industrial automation tasks is not new. For over a decade manufacturers have been seeking for scalable distributed solutions both for manufacturing automation and for collaboration across the manufacturing value chain [3]. Such solutions are driven by future manufacturing requirements, including reduction of costs and time needed to adapt to variable market demand, interoperability across heterogeneous hardware and software elements, integration and interoperability across enterprises (in the manufacturing chain), seamless and cost effective scalability through adding resources without disrupting operations, reusability of devices and production resources, plug-and-play connectivity, as well as better forecasting and predictability of processes and interactions towards meeting real-time demand [4]. These requirements have given rise to distributed decentralized approaches for de-centralizing and virtualization the conventional automation pyramid [5].

One of the most prominent approaches has been the application for intelligent agents in industrial automation, in the scope of in distributed environments where time-critical response, high robustness, fast local reconfiguration, and solutions to complex problems (e.g., production scheduling) are required [6]. Agent-based approaches fall in general devised in the following main categories:

- **Functional decomposition approaches**, where agents correspond to functional modules that are assigned to manufacturing or enterprise processes e.g., order acquisition, planning, scheduling, handling of materials, product distribution and more.
- **Physical decomposition approaches**, where agents are used to represent entities in the physical world (e.g., machines, tools, cells, products, parts, features, operations and more). This decomposition impacts also the implementation of manufacturing processes such as production scheduling. For example, in the case of functional decomposition, scheduling can be implemented as a process that merges local schedules maintained by agents in charge of ordering. Likewise, in the case of physical decomposition scheduling can be implemented based on a negotiation process between agents that represent single resources (e.g., cells, machines, tools, fixtures etc.).

Despite the advantages of agent technology for manufacturing operations (e.g., distribution, autonomy, scalability, reliability), agents are considered inefficient when dealing with low-level control tasks that have very stringent performance requirements. Furthermore, a direct mapping between software agents and manufacturing hardware has not been realized and/or standardized [7].

In addition to software agents' technology, Service Oriented Architecture (SOA) paradigms to decentralized automation have also emerged with a view to exploiting SOA's reusability, autonomy and loose coupling characteristics. SOA approaches to manufacturing automation are based on the identification of operations that can be transformed and exposed as services. Accordingly, these operations are exploited towards implementing service-oriented automation workflows. SOA solutions come with enterprise service bus infrastructures, which decouple producers from consumers, while at the same time facilitating the integration of complex event processing. Furthermore, SOA is a standardized and widely adopted technology, which presents several advantages over software agents, while giving rise to approaches that combine SOA and agents (e.g., [8]). SOA deployments in the shopfloor have also focused on the integration of device level services with enterprise level services, including for example deployments that virtualize Programmable Logic Controllers (PLC) [9], along with implementations of execution environments for Functional Block Instances (FBI), including functional blocks compliant to the IEC 61499 standard [10]. Nevertheless, SOA architectures have been unable to solve the real-time limitations of agent technology, which has given rise to various customizations of the technology (e.g., [11]).

The rise of CPS manufacturing, along with the evolution of the digital technologies that were presented in the previous section (e.g., Cloud Computing, IIoT and Big Data technologies) has led to the emergence of several cloud-based industrial automation platforms, including platforms offered by prominent IT and industrial automation vendors (e.g., IBM, SIEMENS, BOSCH, Microsoft, Software AG, SAP) and platforms developed in the scope of EU projects (e.g., FP7 iMain (http://www.imain-project.eu/), ARTEMIS JU (Joint Undertaking) Arrowhead (http://www.arrowhead.eu), FoF (Factories of the Future) SUPREME (https://www.supreme-fof.eu/) and more). Each of these platforms comes with certain unique value propositions, which aim at differentiating them from competitors.

Acknowledging the benefits of edge computing for industrial automation, Standards Development Organizations (SDOs) have specified relevant reference architectures, while industrial organizations are already working towards providing tangible edge computing implementations. SDOs such as the OpenFog Consortium and the Industrial Internet Consortium (IIC) have produced Reference Architectures (RA). The RA of the OpenFog Consortium prescribes a high-level architecture for internet of things systems, which covers industrial IoT use cases. On the other hand, the RA of the IIC outlines the structuring principles of systems for industrial applications. The IIC RA [12] is not limited to edge computing, but rather based on edge computing principles in terms of its implementation. It addresses a wide range of industrial use cases in multiple sectors, including factory automation. These RAs have been recently released and their reference implementations are still in their early stages.

A reference implementation of the IIC RA's edge computing functionalities [13] for factory automation is provided as part of IIC's edge intelligence testbed. This testbed provides a proof-of-concept implementation of edge computing functionalities on the shopfloor. The focus of the testbed is on configurable edge computing environments, which enable the development and testing of leading edge systems and algorithms for edge analytics. Moreover, Dell-EMC has implemented the EdgeX Foundry framework [14], which is a vendor-neutral open source project hosted by the Linux Foundation that builds a common open framework for IIoT edge computing. The framework is influenced by the above-listed reference architectures and was recently released. Other vendors (e.g., Microsoft and Amazon) are also incorporating support for edge devices and Edge Gateways in their cloud platforms.

The platforms and solutions that are presented in following chapters advance the state of the art in digital automation platforms, based on the implementation of advanced intelligence, resilience and security features, but also through the integration of leading edge technologies (e.g., AI and blockchain technologies). The relevant innovations are presented in the individual chapters that present these solutions. Note, however, that the FAR-EDGE, AUTOWARE and DAEDALUS solutions that are presented in the book fall in the realm of research solutions. Hence, they implement advanced features, yet they lack the maturity for very large scale digital automation deployments.

1.4.2 Outlook Towards a Fully Digital Shopfloor

The digital automation platforms that are listed in the previous paragraphs support the early stage Industry 4.0 deployments, which are characterized by the integration of a limited number of CPS systems and the digitization of selected production processes. As part of the evolution of Industry 4.0 deployments, we will witness a substantial increase of the scope of these deployments in terms of the connected machines and devices, but also in terms of the processes that will be digitized and automated. The ultimate vision is a fully digital shopfloor, where all machines and OT devices will be connected to the IT systems, while acting as CPS systems. This digital shopfloor will support all of the described functionalities and use cases in areas such as automation, simulation, maintenance, quality management, supply chain management and more. Moreover, these functionalities will seamlessly interoperate towards supporting end-to-end processes both inside the factory and across the supply chain. The interaction between these modules will empower more integrated scenarios, where for example information collected by the shopfloor is used to perform a digital simulation and produce outcomes that drive a control operation on the field.

The fully digital shopfloor will enable an autonomous factory, which will be characterized by the following properties:

- **Holistic, Integrated and End-to-End:** The digital shopfloor will deploy digital technologies and capabilities end-to-end, in order to address the digital transformation of all the production processes, rather than of selected processes which is the situation nowadays.
- **Predictive and Anticipatory:** Solutions within the fully digital shopfloor will be able to predict and anticipate important events such as machine failures and occurrence of production defects, as a means of proactively taking action in order to optimize operations.
- **Fast and Real-Time:** Solutions in the digital shopfloor will be fast and able to operate in real-time timescales, which will allow them to remedy potential problems and to perform optimizations on-line (e.g., support on-line defect repairs).
- **Flexible and Adaptive:** In the digital shopfloor of the future, automation solutions will be dynamic and adaptive to changing production requirements and manufacturing contexts. As such their digital capabilities, including their security characteristics, will be flexible and reconfigurable, in order to support dynamic control of production processes and their quality in the system life-cycle.
- **Standards-Based:** The realization of the digital shopfloor could be greatly facilitated based on the integration and use of standards-based solutions, notably solutions that adhere to mainstream digital manufacturing (e.g. RAMI4.0) and Industrial Internet of Things (IIoT) standards

(e.g., OpenFog Consortium). Adherence to such standards will greatly facilitate aspects such as integration and interoperability.

- **Open:** The solutions of the digital shopfloor should be openly accessible through Open APIs (Application Programming Interfaces), which will facilitate their expansion with more features and functionalities.
- **Cost-Effective:** The digital shopfloor will be extremely cost effective in its configuration and operations, based on its flexible, dynamic, reconfigurable and composable nature. In particular, the autonomy of the digital shopfloor solutions will eliminate costs associated with human-mediated error-prone processes, while their composability will lower development and deployment costs.
- **Human-Centric (Human-in-the-Loop):** A fully digital shopfloor shall address human factors end-to-end, including product design aspects, employees' training, proper visualization of production processes, as well as safety of human-in-the-loop processes.
- **Continuous Improvement:** The digital driven production processes will be characterized by a continuous improvement discipline, which will occur at various timescales, including machine, process and end-to-end production levels.

In the scope of the digital shopfloor, products and production processes can be fully virtualized and managed in the digital world (e.g., through their digital twin counterparts). This implies that digital information about the products and the production processes will be collected and managed end-to-end, towards a continuous improvement discipline.

The vision of the digital shopfloor requires development and integration activities across the following complementary pillars:

- **Digitally enhanced manufacturing equipment:** Industry 4.0 hinges on the interconnection of machines and equipment in the cyber world as CPS systems. Currently, legacy machines are augmented based on internet of things protocols in order to become part of Industry 4.0 deployments. At the same time, new machines come with digital interfaces and acts as CPS systems. In the medium and long term, machines will be digitally enhanced in order to provide embedded intelligence functionalities, such as the ability to detect and remedy defects, to identify maintenance parameters and to schedule maintenance activities and more. Such intelligence functionalities will endow machines with flexibility, reconfigurability, adaptability and proactivity properties, which are key enablers of the fully digital shopfloor.

- **Open digital platforms for automation and service engineering:** In the digital shopfloor, digitally enhanced machinery must be interconnected in order to support factory wide processes. To this end, various digital manufacturing platforms shall be integrated based on composable

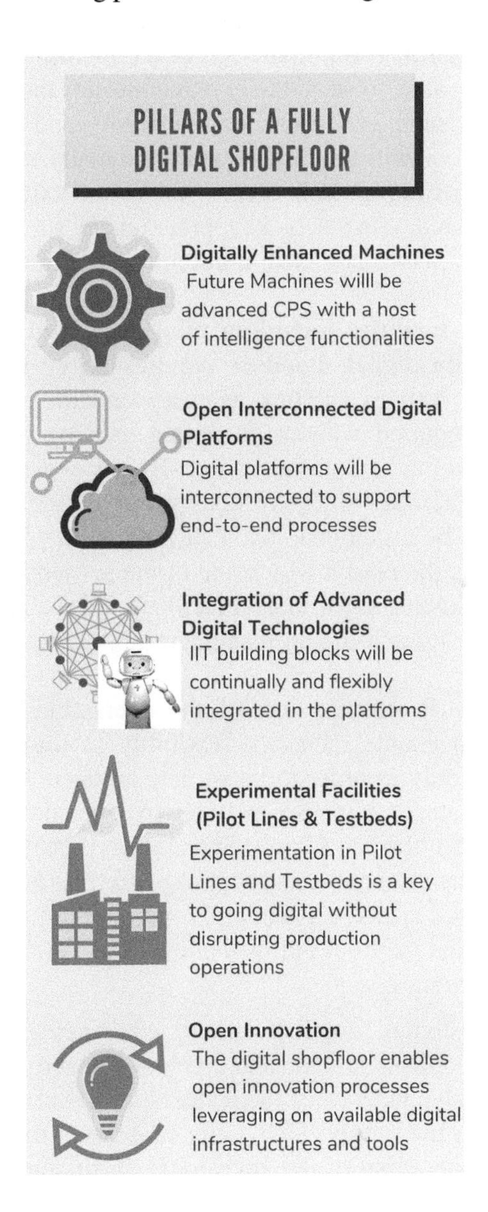
PILLARS OF A FULLY
DIGITAL SHOPFLOOR
Digitally Enhanced Machines
Future Machines willl be
advanced CPS with a host
of intelligence functionalities
Open Interconnected Digital
Platforms
Digital platforms will be
interconnected to support
end-to-end processes
Integration of Advanced
Digital Technologies
IIT building blocks will be
continually and flexibly
integrated in the platforms
Experimental Facilities
(Pilot Lines & Testbeds)
Experimentation in Pilot
Lines and Testbeds is a key
to going digital without
disrupting production
operations
Open Innovation
The digital shopfloor enables
open innovation processes
leveraging on available digital
infrastructures and tools

functionalities. This is also important given that factories and manufacturing chain tend to deploy multiple rather than a single digital automation platform. Hence, the composition of multiple functionalities from different platforms is required in order to support end-to-end production processes as part of the digital shopfloor.

- **Interoperable digital components and technologies:** The digital shopfloor will be able to seamlessly integrate advanced digital and CPS technologies such as sensors, data analytics and AI algorithms. The digital shopfloor will be flexibly and continually upgradable with the best-of-breed of digital technologies for manufacturing as the latter become available. This is a key prerequisite for upgrading the intelligence of the plant floor, with minimum disruption in production operations.
- **Experimentation facilities including pilot lines and testbed**s: The transition to a fully digital shopfloor requires heavy and continuous testing efforts, as well as auditing against standards. Extensive testing is therefore required without disrupting existing operations as a means of guaranteeing smooth migration. To this end, there is a need for experimental facilities and pilot lines where digital manufacturing developments can be tested and validated prior to their deployment in production. This is the reason why some of the subsequent chapters of the book refer to existing experimental facilities and testbeds, as key elements of Industry 4.0 and digital manufacturing platforms ecosystems building efforts.
- **Open Innovation Processes**: One of the overarching objectives of Industry 4.0 is to enable increased flexibility in digital automation deployments, not only in order to boost new production models (such as mass customization), but also in order to ease innovation in digital automation. To this end, open innovation processes should be established over the interconnected digital platforms, leveraging on IT innovation vehicles such as Application Programming Interfaces (APIs) and the experimental facilities outlined above. The latter could serve as a sandbox for innovation.

The road to the fully digital shopfloor is very challenging as a result of the need to develop, establish, validate and combine the above-listed pillars. However, there is already evidence of the benefits of digital technologies in the shopfloor and across the supply chain. Later chapters of this book present this evidence, along with some of the key digital manufacturing platforms that demonstrate the benefits of digital manufacturing platforms and of the related digitally transformed production processes.

1.5 Conclusion

This chapter has introduced Industry 4.0 in general and digital automation platforms in particular, which are at the core of the book. Our introduction to Industry 4.0 has underlined some of the proven and most prominent use cases that are being implemented as part of early deployments. Special emphasis has been given in use cases associated with flexible automation, worker training and safety, predictive maintenance, quality management, digital simulations and more. Basic knowledge about these use cases is a key prerequisite for understanding the automation use cases and applications that are presented in subsequent chapters of the book.

The chapter has also presented the most widely used digital technologies in the scope of Industry 4.0. Emphasis has been put on illustrating the relevance of each technology to Industry 4.0 use cases, but also on presenting how their evolution will impact deployment and adoption of CPS manufacturing systems. This discussion of digital technologies is also a prerequisite for understanding the details of the digital solutions that are presented in subsequent chapters. This is particularly important, given that no chapter of the book presents in detail digital technologies. Rather the emphasis of the book is on presenting advanced manufacturing solutions based on digital automation platforms that leverage the above-listed digital technologies.

Despite early deployments and the emergence of various digital automation platforms, the Industry 4.0 vision is still in the early stages. In the medium- and long-term, different technologies and platforms will be integrated towards a fully digital shopfloor, which supports the digital transformation of industrial processes end-to-end. The vision of a fully digital shopfloor entails the interoperability and interconnection of multiple digitally enhanced machines in-line with the needs of end-to-end automation processes within the factory. As part of this book, we present several automation approaches and functionalities, including field control, data analytics and digital simulations. In the future digital shopfloor, these functionalities will co-exist and seamlessly interoperate in order to enable fully autonomous, intelligent and resource efficient factories. With this wider vision in mind, readers could focus on the more fine-grained descriptions of platforms and technologies presented in subsequent chapters.

References

[1] Alasdair Gilchrist 'Industry 4.0: The Industrial Internet of Things', 1st ed. Edition Apress, June 28, 2016, ISBN-10: 1484220463, ISBN-13: 978-1484220467

[2] F. Bonomi, R. Milito, J. Zhu, S. Addepalli, 'Fog computing and its role in the internet of things', Proceedings of the first edition of the MCC workshop on Mobile cloud computing, MCC '12, pp 13–16, 2012.

[3] F. Jammes and H. Smit, 'Service-Oriented Paradigms in Industrial Automation Industrial Informatics', IEEE Transactions on., pp. 62–70, vol. 1, issue 1, Feb, 2005.

[4] T. Haluška, R. Paulièek, and P. Važan, "SOA as A Possible Way to Heal Manufacturing Industry", International Journal of Computer and Communication Engineering, Vol. 1, No. 2, July 2012

[5] A. W. Colombo (ed.), T. Bangemann (ed.), S. Karnouskos (ed.), J. Delsing (ed.), P. Stluka (ed.), R. Harrison (ed.), F. Jammes (ed.), J. L. Martínez Lastra (ed.), 'Industrial Cloud-Based Cyber-Physical Systems: The IMC-AESOP Approach', Springer. 245 p. 2014.

[6] C Leitão, "Agent-based distributed manufacturing control: A state-of-the-art survey," Engineering Applications of Artificial Intelligence, vol. 22, no. 7, pp. 979–991, Oct. 2009.

[7] P. Vrba, 'Review of Industrial Applications of Multi-agent Technologies', Service Orientation in Holonic and Multi Agent Manufacturing and Robotics, Studies in Computational Intelligence Vol. 472, Springer, pp 327–338, 2013.

[8] Tapia, S. Rodríguez, J. Bajo, and J. Corchado, 'FUSION@, A SOA-Based Multi-agent Architecture', in International Symposium on Distributed Computing and Artificial Intelligence 2008 (DCAI 2008), vol. 50 J. Corchado, S. Rodríguez, J. Llinas, and J. Molina, Eds. Springer Berlin/Heidelberg, 2009, pp. 99–107.

[9] F. Basile, P. Chiacchio, and D. Gerbasio, 'On the Implementation of Industrial Automation Systems Based on PLC', IEEE Transactions on Automation Science and Engineering, Volume: 10, Issue: 4, pp. 9901003, Oct 2013.

[10] W. Dai, V. Vyatkin, J. Christensen, V. Dubinin, 'Bridging Service-Oriented Architecture and IEC 61499 for Flexibility and Interoperability', Industrial Informatics, IEEE Transactions on, Volume: 11, Issue: 3 pp: 771–781, DOI: 10.1109/TII.2015. 2423495, 2015.

[11] T. Cucinotta and Coll, "A Real-Time Service-Oriented Architecture for Industrial Automation," Industrial Informatics, IEEE Transactions on, vol. 5, issue 3, pp. 267–277, Aug. 2009.
[12] Industrial Internet Consortium. 'The Industrial Internet of Things Volume G1: Reference Architecture', version 1.8, [Online], Available from: http://www.iiconsortium.org/IIRA.htm 2018.05.30, 2017
[13] Industrial Internet Consortium 'IIC Edge Intelligence Testbed', Available from: http://www.iiconsortium.org/edge-intelligence.htm 2018.05.30, 2017.
[14] 'EdgeX Foundry Framework' [Online], Available from: https://www.edgexfoundry.org/ 2018.05.30, 2018.

PART I

2

Open Automation Framework for Cognitive Manufacturing

Oscar Lazaro[3], Martijn Rooker[1], Begoña Laibarra[4], Anton Ružić[2], Bojan Nemec[2] and Aitor Gonzalez[3]

[1]TTTech Computertechnik AG, Schoenbrunner Strasse 7,
A-1040 Vienna, Austria
[2]Jožef Stefan Institute, Department of Automatics, Biocybernetics,
and Robotics, Jamova 39, 1000 Ljubljana, Slovenia
[3]Asociacion de Empresas Tecnologicas Innovalia, Rodriguez Arias,
6, 605, 48008-Bilbao, Spain
[4]Software Quality Systems, Avenida Zugazarte 8 1-6, 48930-Getxo, Spain
E-mail: olazaro@innovalia.org; martijn.rooker@tttech.com;
blaibarra@sqs.es; ales.ude@ijs.si; bojan.nemec@ijs.si;
aitgonzalez@innovalia.org;

The successful introduction of flexible, reconfigurable and self-adaptive manufacturing processes relies on evolving traditional automation ISA-95 automation solutions to adopt innovative automation pyramids proposed by CPS vision building efforts behind projects such as PathFinder, ScorpiuS and RAMI 4.0 IEC 62443/ISA99. These evolved automation pyramids demand approaches for the successful integration of data-intensive cloud and fog-based edge computing and communication digital manufacturing processes from the shopfloor to the factory to the cloud. This chapter presents an insight into the business and operational processes and technologies, which motivate the development of a digital cognitive automation framework for collaborative robotics and modular manufacturing systems particularly tailored to SME operations and needs, i.e. the AUTOWARE Operative System (OS).

To meet the requirements of both large and small firms, this chapter elaborates on the proposal of a holistic framework for smart integration of well-established SME-friendly digital frameworks such as the ROS-supported robotic Reconcell framework, FIWARE-enabled data-driven BEinCPPS/MIDIH Cyber Physical Production frameworks and OpenFog [3] compliant open-control hardware frameworks. The chapter demonstrates how AUTOWARE digital abilities are able to support automatic awareness; a first step in the support of autonomous manufacturing capabilities in the digital shopfloor. This chapter also demonstrates how the framework can be populated with additional digital abilities to support the development of advanced predictive maintenance strategies as those proposed by the Zbre4k project.

2.1 Introduction

SMEs are a pillar of the European economy and key stakeholder for a successful digital transformation of the European industry. In fact, manufacturing is the second most important sector in terms of small and medium-sized enterprises' (SMEs) employment and value added in Europe [1]. Over 80% of the total number of manufacturing companies is constituted by SMEs, which represent 59% of total employment in this sector.

In an increasingly global competition arena, companies need to respond quickly and economically feasible to the market requirements. In terms of market trends, a growing product variety and mass customization are leading to demand-driven approaches. Industry, in general, and SMEs, in particular, face significant challenges to deal with the evolution of automation solutions (equipment, instrumentation and manufacturing processes) they should support to respond to demand-driven approaches, i.e. increasing and abrupt changes in market demands intensified by the manufacturing trends of mass customization and individualization, which needs to be coupled with pressure on reduction of production costs, imply that manufacturing configurations need to change more frequently and dynamically.

Current practice is such that a production system is designed and optimized to execute the exact same process over and over again. Regarding the growing dynamics and these major driving trends, the planning and control of production systems has become increasingly complex regarding flexibility and productivity as well as the **decreasing predictability of processes**. It is well accepted that every production system should pursue the following three main objectives: (1) providing capability for rapid responsiveness,

(2) enhancement of product quality and (3) production at low cost. On the one hand, these requirements have been traditionally satisfied through highly stable and repeatable processes with the support of **traditional automation pyramids**. On the other hand, these requirements can be achieved by creating short response times to deviations in the production system, the production process, or the configuration of the product in coherence to overall performance targets. In order to obtain short response times, a high process transparency and reliable provisioning of the required information to the point of need at the correct time and without human intervention are essential.

However, the success of those adaptive and responsive production systems highly depends on real-time and operation-synchronous information from the production system, the production process and the individual product. Nevertheless, it can be stated that the concept of fully automated production systems is no longer a viable vision, as it has been shown that the conventional automation is not able to deal with the ever-rising complexity of modern production systems. Especially, a high reactivity, agility and adaptability required by modern production systems can only be reached by human operators with their immense cognitive capabilities, which enable them to react to unpredictable situations, to plan their further actions, to learn and to gain experience and to communicate with others. Thus, new concepts are required, which apply these cognitive principles to support autonomy in the planning processes and control systems of production systems. Open and smart cyber-physical systems (CPS) are considered to be the next (r)evolution in industrial automation linked to Industry 4.0 manufacturing transformation, with enormous business potential enabling novel business models for integrated services and products. Today, the trend goes towards open CPS devices and we see a strong request for open platforms, which act as computational basis that can be extended during manufacturing operation. **However, the full potential of open CPS has yet to be fully realized in the context of cognitive autonomous production systems.**

In fact, in particular to SMEs, it still seems difficult to understand the driving forces and most suitable strategies behind shopfloor digitalization and how they can increase their competitiveness making use of the vast variety of individualized products and solutions to digitize their manufacturing process, making them cognitive and smart and compliant with Industry 4.0 reference architecture RAMI 4.0 IEC 62443/ISA99. Moreover, as SMEs intend to adopt data-intensive collaborative robotics and modular manufacturing systems, making their advanced manufacturing processes more competitive, they face additional challenges to the implementation of "cloudified" automation processes. While the building blocks for digital automation are available, it is up to the SMEs to align, connect and integrate them to meet the needs of their individual advanced manufacturing processes, leading to difficult and costly digital automation platform adoption.

This chapter presents the AUTOWARE architecture, a concerted effort of a group of European companies under the Digital Shopfloor Alliance (DSA) [12] to provide an open consolidated architecture that aligns currently disconnected open architectural approaches with the European reference architecture for Industry 4.0 (RAMI 4.0) to lower the barrier of small, medium- and micro-sized enterprises (SMMEs) in the development and incremental deployment of cognitive digital automation solutions for next-generation autonomous manufacturing processes. This chapter is organized as follows. Section 2.2 presents the background and state of the art on open digital manufacturing platforms, with a particular focus on European initiatives. Section 2.3 introduces the AUTOWARE open OS building blocks and discusses their mapping to RAMI 4.0, the Reference Architecture for Manufacturing Industry 4.0. Then, Section 2.4 exemplifies how AUTOWARE platform can be tailored and customized to advanced predictive maintenance services. Finally, the chapter concludes with the main features of the AUTOWARE open automation framework.

2.2 State of the Play: Digital Manufacturing Platforms

Industry 4.0 started as a digital transformation initiative with a focus on the digital transformation of European factories towards smart digital production systems through intense vertical and horizontal integration. This resulted in the development by European industry of the RAMI 4.0 reference model built on the strong foundations of the automation European industry. As a consequence, Asian and American countries have also put efforts to define their reference model for the digitization of their manufacturing processes with stronger influences from IT and IoT industries. This has resulted in the development of the IVRA (Industrial Value Chain Reference Architecture) by the Industrial Value Chain Initiative (IVI) in Asia and the Industrial Internet Reference Architecture (IIRA) by the US IIC initiative; see Figure 2.1 below. These initiatives clearly showed the need to consider in the digitalization of European industry not only the Smart Production dimension, but also Smart Product and Smart Supply Chain dimensions.

As a consequence, European industry kicked off complementary efforts to ensure on the one hand RAMI 4.0, IVRA and IIRA interoperability, mapping

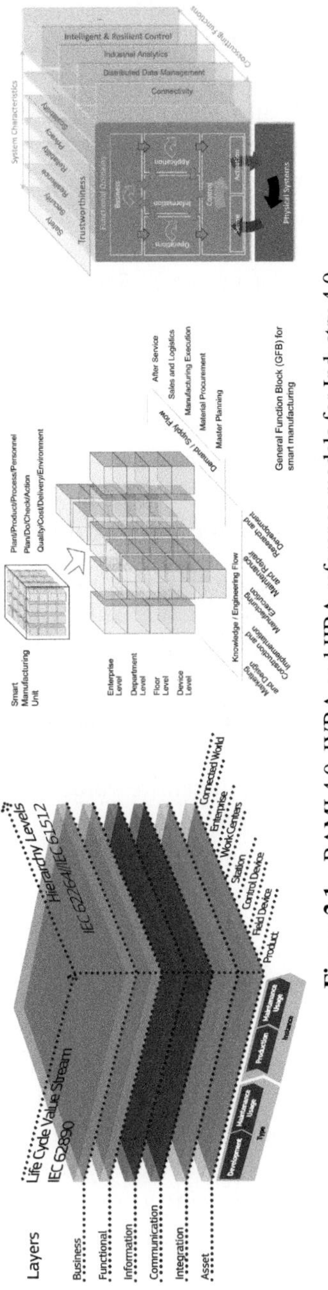

Figure 2.1 RAMI 4.0, IVRA and IIRA reference models for Industry 4.0.

and alignment for global operation of digital manufacturing processes. On the other hand, it has also triggered the need to extend the RAMI 4.0 model with an additional data-driven and digital smart service dimension beyond factory IT/OT integration. This resulted in the development of initiatives such as the Smart Service Welt and the Industrial Data Space to promote the development of smart data spaces as the basis for trusted industrial data exchange. This also derived in a more recent development of a need to support an increased autonomous operation shopfloors in the context of smart data-driven manufacturing processes.

This section provides a state-of-the-art revision of the reference models for factories 4.0 with a focus on RAMI 4.0 and the state of play of digital platforms initiatives developed to address the needs of data-driven operations within Industry 4.0, as the basis and context for the development of a framework for digital automation in industrial SMEs aiming at implementing cognitive and autonomous manufacturing processes.

2.2.1 RAMI 4.0 (Reference Architecture Model Industry 4.0)

The RAMI 4.0 (Reference Architecture Model for Industry 4.0 [34]) specification was published in July 2015. It provided a reference architecture initially for the Industrie 4.0 initiative and later for alignment of European activities and international ones. RAMI 4.0 groups different aspects in a common model and assures the end-to-end consistency of *"...technical, administrative and commercial data created in the ambit of a means of production of the workpiece"* across the entire value stream and their accessibility at all times. Although the RAMI 4.0 is essentially focused on the manufacturing process and production facilities, it tries to focus on all essential aspects of Industry 4.0. The participants (a field device, a machine, a system or a whole factory) can be logically classified in this model and relevant Industry 4.0 concepts are described and implemented.

The RAMI 4.0 3D model (see Figure 2.2) summarizes its objectives and different perspectives and provides relations between individual components. The model adopts the basic ideas of the Smart Grid Architecture Model (SGAM), which was defined by the European Smart Grid Coordination Group (SG-CG) and is worldwide accepted. The SGAM was adapted and modified according to the Industry 4.0 requirements.

The RAMI 4.0 model aims at supporting a common view among different industrial branches like automation, engineering and process engineering. The 3D model combines:

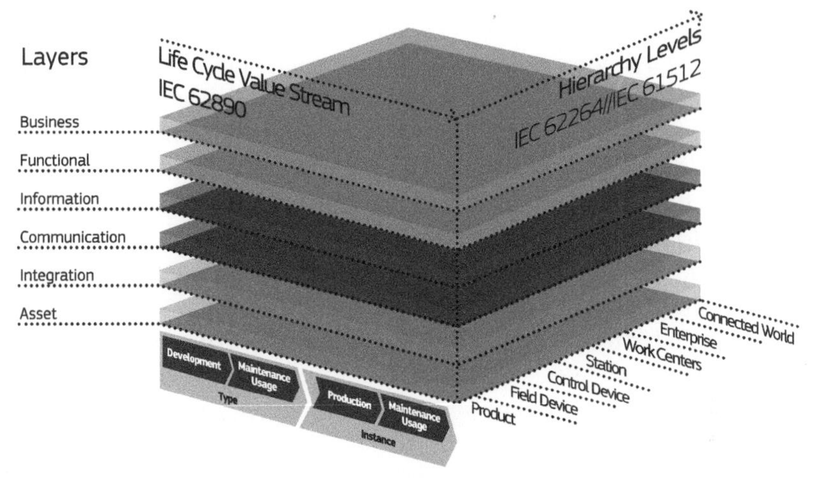

Figure 2.2 RAMI 4.0 3D Model.

- **Hierarchical Levels (Y Axis):** this axis collects the hierarchy levels envisaged by the IEC 62264 international standards on the integration of company computing and control systems;
- **Cycle & Value Stream (X Axis):** the second axis represents the life cycle of facilities and products. The RAMI 4.0 takes the IEC 62890 standard for life cycle management as a reference point to structure the life cycle. This axis focuses on features able to provide a consistent data model during the whole life cycle of an entity.
- **Layers (Z Axis):** the vertical axis, finally, represents the various perspectives from the assets up to the business processes.

The combination of the elements on these three axes represented a quite innovative management of product manufacturing, especially the elements on the X axis. Indeed, the RAMI 4.0 is the only reference architecture to explicitly analyze and take into account entities' life cycles at their time of proposal. Later, other models such as IVRA have also adopted that view.

One of the main objectives of RAMI 4.0 is to provide an end-to-end (i.e. since the inception of the product's idea, until its dismantling or recycling) framework able to connect and consistently correlate all technical, administrative and commercial data so as to create value streams providing added value to the manufacturer.

Many elements are available in RAMI 4.0, e.g. models, types, instances, production lines, factories, etc.). They differentiate between objects, which are elements that have a life cycle and data associated with it. On the other hand, there are the so-called "active" elements inside the different layers and are called Industry 4.0 components (I4.0 component). I4.0 components are also objects, but they have the ability to interact with other elements and can be summarized as follows: (1) they provide data and functions within an information system about an even complex object; (2) they expose one or more end-points through which their data and functions can be accessed and (3) they have to follow a common semantic model.

Therefore, the RAMI 4.0 framework goal is to define how I4.0 components communicate and interact with each other and how they can be coordinated to achieve the objectives set by the manufacturing companies.

2.2.2 Data-driven Digital Manufacturing Platforms for Industry 4.0

The digital convergence of traditional industries is increasingly causing the boundaries between the industrial and service sectors to disappear. In March 2015, Acatech, through the Industry-Science Research Alliance's strategic initiative *"Web-based Services for Businesses"*, has proposed a layered architecture (see Figure 2.3), to facilitate a shift from product-centric to user-centric business models, which extends the Industry 4.0 perspective.

At a technical level, these new forms of cooperation and collaboration will be enabled by new digital infrastructures. **Smart spaces** are the smart environments where smart, Internet-enabled objects, devices and machines (smart products) connect to each other. The term **"smart products"** refers to actual production machines but also encompasses their virtual representations (CPS digital twins). These products are described as "smart" because they know their own manufacturing and usage history and are able to act autonomously. Data generated on networked physical platforms are consolidated and processed on **software-defined platforms**. Providers connect to each other via these service platforms to form **digital ecosystems**.

Digital industrial platforms integrate the different digital technologies into real-world applications, processes, products and services; while new business models re-shuffle value chains and blur boundaries between products and services [16].

In the last few years, a number of initiatives have been announced by the public and private sectors globally dealing with the development of digital manufacturing platforms and multi-sided ecosystems for Industry 4.0 (see Figure 2.4). Vertical initiatives such as AUTOSAR [29] and ISOBUS [28], for instance, in the smart product dimension aim at

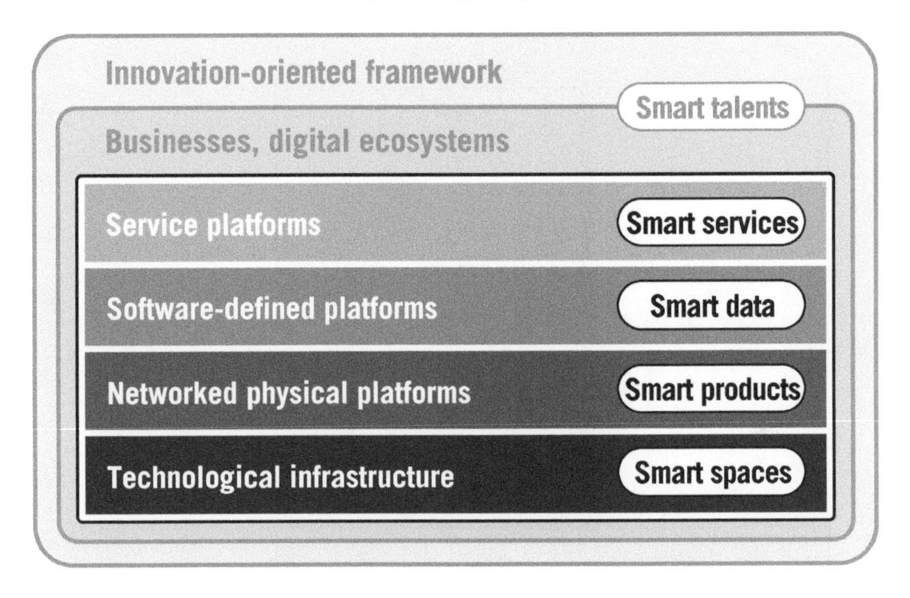

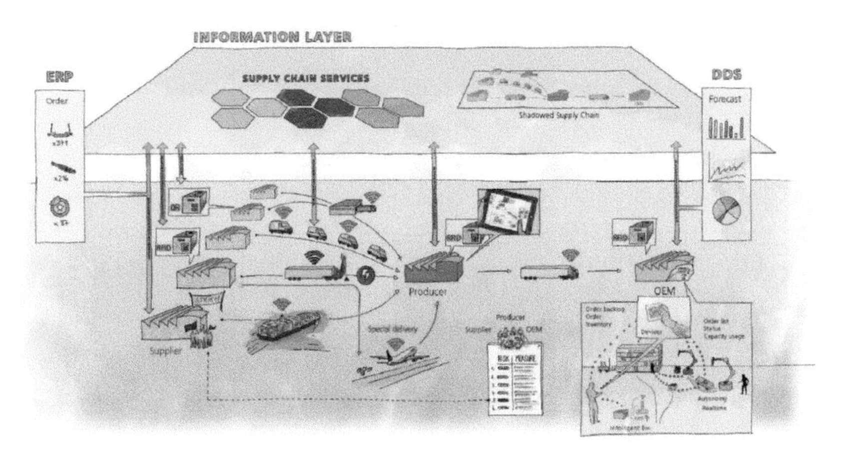

Figure 2.3 Smart Service Welt Reference Model & Vision.

enabling smart products in the automotive and smart agrifood sectors, whereas initiatives such as OPC-UA [31] intend to address manufacturing equipment universal access to a large extent. Similarly, more horizontal open (source) platform initiatives dealing with embedded systems (S3P [27]) or local automation clouds (Arrowhead [26], Productive 4.0 [32]) deal with networked physical product control across vertical industries, e.g. transport, manufacturing, health, energy and agrifood.

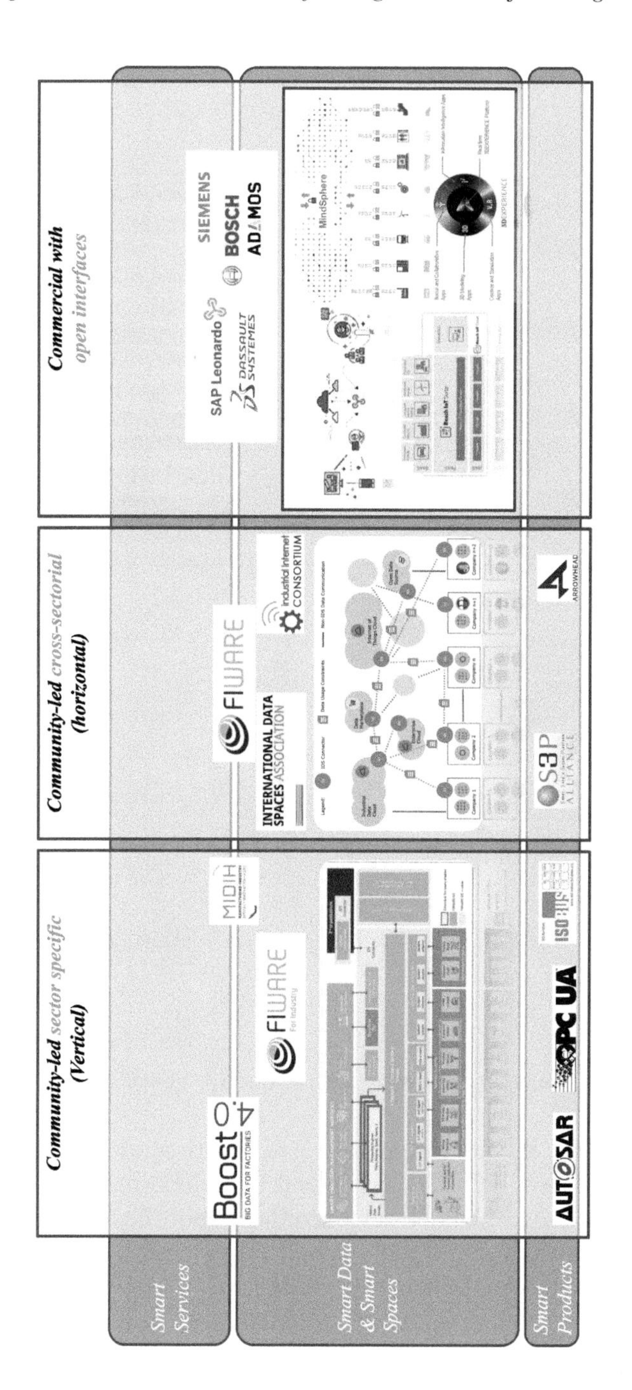

Figure 2.4 Digital manufacturing platform landscape. Adapted from [14] and [15].

However, the largest investment of industry so far has focused on the development of software-defined platforms to leverage smart spaces for smart data; either for vertical industries or for more horizontal approaches. Initiatives such as FIWARE for Smart Industry [22], MIDIH [21] or Boost 4.0 [24] are working to pave the way for the implementation of data-driven smart connected factories. On the other hand, more cross-domain initiatives for smart Internet services (FIWARE [23]), data-sharing sovereignty (International Data Spaces [25]) or Industrial IoT (IIC [30]) are both providing critical general software foundations for the development of vertical solutions such as those mentioned before (FIWARE Smart Industry, Boost 4.0 or MIDIH) and ensuring that interoperability across domains is properly developed as part of the digital transformation supporting the breakup of inter-domain information silos.

Along this line is also worth noting the recent efforts from large industrial software companies to provide commercial solutions with open APIs to respond to the challenge of leveraging digital infrastructures and smart data platforms to support the next generation of digital services. In this area are very relevant initiatives such as Mindsphere by SIEMENS [17], Leonardo by SAP [18], Bosch IoT suite [19] or 3DExperience [20] by Dassault Systems.

2.2.3 International Data Spaces

The **Industrial Data Space initiative** is an initiative driven forward by Fraunhofer together with over 90 key industrial players such as ATOS, Bayer, Boehringer Ingelheim, KOMSA, PricewaterhouseCoopers, REWE, SICK, Thyssen-Krupp, TÜV Nord, Volkswagen, ZVEI, SAP. BOSCH, Audi, Deutche Telekom, Huawei, Rittal and a network of European multipliers (INNOVALIA, TNO, VTT, SINTEF, POLIMI, etc.). **Digital sovereignty over industrial data and trust in data sharing** are key issues in the Industrial Data Space. Data will be shared between certified partners only when it is truly required by the user of that data for a value-added service. The basic principles that form the framework for the technological concept of the Industrial Data Space are summarized as (1) securely sharing data along the entire data supply chain and easily combining own data with publicly available data (such as weather and traffic information, geodata, etc.) and semi-public data, such as from a specific value chain. (2) Sovereignty over data, that is, control over who has what rights in which context, is just as important as legal certainty, to be ensured by certifying participants, data sources and data services. The reference architecture model should be

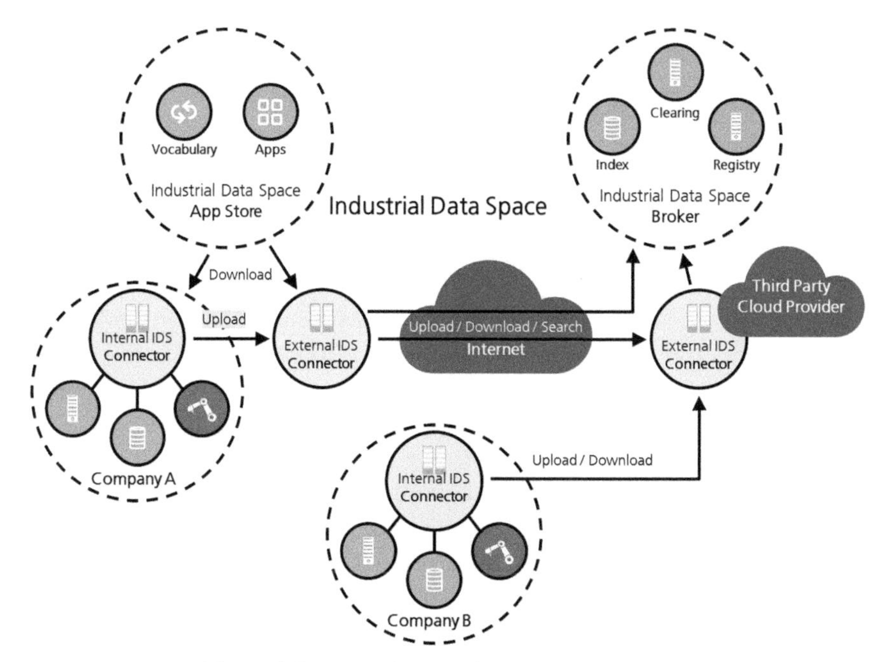

Figure 2.5 Industrial Data Space reference model.

seen as a blueprint for secure data exchange and efficient data combination. Figure 2.5 illustrates the technical architecture of the Industrial Data Space.

The Industrial Data Space fosters secure data exchange among its participants, while at the same time ensures data sovereignty for the participating data owners. The Industrial Data Space Association defines the framework and governance principles for the Reference Architecture Model, as well as interfaces aiming at establishing an international standard which considers the following user requirements: (1) data sovereignty; (2) data usage control; (3) decentralized approach; (4) multiple implementations; (5) standardized interfaces; (6) certification; (7) data economy and (8) secure data supply chains.

In compliance with common system architecture models and standards (such as ISO 42010, 4+1 view model, etc.), the Reference Architecture Model uses a five-layer structure expressing stakeholder concerns and viewpoints at different levels of granularity (see Figure 2.6).

The IDS reference architecture consists of the following layers:

- The **business layer** specifies and categorizes the different stakeholders (namely the roles) of the Industrial Data Space, including their activities and the interactions among them.

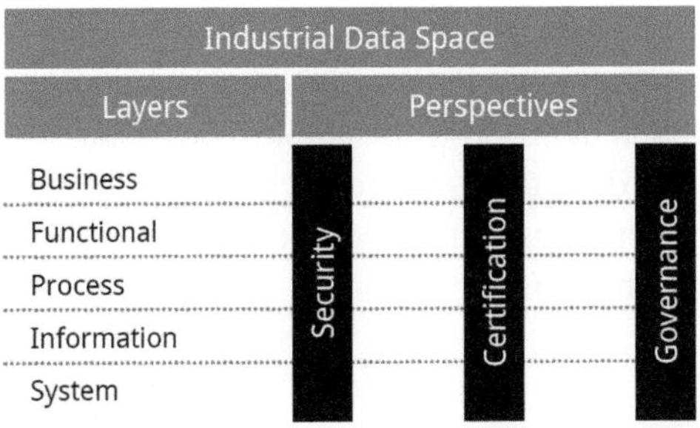

Figure 2.6 General structure of Reference Architecture Model [36].

- The **functional layer** comprises the functional requirements of the Industrial Data Space and the concrete features derived from them (in terms of abstract, technology-agnostic functionalities of logical software components).
- The **process layer** provides a dynamic view of the architecture; using the BPMN notation, it describes the interactions among the different components of the Industrial Data Space.
- The **information layer** defines a conceptual model, which makes use of "linked data" principles for describing both the static and dynamic aspects of the Industrial Data Space's constituents (e.g. participants active, Data Endpoints deployed, Data Apps advertised or datasets exchanged).
- The **system layer** is concerned with the decomposition of the logical software components, considering aspects such as integration, configuration, deployment and extensibility of these components.

In addition, the Reference Architecture Model contains three cross-sectional perspectives:

- **Security:** It provides means to identify participants, protect data communication and control the usage of data.
- **Certification:** It defines the processes, roles, objects and criteria involved in the certification of hardware and software artifacts as well as organizations in IDS.
- **Governance:** It defines the roles, functions and processes from a governance and compliance point of view, defining the requirements to be met by an innovative data ecosystem to achieve corporate interoperability.

System layer: technical components

The most interesting layer for the IDS framework is the system layer, where the roles defined in other layers (business and functional Layers) are now mapped onto a concrete data and service architecture in order to meet the requirements, resulting in what is the technical core of the IDS. From the requirements identified, three major technical components can be derived:

- Connector
- Broker
- App Store

These are supported by four additional components, which are not specific to the IDS:

- Identity provider
- Vocabulary hub
- Update repository (source for updates of deployed connectors)
- Trust repository (source for trustworthy software stacks and fingerprints as well as remote attestation checks).

IDS open source implementation using FIWARE

The most interesting aspect about the IDS business reference architecture is the opportunity to support multiple implementations and to combine it with open source enablers. It is a common goal that a valid open source implementation of the IDS Architecture can be based on FIWARE software components, compatible also with FIWARE architecture principles.

The FIWARE foundation is working towards making sure that core FIWARE Generic Enablers can be integrated together to build a valid open source implementation of the IDS architecture. Both organizations are collaborating on the development of domain data models and communicating about the development of their respective specifications and architectures to keep them compatible.

The way FIWARE software components can be combined to support the implementation of the main IDS architecture components is shown in Figure 2.7. FIWARE technology offers the following features to support IDS implementation:

1. Docker-based tools relying on Docker Hub Services enabling automated deployment and configuration of Data Apps.
2. Standard vocabularies are being proposed at https://www.fiware.org/data-models

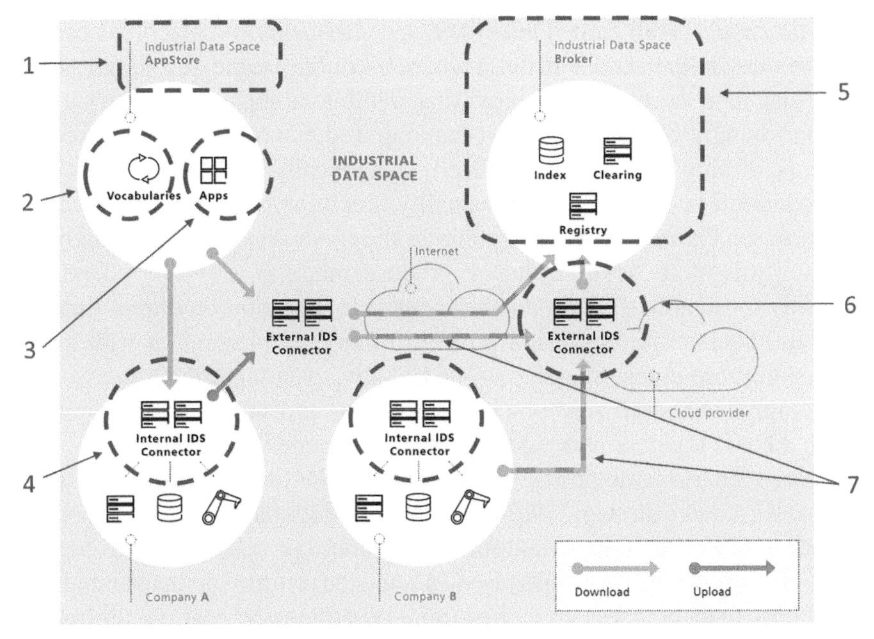

Figure 2.7 Materializing the IDS Architecture using FIWARE.

3. Data Apps map to NGSI adapters or Apps processing context information.
4. Both External and Internal IDS Connectors are implemented using FIWARE Context Broker components.
5. Extended CKAN Data Publication Platform.
6. FIWARE Context Broker components will be used as core component of IDS Connectors.
7. Interface between IDS connectors based on FIWARE NGSI.

2.3 Autoware Framework for Digital Shopfloor Automation

2.3.1 Digital Shopfloor Evolution: Trends & Challenges

The previous section presented the main digital platform and reference architecture work currently in place to deal with data-driven digital transformation in manufacturing. The industrial digitalization supported by Industry 4.0 and its vision of the intelligent networked factory of the future are major talking points as mobile technologies like cloud computing are revolutionizing industrial processes. With embedded systems, components and machines can now talk to one another and self-optimize, self-configure and self-diagnose processes and their current state, providing intelligent support for workers in their increasingly complex decision-making. Today's centrally organized enterprise is turning into a decentralized, dynamically controlled factory whose production is defined by individuality, flexibility and rapidity. As a consequence, see Figure 2.8 below, the digital shopfloor vision is increasingly evolving towards more flexible plug & produce modular assembly islands moving away from more rigid production lines with the ambition of real-time actuation and adaptation **(cognition and autonomy)** of production with an aim of reaching zero defect manufacturing. Equally, manufacturing processes are increasingly collaborative among humans, robots and autonomous mobile systems that come together as needed for mission-oriented tasks.

This new scenario is obviously generating that SMEs face difficulties at various levels to make strategic decisions while building a digital shopfloor, i.e. evolution model to adopt, automation technology selection and cost and time of deployment and operation, associated return on investments that will boost their business strategies (quality, efficiency, cost, flexibility, sustainability, innovation).

Since the 1980s, the IT structure of factories has been ordered hierarchically from field level to the level of factory control. Cloud and edge

Figure 2.8 Digital shopfloor visions for autonomous modular manufacturing, assembly and collaborative robotics.

technologies now make it possible to disengage these hierarchies and link up individual components – from computer numerical control CNC and robot control RC to manufacturing execution systems MES and enterprise resource planning ERP – in flexible networks. The core of this new approach is the virtualization of systems in which software functionality (digital abilities) is decoupled from the specific computer hardware (embedded, edge, cloud, HPC) where it runs. In other words, software used to depend on specific computer or control platforms is now separated from it via virtual machines and transferred to the cloud or the edge based on decision/actuation time scales. In a multitude of ways, **transfer of control functions to the cloud** opens up a whole new dimension of flexibility. First of all, the cloud-edge mechanism *"rapid elasticity"* enables the flexible and mostly automatic distribution of computing capacity. This means that the computing power of a whole group of processor cores in a "private cloud" can be allocated in a few seconds – for instance, between the CPU-intensive processes of the five-axis interpolation of a milling machine or the complex axis control of cooperating robots. Consequently, a much more efficient use of available computing power can be made than was possible with the older, purely decentralized control systems for individual machines and robots. At the same time, further gains in flexibility are given when – with adequate computing power – any number of virtual machine controls VMC or virtual robot controls VRC can be generated. The **cloud-based control** opens the way to upgrading or retrofitting high-quality machines and equipment whose control systems are outdated. The main challenge here is meeting the stringent real-time requirements set by state-of-the-art machine and robot control systems.

AUTOWARE [3], a European initiative under the European Commission initiative for digitizing European Industry, supports the deployment of such autonomous digital shopfloor solutions based on the following three pillars:

- **Pillar1: Harmonized open hardware and software digital automation reference architecture.** From a data-driven perspective for cyber physical production systems (smart products), leverage a reference architecture across open ICT technologies for manufacturing SME (I4MS, www.i4ms.eu) digital transformation competence domains (cloud, edge/OpenFog, BeinCPPS/MIDIH, robotics/ROS-ReconCell). For keeping integration time and costs under control, AUTOWARE framework acts as a glue across manufacturing users and digital automation solution developers in a friendly ecosystem for business development, more efficient service development over harmonized architectures (smart machine, cloudified control, cognitive planning-app-ized operation).

- **Pillar 2: Availability of digital ability technology enablers for digital shopfloor automatic awareness and cloud/edge-based control support.** Leverage a number of SME *digital abilities*, e.g. augmented virtuality, reliable wireless communications, smart data distribution and cognitive planning, to ease the development of automatic awareness capabilities in autonomous manufacturing systems. For ensuring digital shopfloor extendibility, the AUTOWARE framework envisions the development of *usability services* (Cyber Physical Production Systems (CPPS) trusted auto-configuration, programming by demonstration) as well as associated standard compliant *validation & verification services* for digital shopfloor solution.
- **Pillar 3: Digital automation business value model to maximize Industry 4.0 return of investment.** Leverage digital automation investments through a shared SME cognitive manufacturing migration model and an investment assessment platform for incremental brownfield cognitive autonomous solution deployment.

As opposed to other manufacturing environments, digital automation faces an increased challenge in terms of the large diversity of technologies involved. This implies that access to digital technologies or digital services is not enough for Industry 4.0 in general, but SMEs in particular, to leverage the Industry 4.0 business value. In the context of digital automation in general and in the context of cognitive and autonomous systems in particular, safe and secure integration of all technologies involved (robotic systems, production systems, computing platforms, cognitive services and mobile information services) into solutions is the real challenge, as illustrated in Figure 2.9.

Based on these three pillars, AUTOWARE has proposed a framework based on other existing frameworks (e.g. MIDIH, BEinCPPS, FIWARE,

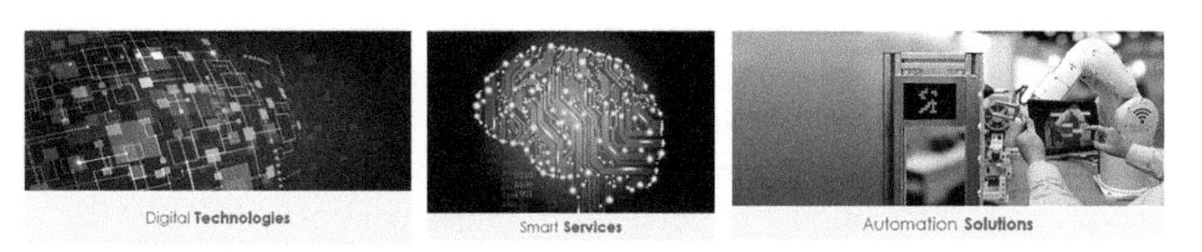

Figure 2.9 AUTOWARE digital automation solution-oriented context.

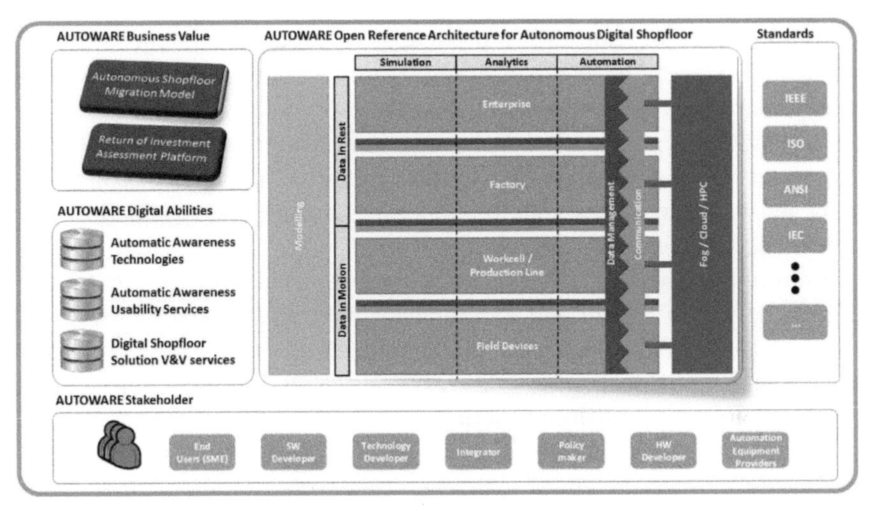

Figure 2.10 AUTOWARE framework.

RAMI 4.0), taking into consideration the industrial requirements from several use cases, aiming to provide a solution-oriented framework for digital shopfloor automation. Figure 2.10 shows the AUTOWARE framework with its main components.

The AUTOWARE framework from a technical perspective offers many features and concepts that are of great importance for cognitive manufacturing in particular to the automatic awareness abilities that AUTOWARE is primarily aiming at:

- **Open platform.** Platforms contain different technology building blocks with communication and computation instances with strong virtualization properties with respect to both safety and security for the cloudification of CPS services.
- **Reference architecture.** Platforms focused on harmonization of reference models for cloudification of CPS services have to make a template style approach for flexible application of an architectural design for suitable implementation of cognitive manufacturing solutions, e.g. predictive maintenance, zero defect manufacturing, energy efficiency.
- **Connectivity to IoT.** Multi-level operation (edge, cloud) and function visualization through open interfaces allow native support for service connection and disconnection from the platform, orchestrating and provisioning services efficiently and effectively.

- **Dynamic configuration.** Software-defined operation of systems allows automatic integration of other systems to connect or disconnect from the system, dynamic configuration including scheduling is implemented. The deployment of new functionalities, new services and new system structures poses new safety and security system requirements; component must be more dynamically configured and validated and finally integrated into these systems.
- **Autonomous controls.** High automation levels and autonomy require a high degree of design and development work in the area of sensors and actuators on the one hand and a high degree of efficient and robust sensor fusion on the other.
- **Virtualization of real-time functions.** Control functions can be virtualized and executed away from machine environments, and machine data can be accessed remotely in real time. This enables a large variety of novel functionalities as it allows the geographical distribution of computationally intensive processes, executed remotely from the location of action.

2.3.1.1 Pillar 1: AUTOWARE open reference architecture for autonomous digital shopfloor

AUTOWARE Reference Architecture (RA) aligns the cognitive manufacturing technical enablers, i.e. robotic systems, smart machines, cloudified control, secure cloud-based planning systems and application platform to provide cognitive automation systems as solutions while exploiting cloud technologies and smart machines as a common system. AUTOWARE leverages a reference architecture that allows harmonization of collaborative robotics, reconfigurable cells and modular manufacturing system control architectures with BEinCPPS and MIDIH data-driven industrial service reference architectures (already fully aligned with ECSEL CRYSTAL and EMC2 CPS design practices) supported by secure and edge-powered reliable industrial (wireless) communication systems (5G, WiFi and OPC-UA TSN) and high-performance cloud computing platforms (CloudFlow) across cognitive manufacturing competence domains (automation, analytics and simulation).

The goal of the AUTOWARE RA is to have a broad industrial applicability, map applicable technologies to different areas and to guide technology and standard development. From a structural perspective, the AUTOWARE RA covers two different areas denoted as domains:

- **Design domain:** it describes the design and development methods, tools and services for designing AUTOWARE CPPS. The components of the design domain enable users to intuitively design the applications (the so-called automatic awareness digital ability usability services).
- **Runtime domain:** it includes all the systems that support the execution and operation of the AUTOWARE autonomous CPPS.

The AUTOWARE RA has four layers/levels (see Figure 2.11), which target all relevant layers for the modeling of autonomous CPPS in the view of AUTOWARE:

- **Enterprise:** The enterprise layer is the top layer of the AUTOWARE reference architecture that encompasses all enterprise's systems, as well as interaction with third parties and other factories.
- **Factory:** At the factory layer, a single factory is depicted. This includes all the various workcells or production lines available for the complete production.
- **Workcell/Production Line:** The workcell layer represents the individual production line of cell within a company. Nowadays, a factory typically contains multiple production lines (or production cells), where individual machines, robots, etc. are located in or become a part of.
- **Field Devices:** The field devices layer is the lowest level of the reference architecture, where the actual machines, robots, conveyer belt, as well as controllers, sensors and actuators are positioned.

To uphold the concept of Industry 4.0 and to move from the old-fashioned automation pyramid (where only communication was mainly possible within a specific layer, and to establish communication between the different layers, complicated interfaces were required), the communication concept is a "pillar" to cover all the mentioned layers. The communication pillar enables direct communication between the different layers. The pillar is named **Fog/Cloud** and uses wired (e.g. IEEE 802.1 TSN) and wireless communication to create direct interaction between the different layers by using Fog/Cloud concepts (blue column in Figure 2.11). In good alignment with this paradigm, this pillar is also responsible for data persistence and potentially distributed transaction management services across the various components of the autonomous digital manufacturing system.

Finally, the last part of the AUTOWARE Reference Architecture focuses on the actual **modeling, programming and configuration** of the different technical components inside the different layers (green column in Figure 2.11). On each layer, different tools or services are applied and

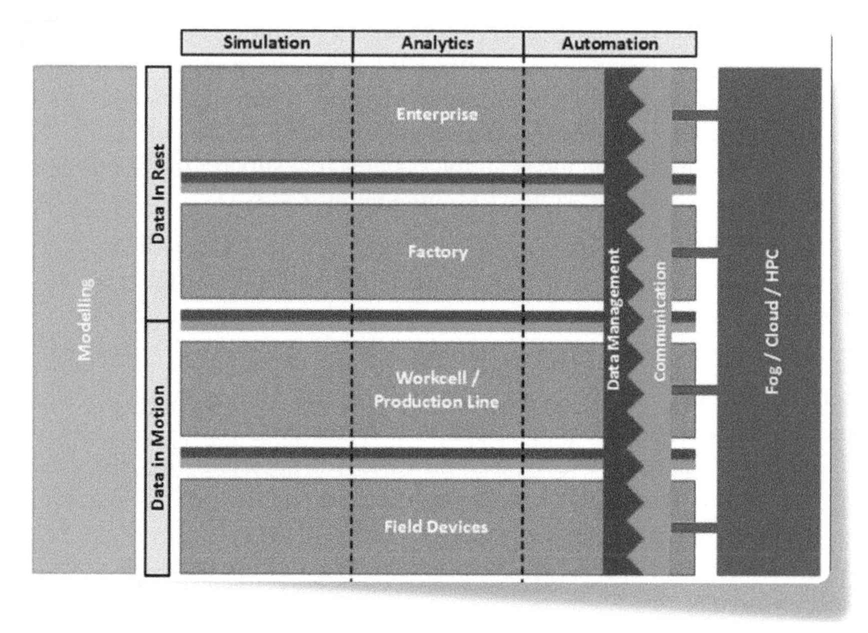

Figure 2.11 AUTOWARE Reference Architecture.

for all of them, different modeling approaches are available. The goal of these modeling approaches is to ease the end user/system developer/system integration developing the tools or technologies for the different levels. Additionally, it could be possible to have modeling approaches that take the different layers into account and make it easier for the users to model the interaction between the different layers.

The AUTOWARE reference architecture also represents the two **data domains** that the architecture anticipates, namely the data in motion and data and rest domains. These layers are also matched in the architecture with the **type of services** automation, analysis and learning/simulation that are also pillars of the RA. The model also represents the layers of the RA where such services could be executed with the support of the fog/cloud computing and persistence services (blue pillar in Figure 2.11).

2.3.1.2 Pillar 2: AUTOWARE digital abilities for automatic awareness in the autonomous digital shopfloor

As an initial and crucial step towards autonomous shopfloor operation, AUTOWARE provides a set of digital technologies and services for setting the foundation **of automatic awareness in a digital shopfloor**. Automatic awareness is the precondition for any form of more advanced autonomous decision and/or self-adaptation process. Autonomous digital shopfloor operations require integration across multiple disciplines. In fact, as discussed in [37] and shown in Figure 2.12, openness and interoperability need to be facilitated across all of those in a harmonized manner to ensure future digital shopfloor extendibility as industry gradually adopts digital abilities and services to build their competitive advantage.

For this purpose, the AUTOWARE framework provides three main components. These AUTOWARE components (technologies, usability services and V&V services) provide a collection of enablers that facilitates the different users of the AUTOWARE framework to interact with the system on different levels. Apart from the enablers developed in the AUTOWARE project, there have been several international projects to promote the creation of new open source enablers for such an architecture. The most interesting ones have come from FIWARE Smart Industry, I4MS and IDS communities and have been integrated into the AUTOWARE framework. Within AUTOWARE, there are three different enablers: technology, usability and verification and validation (V&V), which are crucial to ensure that a particular digital ability (in the specific case of AUTOWARE, automatic awareness) can be effectively and efficiently modeled, programmed, configured, deployed and operated in a digital shopfloor.

On the one hand, within the AUTOWARE framework, there is a collection of technology enablers, which can be identified as the technical tools, methods and components developed or provided within the AUTOWARE framework. Examples of technology enablers within the AUTOWARE project are robotic systems, smart machines, cloudified control systems, fog nodes, secure cloud- and fog-based planning systems as solutions to exploit cloud

Figure 2.12 AUTOWARE harmonized automatic awareness open technologies.

and fog technologies and smart machines as a common system. All these conform to a set of **automatic awareness integrated technologies,** which, as shown in Figure 2.12, adopt i-ROS-ready reconfigurable robotic cell and collaborative robotic bi-manipulation technology, smart product memory technology, OpenFog edge computing and virtualization technology, 5G-ready distributed data processing and reliable wireless mobile networking technologies, OPC-UA compliant Time Sensitive Networking (TSN) technology, Deep object recognition technology and ETSI CIM-ready FIWARE Context Brokering technology.

On the other hand, the AUTOWARE digital ability framework additionally provides **automatic awareness usability services** intended for a more cost-effective, fast and usable modeling, programming and configuration of integrated solutions based on the AUTOWARE enabling automatic digital shopfloor awareness technologies. This includes, for instance, augmented virtuality services, CPPS-trusted auto-configuration services or robot programming by training services.

Through its digital abilities, AUTOWARE facilitates the means for the deployment of completely open digital shopfloor automation solutions for fast data connection across factory systems (from shop floor to office floor) and across value chains (in cooperation with component and machine OEM smart services and knowledge). The AUTOWARE added value is not only to deliver a layered model for the four layers of the digital business ecosystem discussed in Section 2.2 for the digital shopfloor (smart space, smart product, smart data and smart service), but more importantly to provide an open and flexible approach with suitable interfaces to commercial platforms that allows the implementation of collective and collaborative services based on trusted information spaces and extensive exploitation of digital twin capabilities and machine models and operational footprints.

The third element in the AUTOWARE digital ability is the provision of **validation and verification (V&V) services** for digital shopfloor solutions, i.e. CPPS. Although CPPS are defined to work correctly under several environmental conditions, in practice, it is enough if it works properly under specific conditions. In this context, certification processes help to guarantee correct operation under certain conditions, making the engineering process easier, cheaper and shorter for SMEs that want to include CPPS in their businesses. In addition, certification can increase the credibility and visibility of CPPS as it guarantees its correct operation under specific standards. If a CPPS is certified to follow some international or European standards or regulation, then it is not necessary to be certified in each country, so the integration complexity, cost and duration are highly reduced.

2.3.1.3 Pillar 3: AUTOWARE business value

On the one hand, around the world, traditional manufacturing industry is in the throes of a digital transformation that is accelerated by exponentially growing technologies (e.g. intelligent robots, autonomous drones, sensors, 3D printing). Indeed, there are several European initiatives (e.g. I4MS initiative) and interesting platforms that are developing digitalization solutions for manufacturing companies in different areas: robotic solutions, cloudification manufacturing initiatives, CPS platforms implementation, reconfigurable cells, etc. However, all these initiatives were developed in isolation and they act as isolated components.

On the other hand, manufacturing SMEs need to digitalize their processes in order to increase their competitiveness through the adoption of ICT technologies. However, the global competition and the individualized products and solutions that currently exist make it difficult for manufacturing SMEs to access all this potential.

For this reason, AUTOWARE defined a new Autonomous Factory Ecosystem around their AUTOWARE Business Value Pillar allowing manufacturing SMEs to gain a clear competitive advantage for the implementation of their manufacturing processes. This pillar provides access to a set of new generation of tools and decision support toolboxes capable of supporting CPPS and digital services cloudification, robotics systems, reconfigurable cells, thanks to a faster and holistic management of several initiatives and tools into an open ecosystem providing a more seamless transfer of information across physical and digital worlds.

Therefore, AUTOWARE provides an ***open CPPS solution hub ecosystem*** that gathers all resources together, thus enabling SMEs to access all the different components in order to develop digital automation cognitive solutions for their manufacturing processes in a controlled manner and quantifiable business impact.

AUTOWARE reduces the complexity of the access to the different isolated tools significantly and speeds up the process by which multi-sided partners can meet and work together. Indeed, AUTOWARE connects several initiatives for strengthening the European SME offer on cognitive autonomous products and leveraging cognitive autonomous production processes and equipment towards manufacturing SMEs. Thus, AUTOWARE leverages the development of open CPPS ecosystem and joins several stakeholders' needs:

- **End Users (SME):** The main target group of the AUTOWARE project is SMEs (small and medium-sized enterprises) that are looking to change their production according to Industry 4.0, CPPS and Internet of Things (IoT). These SMEs are considered the end user of the AUTOWARE developments, whereby they do not have to use all the developed technologies, but can only be interested in a subset of the technologies.
- **Software Developers:** As the AUTOWARE platform is an open platform, software developers can create new applications that can run on the AUTOWARE system. To support these users in their work, the system provides high usability and intuitiveness level, so that software developers can program the system to their wishes.
- **Technology Developers:** The individual technical enablers can be used as a single technology, but being an open technology, they can also be integrated into different technologies by technology developers. The technology must be open and once again be intuitive to re-use in different applications. Technology developers can then easily use the AUTOWARE technology to develop new technologies for their applications and create new markets for the AUTOWARE results.
- **Integrator:** The integrator is responsible for the integration of the technologies into the whole manufacturing chain. To target this user group, the technologies must support open interfaces, so the system can intuitively be integrated into the existing chain. The advantage of the open interfaces is that the integrator is not bound to a certain brand or vendor.
- **Policy Makers:** Policy makers can make or break a technology. To increase the acceptance rate, the exploitation and dissemination of the technology must be at a professional level, and additionally, the technology must be validated, supporting the right standards and targeting the right problems currently present on the market. Policy makers can push technologies further into the market and act as large catalyst for new technologies.
- **HW Developers:** For hardware developers, it is important to know what kind of hardware is required for the usage of the different technologies. In ideal case, all kind of legacy hardware is capable of interacting with new hardware, but unfortunately, this is not always the case.
- **Automation Equipment Providers:** The technologies developed within the AUTOWARE project can be of interest to other automation equipment providers, e.g. robot providers, industrial controller providers, sensor providers, etc.

2.3.2 AUTOWARE Software-Defined Autonomous Service Platform

Once the complete AUTOWARE framework overview has been presented, this section will focus on the detailed presentation of the software-defined service platform for autonomous manufacturing services. This section extends the main technological blocks underlying the AUTOWARE reference architecture.

Due to the recent development of numerous technical enablers (e.g. IoT, cloud, edge, HPC etc.), it is possible to take a service-based approach for many components of production information systems (IS). When using a service-based approach, instead of developing, deploying and running our own implementations for all production IS tasks, an external service provider can be considered and the end user can rent access to the offered services, reducing the cost and knowledge needed.

AUTOWARE focuses on a service-based approach denoted as software-defined autonomous service platform (in the following, also abbreviated as "service platform") based on open protocols and implementing all the functionalities (physical, control, supervision, MES, ERP) as services. As a result, the components can be reused, the solution can be reconfigured and the technological advanced can be easily followed.

Figure 2.13 includes the reference architecture of the AUTOWARE service platform showing also how all the functionalities are positioned in the overall scheme of production IS. There are different functionalities (and therefore, services) on the different layers depending on the scope, but all of them are interconnected.

2.3.2.1 Cloud & Fog computing services enablers and context management

AUTOWARE considers several cloud services enablers for an easier implementation of the different services or functionalities. Context management and service function virtualization is a critical element to be supported in the delivery of automatic awareness abilities in a digital shopfloor. The use of these open source enablers permits the easier exchange of information and interoperability between different components and services, something really useful for future use cases.

AUTOWARE RA considers FIWARE for Smart Industry technology as the basis to meet AUTOWARE needs of context building management for digital automation information systems with extended support to robotic

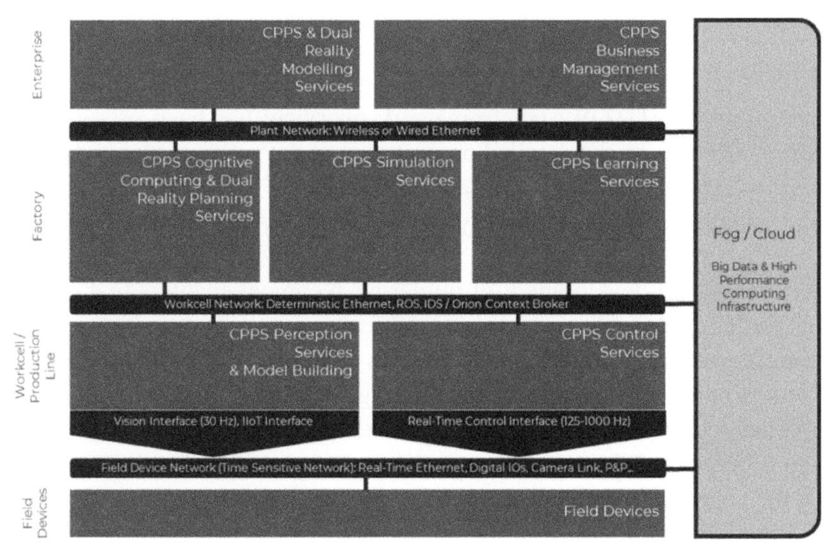

Figure 2.13 AUTOWARE Software-Defined Autonomous Service Platform.

systems. Additionally, AUTOWARE considers OpenFog as the framework for operation of virtualized service functions.

The main features introduced in the cloud & edge computing pillar, beyond those inherent to OpenFog specifications, are the support for automation context information management, processing and visualization. Such functionalities are being provided through edge and cloud support to two main FIWARE components:

- **Backend Device Management – IDAS:** For the translation from IoT-specific protocols to the NGSI context information protocol considered by FIWARE enablers.
- **Orion Context Broker:** It produces, gathers, publishes and consumes context information. This is the main context information communication system throughout the AUTOWARE architecture. It facilitates the exchange of context information between Context Information Producers and Consumers through a Publish/Subscribe methodology (see Figure 2.14). This permits a high decentralized and large-scale context information management and high interoperability between the different components due to the use of a common NGSI protocol. The IDS architecture and connectors permit the use of such a powerful communication tool, making the use of IDS an extension of the AUTOWARE RA through FIWARE support to IDS reference architecture, as described in Section 2.2.

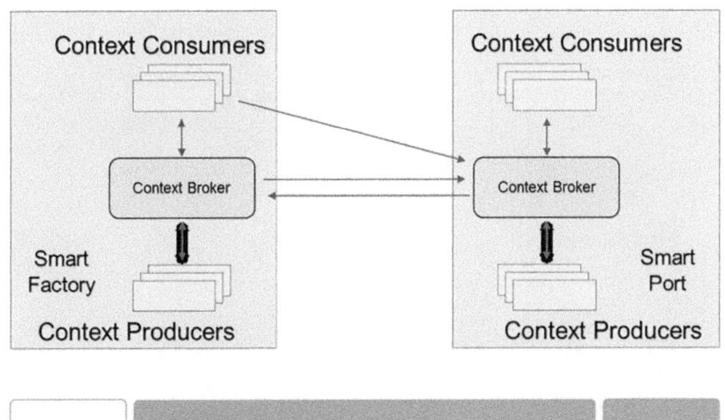

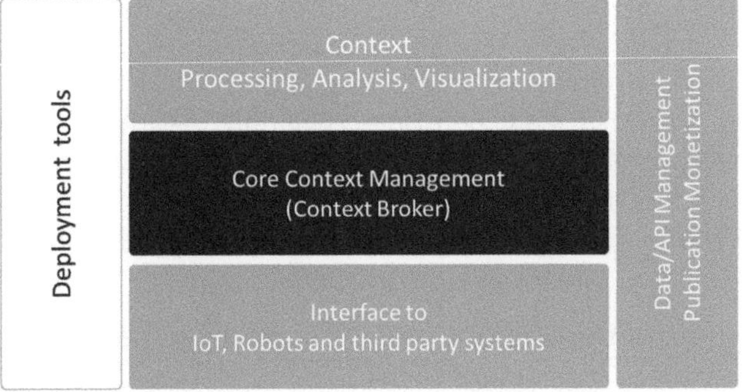

Figure 2.14 Context Broker basic workflow & FIWARE Context Broker Architecture.

- **Backend Device Management – IDAS:** For the translation from IoT-specific protocols to the NGSI context information protocol considered by FIWARE enablers.
- **Cosmos:** For an easier Big Data analysis over context integrated information with most popular Big Data platforms and cloud storage.

AUTOWARE extends a cloud-based architecture to a more flexible and efficient one based on fog computing, which is defined by the OpenFog Consortium as follows: "A horizontal, system-level architecture that distributes computing, storage, control and networking functions closer to the users along a cloud-to-thing continuum". Adding an intermediate layer for data aggregation and computing capabilities at the edge of the network resolves the bottlenecks and disadvantages in complex industrial scenarios: (1) data bottlenecks that occur on the interface between IT and cloud infrastructure; (2) disability to guarantee pre-defined latencies in the communication; (3) sensor data are sent unfiltered to the cloud and (4) limited intelligence on the machine level.

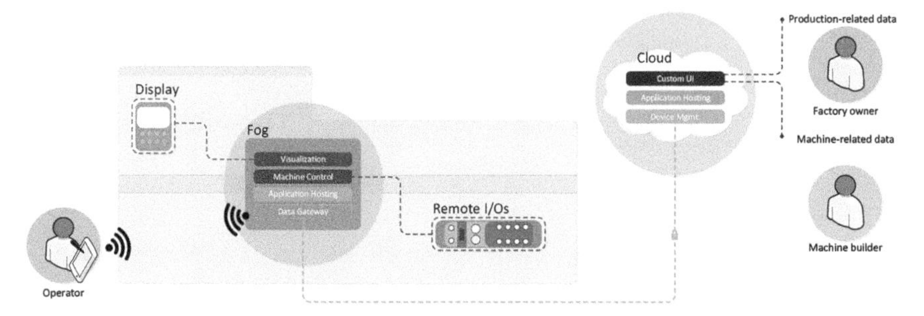

Figure 2.15 Embedding of the fog node into the AUTOWARE software-defined platform as part of the cloud/fog computing & persistence service support.

These drawbacks can be repealed using fog nodes. In addition, strict requirements on timing or even real-time constrains can only be achieved by avoiding long transmission of the data. Thus, the fog computing approach is inherently avoiding the latencies.

Figure 2.15 shows the embedding of the fog node into the AUTOWARE framework. The architecture supports the following aspects:

- **Machine Control Capabilities:** AUTOWARE platform can control the different machines (e.g. robots, machines, etc.) within the plant or the manufacturing cell. It can connect to remote I/Os via an integrated PLC.
- **Device Management Capabilities:** It allows users to perform management of multiple machines in a distributed manner. The device manager is situated in the main office, whereas the devices are distributed over the factories, possible worldwide. The communication between the device manager and the different devices must be implemented over a secure and safe communication channel.
- **Data Gateway:** It enables the communication between other fog nodes, between the fog node and the cloud and with a remote operator.
- **Visualization Capabilities:** The AUTOWARE open platform provides standard interfaces (wired and wireless) to guarantee connectivity via user interfaces to access data via reports, dashboards, etc.
- **Application Hosting Functionality:** It can be located as well in the fog as in the cloud.

The pillars of this architecture, which are common themes of the OpenFog reference architecture, include security, scalability, openness, autonomy, RAS (reliability, availability and serviceability), agility, hierarchy and programmability.

2.3.3 AUTOWARE Framework and RAMI 4.0 Compliance

The overall AUTOWARE Framework and Reference Architecture is also related to the RAMI 4.0, as this is the identified reference architecture for Industry 4.0. The goal of the AUTOWARE project was to keep the developments related to the topics of Industry 4.0 and keep the Reference Architecture and Framework related to the RAMI 4.0 as well as to extend their scope to address the smart service welt data-centric service operations and future autonomous service demands.

To establish this link, the consortium mapped the different concepts and components of the AUTOWARE Framework to the RAMI 4.0 model. In Figure 2.16, the result of such mapping is provided. As it can be observed, the layers of the RAMI 4.0 architecture are well covered by the digital abilities enablers (technologies and service). Moreover, the business value matches with the vision of the business layer of the RAMI 4.0 architecture. On the hierarchical axis, the mapping is provided with the layers of the reference architecture, whereas the lifecycle coverage for type and instance is addressed through the modeling, configuration, programming pillar and the cloud/fog computing and persistence service layers. As discussed in the previous subsection, the data-management services to support at the various layers simulation, learning and knowledge-cognitive capabilities are actually implementing those advanced Industry 4.0 functionalities based on the cloud and edge support. This strict mapping ensures that the AUTOWARE framework not only supports Industry 4.0 scenarios, but also that they can also bring forward more advanced data-driven autonomous operations.

2.4 Autoware Framework for Predictive Maintenance Platform Implementation

In the new Industry 4.0 paradigm, cognitive manufacturing is a fundamental pillar. It transforms manufacturing in three ways:

1. **Intelligent Assets and Equipment:** utilizing interconnected sensors, analytics, and cognitive capabilities to sense, communicate and self-diagnose any type of issues in order to optimize performance and efficiency and reduce unnecessary downtime.
2. **Cognitive Processes and Operations:** analyzing a huge variety of information from workflows, context, process and environment to quality controls, enhance operations and decision-making.

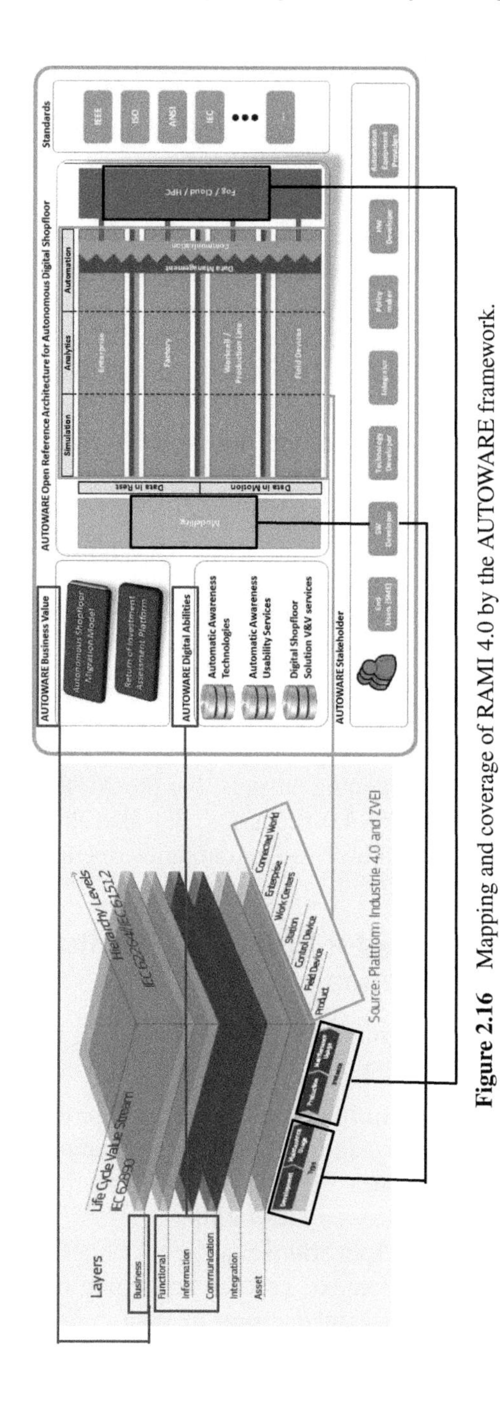

Figure 2.16 Mapping and coverage of RAMI 4.0 by the AUTOWARE framework.

3. **Smarter Resources and Optimization:** combining various forms of data from different individuals, locations, usage and expertise with cognitive insight to optimize and enhance resources such as labor, workforce, and energy, improving in such a way the efficiency of the process.

Predictive maintenance is the prediction of a tool life cycle or other maintenance issues by the use of the information gathered by different sensors and analyzing that information by different types of analytical processes and means. Therefore, predictive maintenance is a clear example of cognitive manufacturing and the focus of the Z-Bre4k project, which employs AUTOWARE Digital Shopfloor reference architecture as its framework for process operation. This section discusses how the AUTOWARE framework can be customized and additional digital abilities and services can be incorporated to implement advanced Industry 4.0 manufacturing processes.

2.4.1 Z-BRE4K: Zero-Unexpected-Breakdowns and Increased Operating Life of Factories

The H2020 project **Z-BRE4K,** https://www.z-bre4k.eu/, looks to implement predictive maintenance strategies to avoid unexpected breakdowns, thus increasing the uptime and overall efficiency of manufacturing scenarios. To this extent, several hardware and software solutions will be implemented in three industrial demonstrators, adapting to the particular needs of each one.

In particular, Z-BRE4K delivers a solution composed of eight scalable strategies at component, machine and system level targeting:

1. **Z-PREDICT.** The prediction occurrence of failure.
2. **Z-DIAGNOSE.** The early detection of current or emerging failure.
3. **Z-PREVENT.** The prevention of failure occurrence, building up or even propagation in the production system.
4. **Z-ESTIMATE.** The estimation of the remaining useful life of assets.
5. **Z-MANAGE.** The management of the strategies through event modeling, KPI monitoring and real-time decision support.
6. **Z-REMEDIATE.** The replacement, reconfiguration, re-use, retirement and recycling of components/assets ().
7. **Z-SYNCHRONISE.** Synchronizing remedy actions, production planning and logistics.
8. **Z-SAFETY.** Preserving the safety, health and comfort of the workers.

The Z-BRE4K solution implementation is expected to have a significant impact, namely (1) increase of the in-service efficiency by 24%, (2) reduced accidents, (3) increased verification according to objectives and (4) 400 new jobs created and over €42M ROI for the consortium.

In order to implement these strategies and reach these impact results, data coming from machine components, industrial lines and shop floors will be fed in the Z-BRE4K platform, which is featured by a communication middleware operative system, a semantic framework module, a dedicated condition monitoring module, a cognitive embedded module, a machine simulator to develop digital twins, an advanced decision support system (DSS), an FMECA module and a predictive maintenance module, together with a cutting-edge vision H/S solution for manufacturing applications associated to advanced HMI.

The General Architecture must be able to support all the components developed under the Z-BRE4K project, which lead to fulfilling the predictive maintenance strategies, being able to keep the information flow constant and well distributed between all the components. At the same time, it must permit an easy implementation in each use case scenario, leading the way towards each particular architecture for each use case and, in the future, different scenarios from other industrial systems. This means that the General Architecture must be highly flexible and easily adapted to new use cases, promoting the predictive maintenance towards its integration in SMEs.

Due to the high flexibility, the architecture requires the main communication middleware operative system to support a high number of different types of data coming from different types of sensors and control software. At the same time, due to the high number of different components, it must also support the need of a continuous communication between all of them, and the interoperability must reach top-notch levels.

2.4.2 Z-Bre4k Architecture Methodology

The Z-Bre4k architecture is designed and developed on the foundations of the AUTOWARE reference architecture and building blocks enabling the convergence of information technology (IT), operational technology (OT), engineering technology (ET) and the leveraging of interoperability of industrial data spaces (IDS), for the support of a factory ecosystem. The objective is to develop a highly adaptive real-time machine (network of components) simulation platform that wraps around the physical equipment for the purpose of predicting uptimes and breakdowns, thus creating intuitive maintenance control and management systems.

The AUTOWARE framework has been selected as open OS for the Z-Bre4k framework for cognitive CPPS service development and strategy implementation. The AUTOWARE open framework is particularly well suited for integration of Z-Bre4k strategies over legacy machines and IT systems with minimum interference and that even SMEs are able to easily integrate advanced predictive maintenance strategies in the very same IT framework used to deal with production optimization or zero defect manufacturing processes.

2.4.3 Z-BRE4K General Architecture Structure

The Z-BRE4K General Architecture will be a combination of the AUTOWARE RA from Figure 2.17 with a vertical separation definition included in the Digital Shopfloor Alliance Reference Architecture and the integration of the IDS General Architecture from Figure 2.6 by using FIWARE Generic Enablers as IDS core components. The main result is shown in Figure 2.18, where the Z-BRE4K Automation, Z-BRE4K Analytics and Z-BRE4K Simulation are presented following the Far-Edge Architecture principles envisioned in the DSA Reference Architecture.

2.4.4 Z-BRE4K General Architecture Information Workflow

Since the predictive maintenance Z-BRE4K is aiming at has been envisioned as a service, the General Architecture will adapt AUTOWARE Service Platform Reference Architecture to the Z-BRE4K structure as shown in Figure 2.19. Figure 2.19 shows the different services divided into the AUTOWARE different blocks and layers, all of them interconnecting through suitable data-buses constructed across information contexts. The main work cell and plant network will be done through the IDS Connector and FIWARE Orion Context Broker principally, but not necessary, so other communication methodologies are also supported, to be able to adapt the architecture to any future use case implementation. The Fog/Cloud interconnection is always available through the fog nodes described in Section 2.3. This will permit the use of storage, HPC and Deep Learning FIWARE Generic Enablers for better computing and calculating processes.

The information captured by the field devices (sensors, machines, etc.) is sent through the Time Sensitive Network (TSN) located in the end users facilities to the Control Services and Perception Services & Model Building components in Real Time.

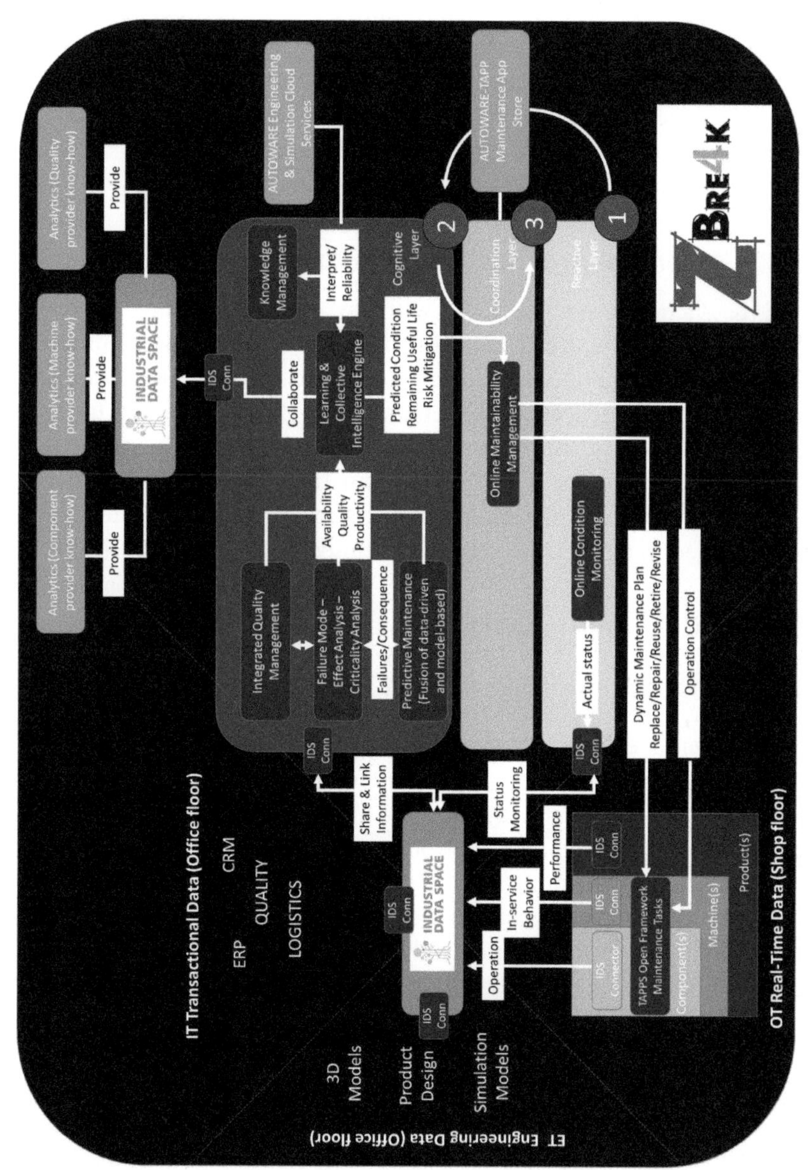

Figure 2.17 Z-Bre4k zero break down workflow & strategies.

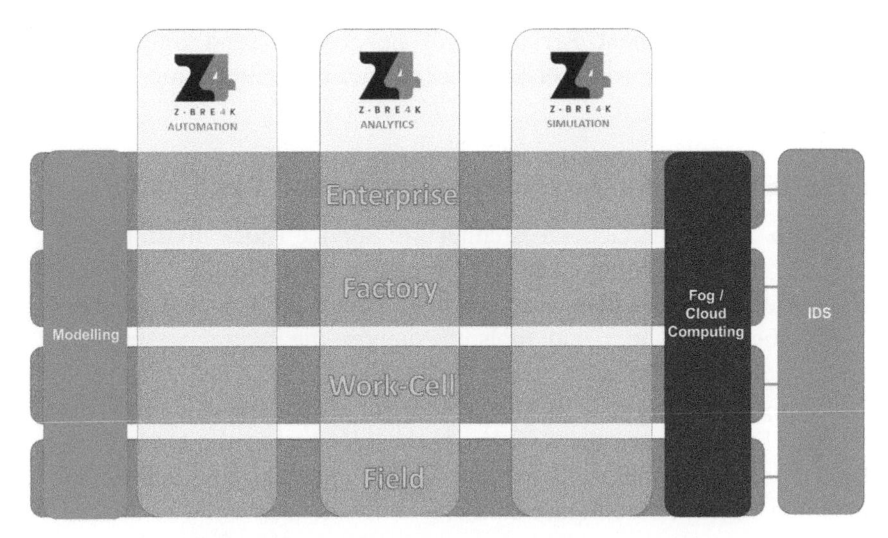

Figure 2.18 Z-BRE4K General Architecture Structure.

The next step is, through the IDS Connectors connected to the Work-cell layer components, the data (normally preprocessed by the Workcell components) is sent to (published) the Orion Context Broker. The different components from the factory layer that are subscribed to each data set will receive it for their analysis and processing. The factory services components, which are divided into Learning, Simulating and Cognitive Computing Services, may require processed data from another factory layer service. The outputs from factory layer components that are required as inputs by other factory layer components will be published once again in the Orion Context Broker in the Workcell. The factory layer components that need those outputs as inputs will be subscribed to that data and will receive it. That is how the communication and information flow will be carried out through the different hierarchical levels.

The Learning, Simulating and Cognitive Computing Services will end up creating valuable information as outputs that will be published in the Plant Network's Orion Context Broker. The different Business Management Services will recollect the information required as inputs for their processing and will elaborate reports, actions, alerts, decision support actions, etc. Dual Reality and Modelling Services will also gather information and will process it to give extra support information for business management decision making and user interfaces by publishing it back in the Plant's Orion Context Broker.

The Business Management Services will be able to send information to the Control Services for user interface issues or optimization actions if necessary.

2.4.5 Z-BRE4K General Architecture Component Distribution

Following the Z-BRE4K General Architecture Service Block division from Figure 2.19 and the component for predictive maintenance, the final Z-BRE4K General OS will be as shown in Figure 2.20, where the specific technologies, services and tools to support the required predictive maintenance digital ability are actually illustrated.

The strength of the AUTOWARE RA to serve the Z-Bre4k predictive maintenance lies that once the data has been published in the Orion Context Broker in any of the scenarios considered, they can consider similar information workflows (see Figure 2.21).

The information in the particular use cases, presented in Figure 2.21, for the predictive maintenance will go as follows: (1) The information is gathered by the field devices, pre-processed if necessary by the control and model building services and published in the Orion Context Broker through each use cases' IDS Connector. (2) The data is collected by subscription by the C-03 Semantic Framework, where it is given the semantic structure and stored in a DB (fog/cloud computing most probable). Then, it is published again in the Context Broker. (3) Data is used to feed the C-08 Machine Simulators. (4) Prediction algorithms (from the C-07 Predictive Maintenance) are run through the C-08 outputs. (5) The C-04 DSS gathers information from the C-07 Predictive Maintenance and analyzes it, giving as an output the failure mode. (6) The C-05 FMECA gets the failure mode from the DSS through the context broker. (7) FMECA returns criticality, risk, redundancy, etc. for the specific failure mode to the DSS through the Context Broker. (8) The DSS, based on the Rules set, provides Recommendations to the Technicians through a common User Interface and control services. (9) The Technicians can use the C-06 VRfx for the better understanding of the information. (10) The Technicians take Actions on the assets through the control services based on the recommendations given. (11) The Technicians provide Feedback on the accuracy of the Recommendations given by the DSS. (12) The DSS improves its Rules and Recommendations based on the Feedback received.

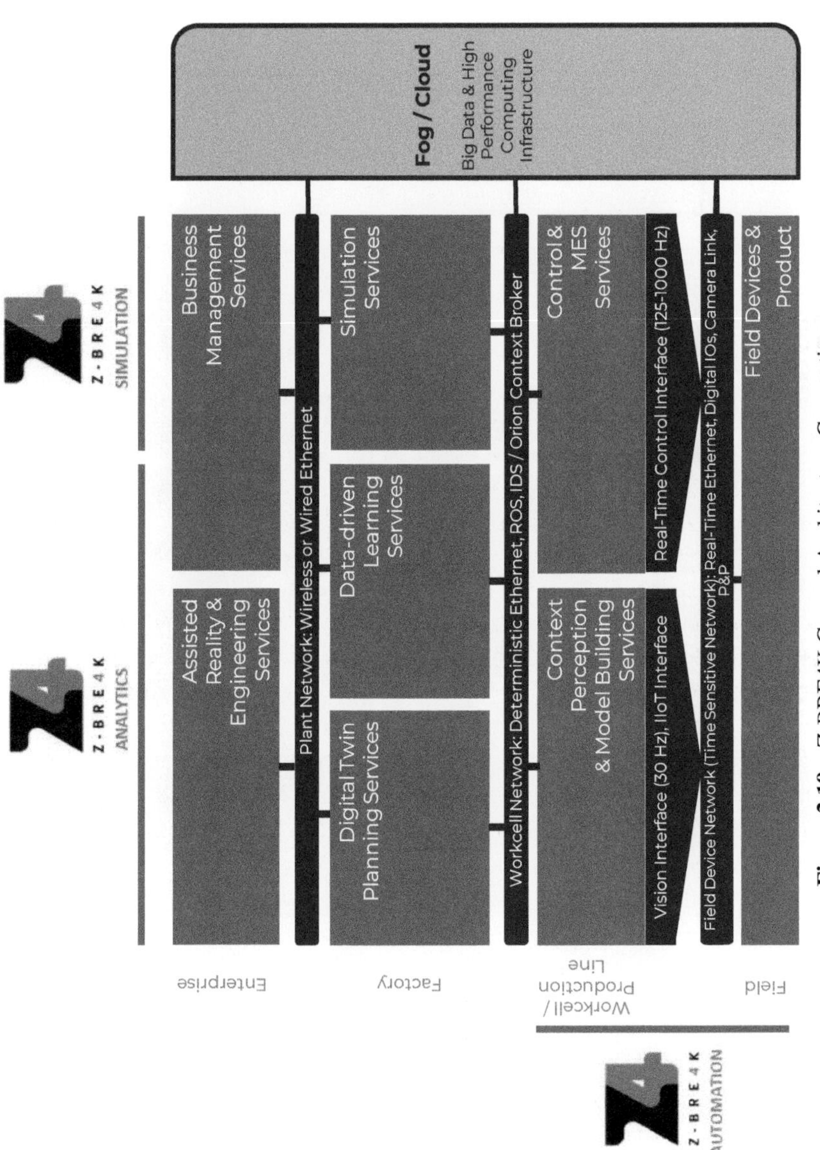

Figure 2.19 Z-BRE4K General Architecture Connections.

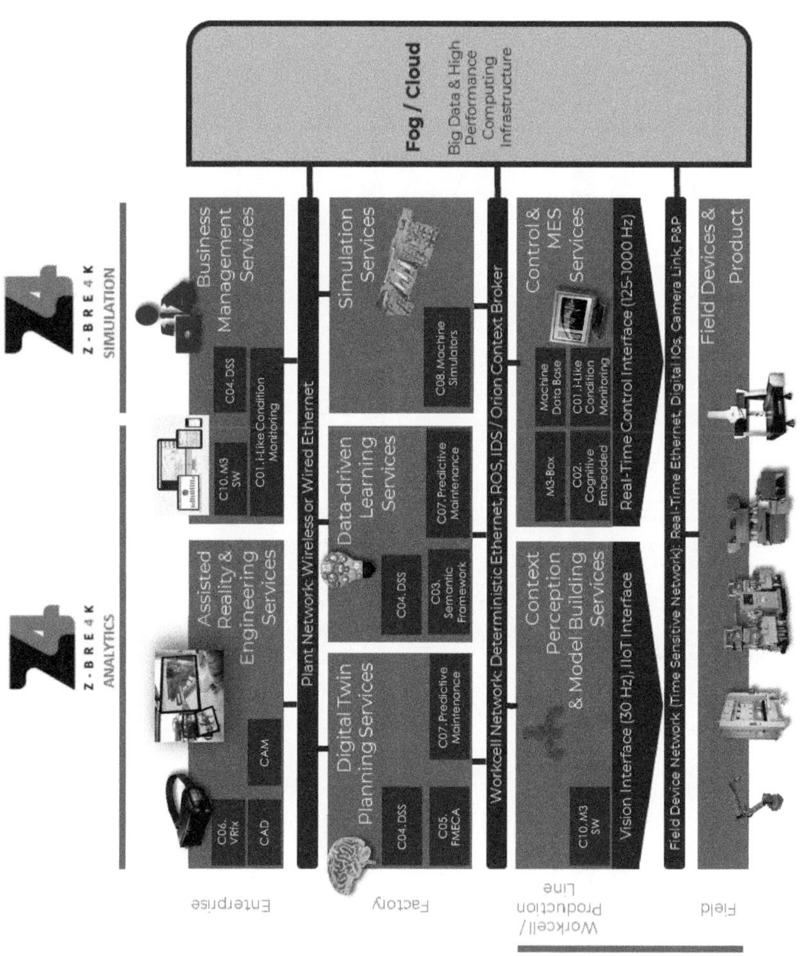

Figure 2.20 Z-BRE4K General OS.

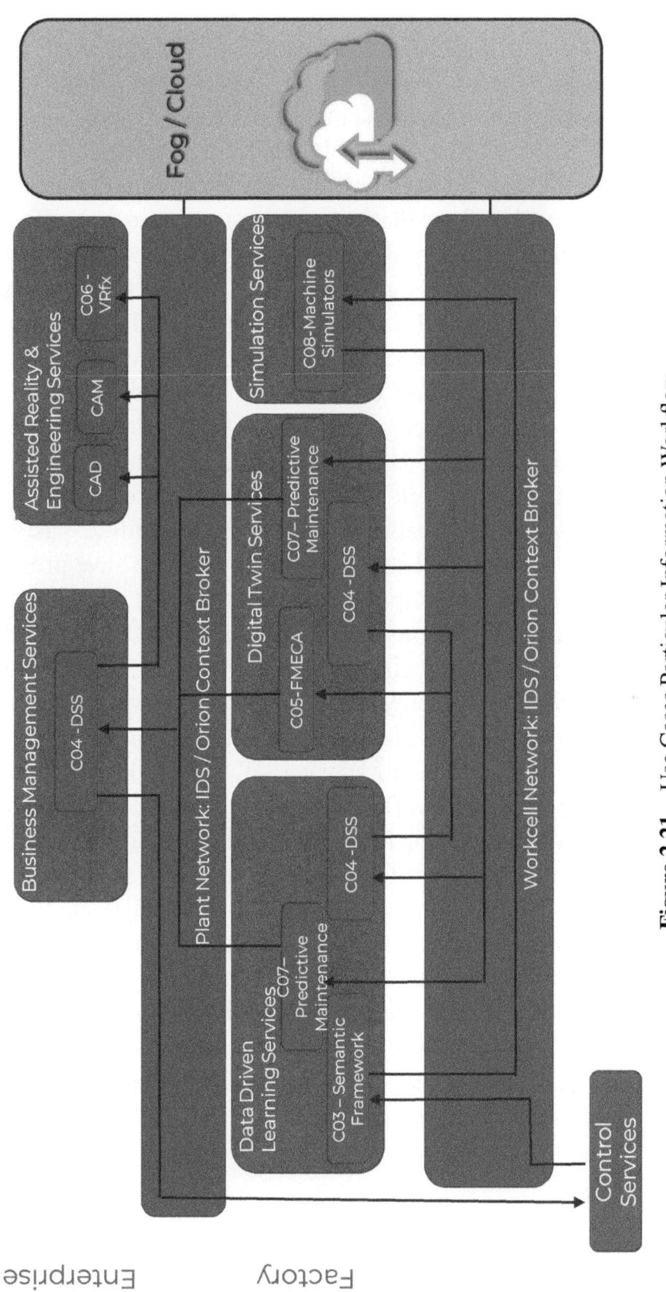

Figure 2.21 Use Cases Particular Information Workflow.

2.5 Conclusions

In this chapter, we have discussed the needs for development of a digital automation framework for the support of autonomous digital manufacturing workflows. We have also presented how various open platforms (i-ROS, OpenFog, IDS, FIWARE, BeinCPPS, MIDIH, ReconCell, Arrowhead, OPC-UA/TSN, 5G) can be harmonized through open APIs to deliver a software-defined digital shopfloor platform enabling a more cost-effective, control and extendable deployment of digital abilities in the shopfloor in close alignment with business strategies and investments available. This chapter has also presented how AUTOWARE is also bringing forward the technology enablers (connectivity, data distribution, edge extension of automation and control equipment for app-ized smart open control hardware (open trusted platforms) operation, deep object recognition), usability services (augmented virutality, CPPS autoconfiguration, robotic programming by training) and verification and validation framework (safety & standard compliant) to the deployment and operation of automatic awareness digital abilities, as a first step in cognitive autonomous digital shopfloor evolution. We have presented how open platforms for fog/edge computing can be combined with cloudified control solutions and open platforms for collaborative robotics, modular manufacturing and reconfigurable cells for delivery of advanced manufacturing capabilities in SMEs. Moreover, we have also presented how the AUTOWARE framework is flexible enough to be adopted and enriched with additional digital capability services to support advanced and collaborative predictive maintenance decision workflows. AUTOWARE is adapted for operation of predictive maintenance strategies in high diversity of machinery (robotic systems, inline quality control equipment, injection molding, stamping press, high-performance smart tooling/dies and fixtures), very challenging and sometimes critical manufacturing processes (highly automated packaging industry, multi-stage zero-defect adaptive manufacturing of structural lightweight component for automotive industry, short-batch mass customized production process for consumer electronics and health sector) and key economic European sectors with the strongest SME presence (automotive, food and beverage, consumer electronics).

Acknowledgments

This work was funded by the European Commission through the FoF-RIA Project AUTOWARE: Wireless Autonomous, Reliable and Resilient Production Operation Architecture for Cognitive Manufacturing

(No. 723909) and through the FoF-IA Project Zbre4k: Strategies and Predictive Maintenance models wrapped around physical systems for Zero-unexpected-Breakdowns and increased operating life of Factories (No. 768869).

References

[1] P. Muller, J. Julius, D. Herr, L. Koch, V. Peycheva, S. McKiernan, Annual Report on European SMEs 2016/2017 European Commission, Directorate-General for Internal Market, Industry, Entrepreneurship and SMEs; Directorate H. https://ec.europa.eu/growth/smes/business-friendly-environment/performance-review_es#annual-report

[2] AUTOWARE http://www.AUTOWARE-eu.org/

[3] OpenFog Consotium; https://www.openfogconsortium.org/#consortium

[4] B. Laibarra "Digital Shopfloor Alliance", EFFRA ConnectedFactories Digital Platforms for Manufacturing Workshop, Session 2 - Integration between projects' platforms – standards & interoperability, Brussels, 5, 6 February 2018 https://cloud.effra.eu/index.php/s/2tlFxI811TOjCOp

[5] Acatech, "Recommendations for the Strategic Initiative Web-based Services for Businesses. Final report of the Smart Service Working Group", 19 August 2015. https://www.acatech.de/Publikation/recommendations-for-the-strategic-initiative-web-based-services-for-businesses-final-report-of-the-smart-service-working-group/

[6] M. Lemke "Digital Industrial Platforms for the Smart Connected Factory of the Future", Manufuture – Tallinn 24 October 2017. http://manufuture2017.eu/wp-content/uploads/2017/10/pdf-Max-Lemke-24.10.pdf

[7] A. Zwegers, Workshop on Digital Manufacturing Platforms for Connected Smart Factories, Brussels, 19 October 2017. https://ec.europa.eu/digital-single-market/en/news/workshop-digital-manufacturing-platforms-connected-smart-factories

[8] EC, "Digitising European Industry: progress so far, 2 years after the launch". March 2018. https://ec.europa.eu/digital-single-market/en/news/digitising-european-industry-2-years-brochure

[9] SIEMENS Mindsphere https://siemens.mindsphere.io/

[10] SAP Leonardo https://www.sap.com/products/leonardo.html

[11] Bosch IoT Suite https://www.bosch-si.com/iot-platform/bosch-iot-suite/homepage-bosch-iot-suite.html

[12] Dassault 3D experience platform https://www.3ds.com/

[13] Manufacturing Industry Digital Innovation Hubs (MIDIH) http://www.midih.eu/
[14] Fiware Smart Industry https://www.fiware.org/community/smart-industry/
[15] Fiware https://www.fiware.org/
[16] Big data for Factories 4.0 http://boost40.eu/
[17] International Data Spaces Association (IDSA) https://www.internationaldataspaces.org/en/
[18] Arrowhead framework http://www.arrowhead.eu/
[19] Smart, Safe & Secure Platform http://www.esterel-technologies.com/S3P-en.html
[20] ISOBUS https://www.aef-online.org/the-aef/isobus.html
[21] AUTOSAR (Automotive Open System Architecture) https://www.autosar.org/
[22] Industrial Interner Consortium (IIC) https://www.iiconsortium.org/
[23] OPC-UA https://opcfoundation.org/
[24] Productive 4.0 https://productive40.eu/
[25] Industrial Value Chain Initiative (IVI) https://iv-i.org/wp/en/
[26] RAMI 4.0 https://www.zvei.org/en/subjects/industry-4-0/the-reference-architectural-model-rami-40-and-the-industrie-40-component/
[27] IIRA https://www.iiconsortium.org/wc-technology.htm
[28] Otto B., Lohmann S., Steinbuss S. IDS Reference Architecture Model Version 2.0, April 2018.
[29] E. Molina, O. Lazaro, et al. "The AUTOWARE Framework and Requirements for the Cognitive Digital Automation", In: Camarinha-Matos L., Afsarmanesh H., Fornasiero R. (eds) Collaboration in a Data-Rich World. PRO-VE 2017. IFIP Advances in Information and Communication Technology, vol. 506. Springer, Cham.

3

Reference Architecture for Factory Automation using Edge Computing and Blockchain Technologies

Mauro Isaja

Engineering Ingegneria Informatica SpA, Italy
E-mail: mauro.isaja@eng.it

This chapter will introduce the reader to the FAR-EDGE Reference Architecture (RA): the conceptual framework that, in the scope of the FAR-EDGE project, was used as the blueprint for the proof-of-concept implementation of a novel *edge computing platform for factory automation*: the FAR-EDGE Platform. Such platform is going to prove edge computing's potential to increase flexibility and lower costs, without compromising on production time and quality. The FAR-EDGE RA exploits best practices and lessons learned in similar contexts by the global community of system architects (e.g., Industrie 4.0, Industrial Internet Consortium) and provides a terse representation of concepts, roles, structure and behaviour of the system under analysis. Its unique approach to edge computing is centered on the use of *distributed ledger technology (DLT)* and *smart contracts* – better known under the collective label of *Blockchain*. The FAR-EDGE project is exploring the use of Blockchain as a key enabling technology for industrial automation, analytics and virtualization, with validation use cases executed in real-world environments that are briefly described at the end of the chapter.

3.1 FAR-EDGE Project Background

FAR-EDGE's main goal is to provide a novel edge computing solution for the virtualization of the factory automation pyramid. The idea of decentralizing factory automation is not new. Rather, for over a decade, several initiatives, including background projects of the consortium partners, have introduced decentralized factory automation solutions based on various technologies like intelligent agents and service-oriented architectures (SOA). These background initiatives produced proof-of-concept implementations that highlighted the benefits of decentralized automation in terms of flexibility; yet they are still not being widely deployed in manufacturing plants. Nevertheless, the vision is still alive, as this virtualization can make production systems more flexible and agile, increase product quality and reduce cost, e.g., enable scalable, fast-configurable production lines to meet the global challenges of *mass-customization* and *reshoring*.

With the advent of the Industrie 4.0 and the Industrial Internet of Things (IIoT), such solutions are revisited in the light of the integration of Cyber-Physical Systems (CPS) within cloud computing infrastructures. Therefore, several cloud-based applications are deployed and used in factories, which leverage the capacity and scalability of the cloud, while fostering supply chain collaboration and virtual manufacturing chains. Early implementations have also revealed the limitations of the cloud in terms of efficient bandwidth usage and its ability to support real-time operations, including operations close to the field. In order to alleviate these limitations, edge computing architectures have recently introduced. Edge computing architectures introduce layers of edge nodes between the field and the cloud, as a means of:

- **Saving bandwidth and storage**, as edge nodes can filter data streams from the field in order to get rid of information that does not provide value for industrial automation.
- **Enabling low-latency and proximity processing**, since information can be processed close to the field, rather in a remote (back-end) cloud infrastructure.
- **Providing enhanced scalability**, given that edge computing supports decentralized storage and processing that scale better when compared to conventional centralized cloud processing. This is especially the case when interfacing to numerous devices is needed.
- **Supporting shopfloor isolation and privacy-friendliness**, since edge nodes deployed at the shopfloor can be isolated from the rest of the edge network. This can provide increased security and protection of manufacturing dataset in cases required.

These benefits make edge computing suitable for specific classes of use cases in factories, including:

- **Large-scale distributed applications**, typically applications that involve multiple plants or factories, which collect and process streams from numerous distributed devices in large scale.
- **Nearly real-time applications**, which need to analyze data close to the field or even control CPS such as smart machines and industrial robots. A special class of such real-time applications involves edge analytics applications.

As a result, the application of edge computing for factory automation is extremely promising, since it can support decentralized factory automation in a way that supports real-time interactions and analytics in large scale. FAR-EDGE researches have explored the application of the edge computing paradigm in factory automation, through designing and implementing reference implementations in line with recent standards for edge computing in industrial automation applications. Note that FAR-EDGE was one of the first initiatives to research and experiment with edge computing in the manufacturing shopfloor, as relevant activities were in their infancy when FAR-EDGE project was approved. However, the state of the art in factory automation based on edge computing has evolved and FAR-EDGE efforts are taking into account this evolution.

3.2 FAR-EDGE Vision and Positioning

FAR-EDGE's vision is to research and provide a proof-of-concept implementation of an *edge computing platform for factory automation*, which will prove edge computing's potential to increase automation flexibility and lower automation costs, without however compromising production time and quality. The FAR-EDGE architecture is aligned to the IIC RA, while exploiting concepts from other RAs and standards such as the OpenFog RA and RAMI 4.0 (see below for more details). Hence, the project will be providing one of the world's first reference implementation of edge computing for factory automation. Within this scope, FAR-EDGE will offer a host of functionalities that are not addressed by other implementations, such as IEC-61499 compliant automation and simulation.

Beyond its functional uniqueness, FAR-EDGE is also unique from a research perspective. In particular, the project is researching the applicability of disruptive KETs: *distributed ledger technology (DLT)* and *smart contracts* – better known under the collective label of *Blockchain*. The Blockchain concept, while being well understood and thoroughly tested in mission-critical areas like digital currencies (e.g., Bitcoin, Ethereum), has never been applied before to industrial systems. FAR-EDGE aims at demonstrating how a pool of services built on a generic Blockchain platform can enable decentralized factory automation in an effective, reliable, scalable and secure way. In FAR-EDGE, such services are responsible for sharing process state and enforcing business rules across the computing nodes of a distributed system, thus permitting virtual automation and analytics processes that span multiple nodes – or, from a bottom-up perspective, autonomous nodes that cooperate to a common goal.

3.3 State of the Art in Reference Architectures

A *reference architecture* (RA) is often a synthesis of best practices having their roots in past experience. Sometimes it may represent a "vision", i.e., a conceptual framework that aims more at shaping the future and improving over state-of-the-art design rather than at building systems faster and with lower risk. The most successful RAs – those that are known and used beyond the boundaries of their native ground – are those combining both approaches. Whatever the strategy, an RA is for teamwork: its major contribution to development is to set a common context, vocabulary and repository of patterns for all stakeholders.

In FAR-EDGE, where we explore the business value of applying innovative computing patterns to the smart factory, starting from an effective RA is of paramount importance. For this reason, the FAR-EDGE Reference Architecture was the very first outcome of the project's platform development effort.

In our research, we considered some well-known and accepted *generic RAs* (see sub-section below) as sources of inspiration. The goal was twofold: on the one hand, to leverage valuable experience from large and respected communities; on the other hand, to be consistent and compatible with the mainstream evolution of the smart factory, e.g., Industrial IoT and Industry 4.0. At the end of this journey, we expect the FAR-EDGE RA to become an asset not only in the scope of the project (as the basis for the FAR-EDGE Platform's design), but also in the much wider one of factory automation, where it may guide the design of ad-hoc solutions having edge computing as their main technology driver.

3.3.1 Generic Reference Architectures

A generic RA is one that, while addressing a given field of technology, is not targeting any specific application, domain, industry or even (in one case) sector. Its value is mainly in communication: lowering the impedance of information flow within the development team and possibly also towards the general public. As such, it is basically an ontology and/or a mind mapping tool. However, as we will see further on in this analysis, sometimes the ambition of a generic RA is also to set a standard for runtime interoperability of systems and components, placing some constraints on implementation choices. Obviously, for this approach to make sense, it should be backed by a critical mass of solution providers, all willing to give up the vendor-lock-in competitive factor in exchange for the access to a wider market.

We have identified three generic RAs that have enough traction to influence the "technical DNA" of the FAR-EDGE Platform: RAMI 4.0, IIRA and OpenFog RA. In the following sub-sections, each of them is briefly analysed and, when it is the case, some elements that are relevant to FAR-EDGE are extracted to be reused later on.

3.3.2 RAMI 4.0

The Reference Architectural Model for Industrie 4.0 (RAMI 4.0)[1] is a generic RA addressing the manufacturing sector. As its name clearly states, it is the outcome of Platform Industrie 4.0,[2] the German public–private initiative addressing the fourth industrial revolution, i.e., merging the digital, physical and biological worlds into CPS.

According to some experts [1], the expected benefits of the adoption of CPS in the factory are:

- higher quality
- more flexibility
- higher productivity
- standardization in development
- products can be launched earlier
- continuous benchmarking and improvement
- global competition among strong businesses
- new labour market opportunities
- creation of appealing jobs at the intersection of mechanical engineering, automation and IT
- new services and business model

[1]https://www.zvei.org/en/subjects/industry-4-0/

[2]http://www.plattform-i40.de/I40/Navigation/EN/Home/home.html

To ensure that all participants involved in discussions understand each other, RAMI 4.0 defines a 3D structure for mapping the elements of production systems in a standard way.

RAMI 4.0, however, is also a standard-setting effort. While still a work in (slow) progress at the time of writing [2], its roadmap includes the definition of a globally standardized communication architecture that should enable the plug-and-play of *Things* (e.g., field devices, connected factory tools and equipment, smart machines, etc.) into composite CPS. Currently, only the general concept of *I4.0 Component* has been introduced: any Thing that is wrapped inside an *Administration Shell*, which provides a standard interface for communication, control and management while hiding the internals of the actual physical object. Future work will identify standard languages for the exchange of information, define standard data and process models and include recommendations for implementation – communication protocols in the first place.

With respect to the latter point, OPC UA is central to the RAMI 4.0 strategy. It is the successor of the much popular (in Microsoft-based shopfloors) OPC machine-to-machine communication protocol for industrial automation. As opposed to OPC, OPC UA is an open, royalty-free cross-platform and supports very complex information models. I4.0 Components will be required to adopt OPC UA as their interfacing mechanism, while also relying on several IEC standards (e.g., 62832, 61804, etc.) for information sharing.

RAMI 4.0 has gained a significant traction in Germany and is also driving the discussion around Industry 4.0 solutions and platforms in Europe. In particular, its glossary and its 3D structure for element mapping are increasingly used in sector-specific projects (in particular, platform-building ones) and working groups as a common language. The FAR-EDGE RA will adopt some of the RAMI 4.0 conceptual framework as its own, simplifying communication with the external communities of developers and users.

3.3.3 IIRA

The Industrial Internet Reference Architecture (IIRA)[3] has been developed and is actively maintained by the Industrial Internet Consortium (IIC), a global community of organizations (>250 members, including IBM, Intel, Cisco, Samsung, Huawei, Microsoft, Oracle, SAP, Boeing, Siemens, Bosch and General Electric) committed to the wider and better adoption of the

[3]http://www.iiconsortium.org/IIRA.htm

Internet of Things by the industry at large. The IIRA, first published in 2015 and since evolved into version 1.8 (Jan 2017), is a standards-based architectural template and methodology for the design of Industrial Internet Systems (IIS). Being an RA, it provides an ontology of IIS and some architectural patterns, encouraging the reuse of common building blocks and promoting interoperability. It is worth noting that a collaboration between the IIC and Platform Industrie 4.0, with the purpose of harmonizing RAMI 4.0 and IIRA, has been announced.[4]

IIRA has four separate but interrelated *viewpoints*, defined by identifying the relevant stakeholders of IIoT use cases and determining the proper framing of concerns. These viewpoints are: business, usage, functional and implementation.

- The *business viewpoint* attends to the concerns of the identification of stakeholders and their business vision, values and objectives. These concerns are of particular interest to decision-makers, product managers and system engineers.
- The *usage viewpoint* addresses the concerns of expected system usage. It is typically represented as sequences of activities involving human or logical users that deliver its intended functionality in ultimately achieving its fundamental system capabilities.
- The *functional viewpoint* focuses on the functional components in a system, their interrelation and structure, the interfaces and interactions between them and the relation and interactions of the system with external elements in the environment.
- The *implementation viewpoint* deals with the technologies needed to implement functional components, their communication schemes and their life cycle procedures.

In FAR-EDGE, which deals with platforms rather than solutions, the functional and implementation viewpoints are the most useful.

The functional viewpoint decomposes an IIS into functional domains, which are, following a bottom-up order, *control*, *operations*, *information*, *application* and *business*. Of particular interest in FAR-EDGE are the first three.

The *control domain* represents functions that are performed by industrial control systems: reading data from sensors, applying rules and logic and exercising control over the physical system through actuators. Both accuracy and

[4]http://www.iiconsortium.org/iic-and-i40.htm – to date, no concrete outcomes of such collaboration have been published.

resolution in timing are critical. Components implementing these functions are usually deployed in proximity to the physical systems they control, and may therefore be distributed.

The *operations domain* represents the functions for the provisioning, management, monitoring and optimization of the systems in the control domain.

The *information domain* represents the functions for gathering and analysing data to acquire high-level intelligence about the overall system. As opposed to their control domain counterparts, components implementing these functions have no timing constraints and are typically deployed in factory or corporate data centres, or even in the cloud as a service.

Overall, the functional viewpoint tells us that control, management and data flow in IIS are three separate concerns having very different non-functional requirements, so that implementation choices may also differ substantially.

The implementation viewpoint describes some well-established architectural patterns for IIS: the Three-tier, the Gateway-mediated Edge Connectivity and Management and the Layered Databus. They are of particular interest in FAR-EDGE, as they all deal with edge computing, although in different ways.

The *Three-tier architectural pattern* distributes concerns to separate but connected tiers: Edge, Platform and Enterprise. Each of them play a specific role with respect to control and data flows. Consistently with the requirements stemming from the functional viewpoint, control functionality is positioned in the Edge Tier, i.e., in close proximity to the controlled systems, while data-related (information) and management (operations) services are part of the Platform. However, the IIRA document v1.8 also states that in real systems, some functions of the information domain may be implemented in or close to the edge tier, along with some application logic and rules to enable *intelligent edge computing*. Interestingly enough, though, the opposite – edge computing as part of Platform functionality – is not contemplated by IIRA, probably because intelligent edge nodes (i.e., connected factory equipment with on-board computing capabilities) are deemed to be an OEM's (Original Equipment Manufacturer) concern. However, there is a component in the IIRA diagram suggesting that such boundaries may be blurred: the Gateway, which is part of the Edge Tier, connects it to both the Platform and Enterprise ones.

The Edge Gateway (EG) is in fact the focus point of another IIRA architectural pattern: the *Gateway-mediated Edge Connectivity and Management*. It allows for localizing operations and controls (edge analytics and computing). Its main benefit is in breaking down the complexity of the IIS, so that it may scale up in both numbers of managed assets and networking. The EG acts as an endpoint for the wide-area network while isolating the individual local networks of edge nodes. It may be used as a management point for devices and as an aggregation hub where some data processing and control logic is deployed.

The implementation viewpoint indeed provides some very relevant building blocks for the FAR-EDGE platform. What we see as a gap in the IIRA approach, up to this point, is the lack of such a block for addressing *distributed computing,* which is implied in the very notion of edge computing when used as a load-distribution technique for systems that are still centralized in their upper tiers. A partial answer to this question is given by the third and last IIRA architectural pattern: the *Layered Databus*. According to this design, an IIS can be partitioned into multiple horizontal layers that together define a hierarchy of scopes: machine, system, system of systems and Internet. Within each layer, components communicate with each other in a *peer-to-peer* (P2P) fashion, supported by a layer-specific databus. A databus is a logical connected space that implements a common data model, allowing interoperable communications between endpoints at that layer. For instance, a databus can be deployed within a smart machine to connect its internal sensors, actuators, controls and analytics. At the system level, another databus can be used for communications between different machines. At the system of systems level, still another databus can connect together a series of systems for coordinated control, monitoring and analysis.

In FAR-EDGE, the concept of cross-node P2P communication is going to play a key role as the enabling technology for edge computing in the three functional domains of interest: control, operations and information.

3.3.4 OpenFog RA

The OpenFog Consortium[5] is a public–private initiative, which was born in 2015 and shares similarities to the IIC: both consortia share big players like IBM, Microsoft, Intel and Cisco as their founding members and both use the ISO/IEC/IEEE 42010:2011 international standard[6] for communicating architecture descriptions to stakeholders. However, the OpenFog initiative is not constrained to any specific sector: it is a technology-oriented ecosystem that fosters the adoption of *fog computing* in order to solve the bandwidth, latency

[5]https://www.openfogconsortium.org/
[6]https://www.iso.org/standard/50508.html

and communications challenges of IoT, AI, robotics and other advanced concepts in the digitized world. Fog computing is a term first introduced by Cisco, and is basically a synonym for edge computing[7]: both refer to the practice of moving computing and/or storage services towards the edge nodes of a networked system.

The OpenFog RA was first released at the beginning of 2017, and as such it is the most recent contribution to the mainstream world of IoT-related architectures. The technical paper that describes it[8] is quite rich in content. As in IIRA, *viewpoints* are used to frame similar concerns, which in OpenFog RA are restricted to *functional* and *deployment* (the latter being roughly equivalent of IIRA's *implementation* viewpoint). However, these topics are not discussed in much detail. In particular, the functional viewpoint is nothing more than a placeholder, for example, use cases (one of them provided as an annex to the document), while the deployment viewpoint just skims the surface, introducing the concept of multi-tier systems. With respect to this, however, a very interesting example is made, which shows how the OpenFog approach to deployment is close to IIRA's Layered Databus pattern: it is a hierarchy of layers where nodes on the same level can interact with each other – in what is called "east–west communication" – without the mediation of higher-level entities. The layers themselves, although more relevant to a smart city context, are quite consistent with the IIRA ones. The means by which P2P communication should be implemented are not specified (no databus, in this case).

Besides viewpoints, two additional kinds of frames are used to organize concepts: *views* and *perspectives*. The former include aspects (i.e., *node*, *system* and *software*) that have a clear positioning in the structure of a system, and are further articulated into sub-aspects (e.g., the node view includes security, management, network, accelerators, compute, storage, protocol abstraction and sensors/actuators); the latter are crosscutting concerns (e.g., performance, security, etc.).

Overall, the OpenFog RA gives the impression of being an ambitious exercise, having the main goal of creating a universal conceptual framework that is at the same time generic, comprehensive and detailed. The mapping of a large scale, complex and critical use case (airport visual security), as provided in the document, is impressive, but this comes as no surprise because that was obviously the case study on which the RA itself was fine-tuned. The reverse path – designing a new system using OpenFog RA as the blueprint –

[7]The term conveys the concept of cloud computing moved at the ground level

[8]https://www.openfogconsortium.org/ra/

appears to be a daunting task, in particular in industrial scenarios where a very pragmatic approach is the norm. In FAR-EDGE, the value that we see in OpenFog RA is – again, as it was also introduced in IIRA – the concept of a hierarchy of geo-scoped layers that use P2P communication internally.

3.4 FAR-EDGE Reference Architecture

The FAR-EDGE Reference Architecture is the conceptual framework that has driven the design and the implementation of the FAR-EDGE Platform. As an RA, its first goal is communication: providing a terse representation of concepts, roles, structure and behaviour of the system under analysis both internally for the benefit of team members and externally for the sake of dissemination and ecosystem-building. There is a second goal, too, which is reuse: exploiting best practices and lessons learned in similar contexts by the global community of system architects.

The FAR-EDGE RA is described from two architectural viewpoints: the *functional viewpoint* and the *structural viewpoint*. In the sections that follow, they are described in detail. A partial *implementation viewpoint* is also provided further on, with its scope limited to the Ledger Tier. Figure 3.1 provides an overall architecture representation that includes all elements.

3.4.1 Functional Viewpoint

According to the FAR-EDGE RA, the functionality of a factory automation platform can be decomposed into three high-level *Functional Domains* – *Automation*, *Analytics* and *Simulation* – and four *Crosscutting*

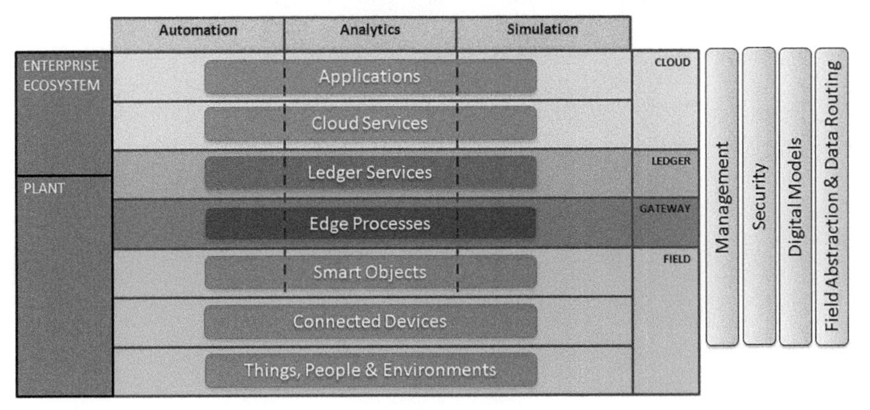

Figure 3.1 FAR-EDGE RA overall view.

(XC) Functions – Management, *Security*, *Digital Models* and *Field Abstraction & Data Routing*. To better clarify the scope of such topics, we have tried to map them to similar IIRA concepts. However, the reader should be aware that the overall scope of the IIRA is wider, as it aims at modelling entire Industrial Internet Systems, while the FAR-EDGE RA is more focused and detailed: oftentimes, concept mapping is partial or even impossible.

Functional Domains and XC Functions are orthogonal to structural Tiers (see next section): the implementation of a given functionality may – but is not required to – span multiple Tiers, so that in the overall architecture representation (Figure 3.1), Functional Domains appear as vertical lanes drawn across horizontal layers. In Figure 3.2 the relationship between Functional Domains, their users and the factory environment is highlighted by arrows showing the flow of data and of control.

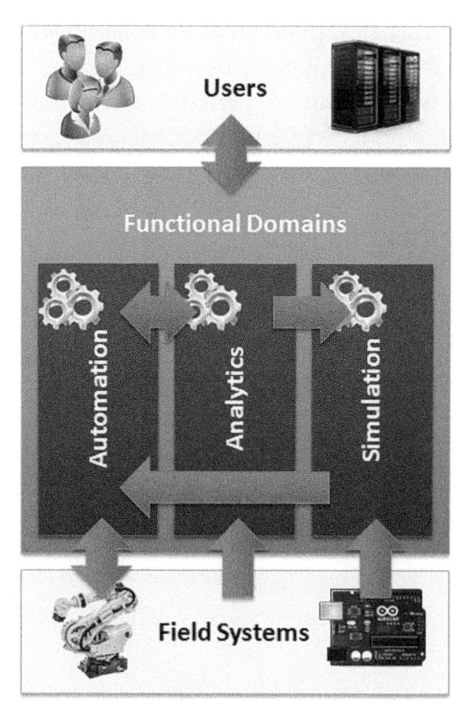

Figure 3.2 FAR-EDGE RA Functional Domains.

3.4.1.1 Automation domain

The FAR-EDGE Automation domain includes functionalities supporting automated control and automated configuration of physical production processes. While the meaning of "control" in this context is straightforward, "configuration" is worth a few additional words. Automated configuration is the enabler of plug-and-play factory equipment – better known as *plug-and-produce* – which in turn is a key technology for mass-customization, as it allows a faster and less expensive adjustment of the production process to cope with a very dynamic market demand. The Automation domain requires a bidirectional monitoring/control communication channel with the Field, typically with low bandwidth but very strict timing requirements (tight control loop). In some advanced scenarios, Automation is controlled – to some extent – by the results of Analytics and/or Simulation (see below for more details on this topic).

The Automation domain partially maps to the Control domain of the IIRA. The main difference is that Control is also responsible for decoupling the real word from the digital world, as it includes the functionality for Field communication, entity abstraction, modelling and asset management. In other words, Control mediates all Field access from other domains like Information, Operations, etc. In the FAR-EDGE RA, instead, the Automation domain is only focused on its main role, while auxiliary concerns are dealt with by Data Models and by Field Abstraction & Data Routing, which are XC Functions.

3.4.1.2 Analytics domain

The FAR-EDGE Analytics domain includes functionalities for gathering and processing Field data for a better understanding of production processes, i.e., a factory-focused business intelligence. This typically requires a high-bandwidth Field communication channel, as the volume of information that needs to be transferred in a given time unit may be substantial. On the other hand, channel latency tends to be less critical than in the Automation scenario. The Analytics domain provides intelligence to its users, but these are not necessarily limited to humans or vertical applications (e.g., a predictive maintenance solution): the Automation and Simulation domains, if properly configured, can both make direct use of the outcome of data analysis algorithms. In the case of Automation, the behaviour of a workflow might change in response to changes detected in the controlled process, e.g., a process drift caused by the progressive wear of machinery or by the quality of assembly components being lower than usual. In the case of Simulation, data analysis can be used to update the parameters of a digital model (see the following section).

The Analytics domain matches perfectly the Information domain of the IIRA, except that the latter is receiving data from the Field through the mediation of Control functionalities.

3.4.1.3 Simulation domain

The FAR-EDGE Simulation domain includes functionalities for simulating the behaviour of physical production processes for the purpose of optimization or of testing what/if scenarios at minimal cost and risk and without any impact of regular shop activities. Simulation requires digital models of plants and processes to be in-sync with the real-world objects they represent. As the real world is subject to change, models should reflect those changes. For instance, the model of a machine assumes a given value of electric power/energy consumption, but the actual values will diverge as the real machine wears down. To detect this gap and correct the model accordingly, raw data from the Field (direct) or complex analysis algorithms (from Analytics) can be used. However, it is important to point out that model synchronization functionality is *not* part of the Simulation domain, which acts just as a consumer of the Digital Models XC Functions.

There is no mapping between the Simulation domain and any functional domain of the IIRA: in the latter, simulation support is not considered as an integral part of the infrastructure.

3.4.1.4 Crosscutting functions

Crosscutting Functions address, as the name suggests, common specific concerns. Their implementation tends to be pervasive, affecting several Functional Domains and Tiers. They are briefly listed and described here.

- **Management:** Low-level functions for monitoring and commissioning/decommissioning of individual system modules, i.e., factory equipment and IT components that expose a management interface. They partially correspond to IIRA's Operations functional domain, with the exclusion of its more high-level functions like diagnostics, prognostics and optimization.
- **Security:** Functions securing the system against the unruly behaviour of its user and of connected systems. These include digital identity

management and authentication, access control policy management and enforcement, communication and data encryption. They partially correspond to the Trustworthiness subset of System Characteristics from IIRA.

- **Digital Models:** Functions for the management of digital models and their synchronization with the real-world entities they represent. Digital modes are a shared asset, as they may be used as the basis for automated configuration, simulation and field abstraction, e.g., semantic interoperability of heterogeneous field systems. They correspond to the Modeling and Asset Management layers of IIRA's Control functional domain.
- **Field Abstraction & Data Routing:** Functions that ensure the connectivity of business logic (FAR-EDGE RA Functional Domains) to the Field, abstracting away the technical details, like device discovery and communication protocols. Data routing refers to the capability of establishing direct producer–consumer channels on demand, optimized for unidirectional massive data streaming, e.g., for feeding Analytics. They correspond to the Communication and Entity Abstraction layers of IIRA's Control functional domain.

3.4.2 Structural Viewpoint

The FAR-EDGE RA uses two classes of concepts for describing the structure of a system: *Scopes* and *Tiers*.

Scopes are very simple and straightforward: they define a coarse mapping of system elements to either the factory – *Plant Scope* – or the broader world of corporate IT – *Enterprise Ecosystem Scope*. Examples of elements in Plant Scope are machinery, field devices, workstations, SCADA and MES systems, and any software running in the factory data centre. To the Enterprise Ecosystem Scope belong ERP and PLM systems and any application or service shared across multiple factories or even companies, e.g., supply chain members.

Tiers are a more detailed and technically oriented classification of deployment concerns: they can be easily mapped to scopes, but they provide more insight into the relationship between system components. Not surprisingly, FAR-EDGE being inspired by edge and distributed computing paradigms, this kind of classification is quite similar to the OpenFog RA's deployment viewpoint, except for the fact that FAR-EDGE Tiers are industry-oriented whereas OpenFog ones are not. That said, FAR-EDGE Tiers are one of the most innovative traits of its RA, and they are individually described here.

3.4.2.1 Field Tier

The Field Tier (see Figure 3.3) is the bottom layer of the FAR-EDGE RA and is populated by *Edge Nodes (EN)*: any kind of device that is connected to the *digital world* on one side and to the *real world* to the other. ENs can have embedded intelligence (e.g., a smart machine) or not (e.g., an IoT sensor or actuator); the FAR-EDGE RA honours this difference: *Smart Objects* are ENs with on-board computing capabilities, and *Connected Devices* are those without. The Smart Object is where local control logic runs: it is a semi-autonomous entity that does not need to interact too frequently with the upper layers of the system.

The Field is also populated by entities of the real world, i.e., those physical elements of production processes that are not directly connected to the network, and as such are not considered as ENs: *Things*, *People* and *Environments*. These are represented in the digital world by some kind of EN "wrapper". For instance, room temperature (Environment) is measured by an IoT sensor (Connected Device), the proximity of a worker (People) to a physical checkpoint location is published by an RFID wearable and detected by an RFID Gate (Connected Device) and a conveyor belt (Thing) is operated by a PLC (Smart Object).

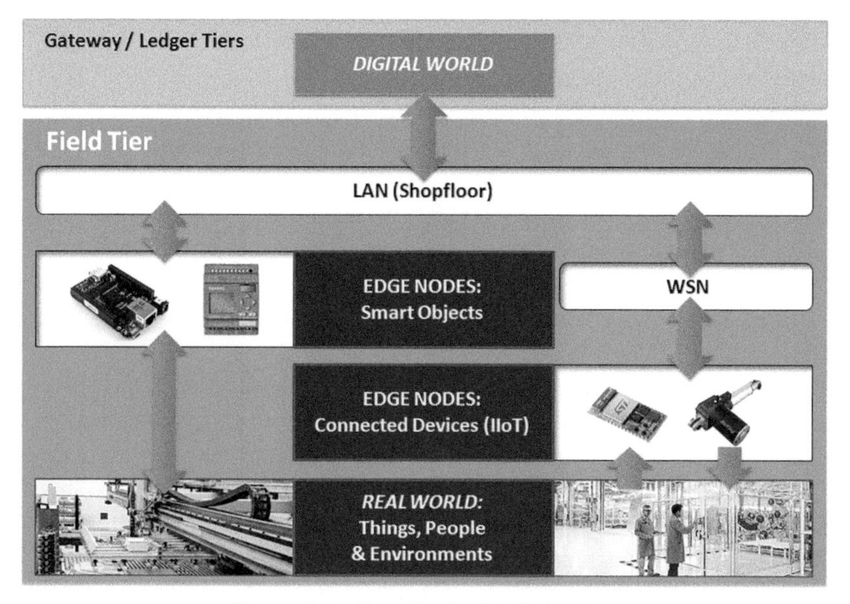

Figure 3.3 FAR-EDGE RA Field Tier.

The Field Tier is in Plant Scope. Individual ENs are connected to the digital world in the upper Tiers either directly by means of the shopfloor's LAN, or indirectly through some special-purpose local network (e.g., WSN) that is bridged to the former.

From the RAMI 4.0 perspective, the FAR-EDGE Field Tier corresponds to the **Field Device** and **Control Device** levels on the **Hierarchy** axis (IEC-62264/IEC-61512), while the entities there contained are positioned across the **Asset** and **Integration Layers**.

3.4.2.2 Gateway Tier

The Gateway Tier (see Figure 3.4) is the core of the FAR-EDGE RA. It hosts those parts of Functional Domains and XC Functions that can leverage the edge computing model, i.e., software designed to run on multiple, distributed computing nodes placed close to the field, which may include resource-constrained nodes. The Gateway Tier is populated by *Edge Gateways (EG)*: computing devices that act as a digital world gateway to the real world of the Field. These machines are typically more powerful than the average intelligent EN (e.g., blade servers) and are connected to a fast LAN. Strategically positioned close to physical systems, the EG can execute *Edge*

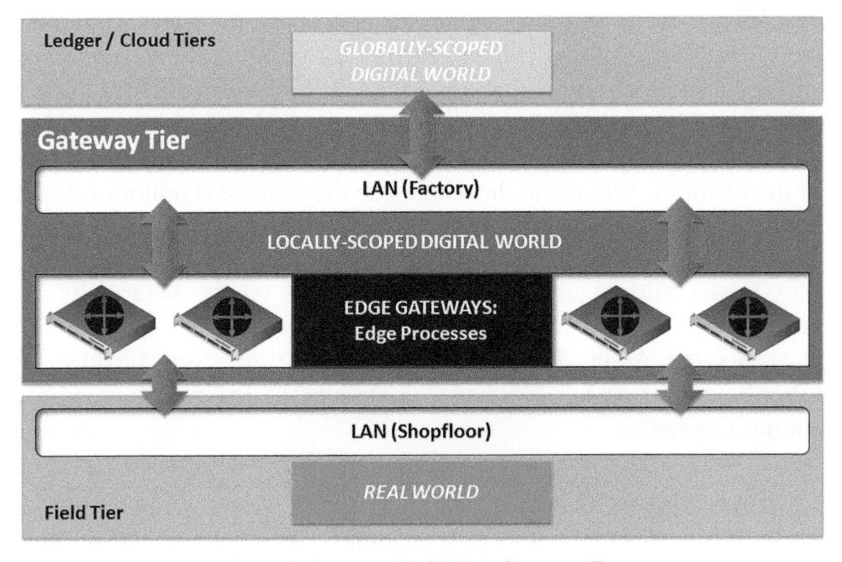

Figure 3.4 FAR-EDGE RA Gateway Tier.

Processes: time- and bandwidth-critical functionality having *local scope*. For instance, the orchestration of a complex physical process that is monitored and operated by a number of sensors, actuators (Connected Devices) and embedded controllers (Smart Objects); or the real-time analysis of a huge volume of live data that is streamed from a nearby Field source.

By itself, the Gateway Tier does not introduce anything new: deploying computing power and data storage in close proximity to where it is actually used is a standard best practice in the industry, which helps reduce network latency and traffic. However, this technique basically requires that the scope of individual subsystems is narrow (e.g., a single work station). If instead the critical functionality applies to a wider scenario (e.g., an entire plant or enterprise), it must be either deployed at a higher level (e.g., the Cloud) – thus losing all benefits of proximity – or run as multiple parallel instances, each focused on its own narrow scope. In the latter case, new problems may arise: keeping global variables in-sync across all local instances of a given process, reaching a consensus among local instances on a "common truth", collecting aggregated results from independent copies of a data analytics algorithm, etc. These problems are well known: the need for peer nodes of a distributed system to mutually exchange information is recognized by the OpenFog RA. The innovative approach in FAR-EDGE is to define a specific system layer – the Ledger Tier – that is responsible for the implementation of such mechanisms and to guarantee an appropriate Quality of Service level.

The Gateway Tier is in Plant Scope, located above the Field Tier and below the Cloud Tier – in this context, we do not consider the Ledge Tier as part of the north-south continuum, due to its very specific role of support layer. Individual EGs are connected with each other and with the north side of the system – i.e., the globally scoped digital world in the Cloud Tier – by means of the factory LAN, and to the south side through the shopfloor LAN.

From the RAMI 4.0 perspective, the FAR-EDGE Gateway Tier corresponds to the **Station** and **Work Centre** levels on the **Hierarchy** axis (IEC-62264/IEC-61512), while the EGs there contained are positioned across the **Asset**, **Integration** and **Communication Layers**. Edge Processes running on EGs, however, map to the **Information** and **Functional Layers**.

3.4.2.3 Ledger Tier

The Ledger Tier (see Figure 3.5) is a complete abstraction: it does not correspond to any physical deployment environment, and even the entities

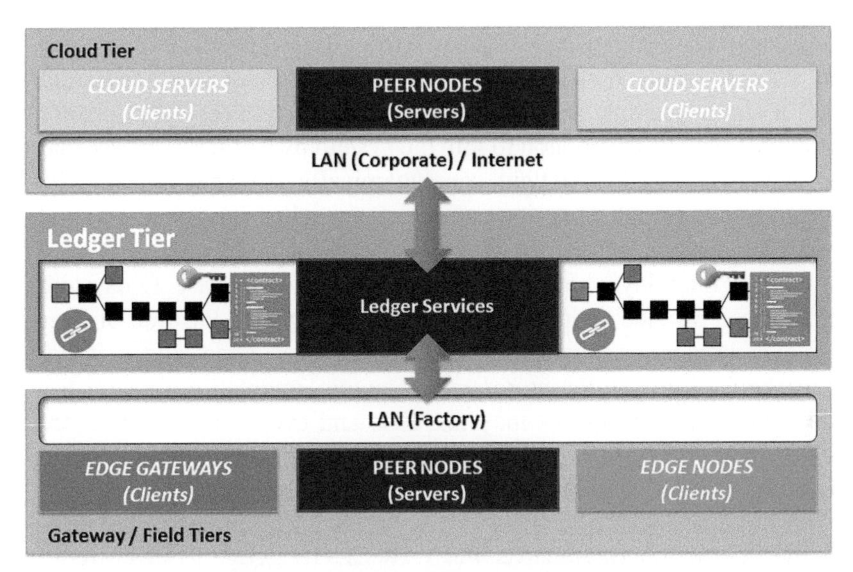

Figure 3.5 FAR-EDGE RA Ledger Tier.

that it "contains" are conventional abstractions. Such entities are *Ledger Services*, which implement decentralized business logic as *smart contracts* executed on a Blockchain platform (see next section for an in-depth technical analysis).

Ledger Services are transaction-oriented: each service call that needs to modify the shared state of a system must be evaluated and approved by *Peer Nodes* before taking effect. Similarly to "regular" services, Ledger Services are implemented as executable code; however, they are not actually executed on any specific computing node: each service call is executed in parallel by all Peer Nodes that happen to be online at the moment, which then need to reach a consensus on its validity. Most importantly, even the executable code of Ledger Services can be deployed and updated online by means of a distributed ledger transaction, just like any other state change.

Ledger Services implement the part of Functional Domains and/or XC Functions that enable the edge computing model, through providing support for their Edge Service counterpart. For example, the Analytics Functional Domain may define a local analytics function (Edge Service) that must be executed in parallel on several EGs, and also a corresponding service call (Ledger Service) that will be invoked from the former each time new or updated local results become available, so that all results can converge into an aggregated dataset. In this case, aggregation logic is included in the

Ledger Service. Another use case may come from the Automation Functional Domain, demonstrating how the Ledger Tier can also be leveraged from the Field: a smart machine with embedded *plug-and-produce* functionality (Smart Object) can ask permission to join the system by making a service call and then, having received green light, can dynamically deploy its own specific Ledger Service for publishing its current state and/or receiving external high-level commands.

The Ledger Tier lays across the Plant and the Enterprise Ecosystem Scopes, as it can provide support to any Tier. The physical location of Peer Nodes, which implement smart contracts and the distributed ledger, is not defined by the FAR-EDGE RA as it depends on implementation choices. For instance, some implementations may use EGs and even some of the more capable ENs in the role of Peer Nodes; others may separate concerns, relying on specialized computing nodes that are deployed on the Cloud.

From the RAMI 4.0 perspective, the FAR-EDGE Ledger Tier corresponds to the **Work Centre**, **Enterprise** and **Connected World** levels on the **Hierarchy** axis (IEC-62264/IEC-61512), while the Ledger Services there contained are positioned across the **Information** and **Functional Layers**.

3.4.2.4 Cloud Tier

The Cloud Tier (see Figure 3.6) is the top layer of the FAR-EDGE RA, and also the simplest and more "traditional" one. It is populated by *Cloud Servers (CS)*: powerful computing machines, sometimes configured as clusters, that are connected to a fast LAN internally to their hosting data centre, and made accessible from the outside world by means of a corporate LAN or the Internet. On CSs runs that part of the business logic of Functional Domains and XC Functions that benefits from having the widest of scopes over production processes, and can deal with the downside of being physically deployed far away from them. This includes the planning, monitoring and management of entire factories, enterprises and supply chains (e.g., MES, ERP and SCM systems). The Cloud Tier is populated by *Cloud Services* and *Applications*. The difference between them is straightforward: Cloud Services implement specialized functions that are provided as individual API calls to Applications, which instead "package" a wider set of related operations that are relevant to some higher-level goal and often – but not necessarily – expose an interactive human interface.

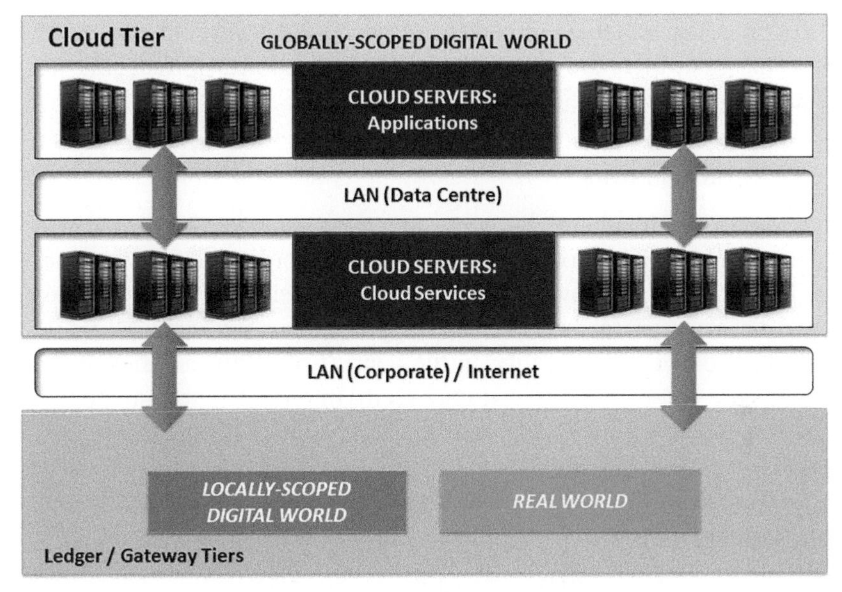

Figure 3.6 FAR-EDGE Cloud Tier.

The Cloud Tier is in Enterprise Ecosystem scope. The "Cloud" term in this context implies that Cloud Services and Applications are visible from all Tiers, wherever located. It does *not* imply that CSs should be actually hosted on some commercial ISP's infrastructure. More often, in particular in large enterprises, the Cloud Tier corresponds to one or more corporate data centres (private cloud), ensuring that the entire system is fully under the control of its owner.

From the RAMI 4.0 perspective, the FAR-EDGE Cloud Tier corresponds to the **Work Centre**, **Enterprise** and **Connected World** levels on the **Hierarchy** axis (IEC-62264/IEC-61512), while the Cloud Services and Applications there contained are positioned across the **Information**, **Functional** and **Business Layers**.

3.5 Key Enabling Technologies for Decentralization

In this section, our main concern is the use of Blockchain and smart contracts as the key enabling technologies of Ledger Services (see the Ledger Tier section above). In FAR-EDGE, the baseline Blockchain platform is an off-the-shelf product, which is enriched by application-specific smart contract software. That said, there are some Blockchain-related basic issues that we need to account for.

3.5.1 Blockchain Issues

For those familiar with the technology, the main question is: how can a Blockchain fit industrial automation scenarios? According to conventional wisdom, Blockchains are slow and cumbersome systems with limited scalability and an aversion to data-intensive applications. Nevertheless, while this vision has solid roots in reality, in the context of smart factories, these shortcomings are not as relevant as it may seem. In order to substantiate this claim, though, we first need to explain some key points of the technology.

First and foremost, the Blockchain is a log of all transactions (i.e., state changes) executed in the system. The log, which is basically a witness of past and current system states, is replicated and kept in-sync across multiple nodes. All nodes are peers, so that no "master node" or "master copy" of the log exists anywhere at any time. Internally, the log is a linear sequence of records (i.e., *blocks* containing transactions) that are individually immutable and time-stamped. The sequence itself can only be modified by appending new records at the end. The integrity of both records and sequence is protected by means of strong cryptographic algorithms [3]. Moreover, all records must be approved by consensus among peers, using some sort of *Byzantine Fault Tolerance (BFT)* mechanism as a guarantee that an agreement on effective system state can always be reached, even if some peers are unavailable or misbehaving (in good faith or for malicious purposes) [4, 5].

The process described above is all about trust: the consensus protocol guarantees that all approved transactions conform to the business logic that peers have agreed on, while the log provides irrefutable evidence of transactions. For this to work in a zero-trust environment, where peers that do not know (let alone trust) each other and are not subject to a higher authority, there is yet another mechanism in place: an economic incentive that rewards "proper" behaviour and makes the cost of cheating much higher than the profit. Given that the whole system must be self-contained and autonomous, such incentive is based on native digital money: a *cryptocurrency*. This closes the loop: all public Blockchain networks need *cryptoeconomics* to make their BFT mechanism work. For some of them (e.g., Bitcoin), the cryptocurrency itself is the main goal of the system: transactions are only used to exchange value between users. Other systems (e.g., Ethereum) are much more flexible, as we will see further on. That said, cryptocurrencies are problematic for many reasons, including regulatory compliance, and hinder the adoption of the Blockchain in the corporate world.

Another key point of Blockchain technology that is worth mentioning is the problem of *transaction finality*. Most BFT implementations rely on *forks* to resolve conflicts between peer nodes: when two incompatible opinions on the validity of some transaction exist, the log is split into two branches, each corresponding to one alternate vision of reality, i.e., of system state. The other nodes of the network will then have to choose which branch is the valid one, and will do this by appending their new blocks to the "right" branch only. Over time, consensus will coalesce on one branch (the one having more new blocks appended), and the losing branch will be abandoned. While this scheme is indeed effective for achieving BFT in public networks, it has one important consequence: there is no absolute guarantee that a committed transaction will stay so, because it may be deemed invalid *after* it is written to the log. In other words, it may appear only on the "bad" branch of a fork and be reverted when the conflict is resolved. Clearly enough, this behaviour of the Blockchain is not acceptable in scenarios where a committed transaction has side effects on other systems.

This is how first-generation Blockchains work. For all these reasons, public Blockchains are, at least to date, extremely inefficient for common online transaction processing (OLTP) tasks. This is most unfortunate, because second-generation platforms like Ethereum have introduced the smart contract concept. Smart contracts were initially conceived as a way for users to define their custom business logic for transaction, i.e., making the Blockchain "smarter" by extending or even replacing the built-in logic of the platform. It then became clear that smart contracts, if properly leveraged, could also turn a Blockchain into a distributed computing platform with unlimited potential. However, distributed applications would still have to deal with the scalability, responsiveness and transaction finality of the underlying BFT engine, which significantly limits the range of possible use cases.

To tackle this problem, the developer community is currently treading two separate paths: upgrading the BFT architecture on the one hand and relax functional requirements on the other hand. The former approach is ambitious but slow and difficult: it is followed by a third generation of Blockchain platforms that are proposing some innovative solution, although transaction finality still appears to be an open point nearly everywhere. The latter is much easier: if we can assume some limited degree of trust between parties, we can radically simplify the BFT architecture and thus remove the worst bottlenecks. From this reasoning, an entirely new species was born in recent years: *permissioned* Blockchains. Given their simpler architecture, commercial-grade permissioned Blockchains are already available today (e.g., Hyperledger, Corda), as opposed to third-generation ones (e.g., EOS, NEO) which are still experimental.

3.5.2 Permissioned Blockchains

Permissioned Blockchains are second-generation architectures that do not support anonymous nodes and do not rely on cryptoeconomics. Basically, they are meant to make the power of Blockchain and smart contracts available to the enterprise, at least to some extent. Their BFT is still a decentralized process executed by peer nodes; however, the process runs under the supervision of a central authority. This means that all nodes must have a strong digital identity (no anonymous parties) and be trusted by the authority in order to join the system. Trust, and thus access to the Blockchain, can be revoked at any time. The BFT protocol can then rely on some basic assumptions and perform much faster, narrowing the distance from OLTP standards in terms of both responsiveness and throughput. Some BFT implementation also support final transactions, as consensus on transaction validity can be reached in near-real-time *before* anything is written to the log.

The key point of permissioned Blockchains is that they are only partially decentralized, leaving governance and administration roles in the hands of a leading entity – be it a single organization or a consortium. This aspect is a boon for enterprise adoption, for obvious reasons. Typically, these networks are also much smaller than public ones, with the positive side effect of limiting the inefficiency of data storage caused by massive data replication across peer nodes. Overall, we can argue that permissioned Blockchains are a viable compromise between the original concept and legacy OLTP systems. But then, to what extent? Can we identify some use cases that a state-of-the-art permissioned Blockchain can effectively support? This is exactly what the FAR-EDGE project aims at, with the added goal of validating claims on the field, by means of pilot applications deployed in real-world industrial environments.

3.5.3 The FAR-EDGE Ledger Tier

The first problem that FAR-EDGE had to face was to define the *performance envelope* of current Blockchain implementations, so that validation cases could be shaped according to the sustainable workload. The idea was to set the benchmark for a Blockchain *comfort zone* in terms of a few objective and measurable Key Performance Indicators (KPI), targeting the known weak points of the technology:

- *Transaction Average Latency (TrxAL)* – The average waiting time for a client to get confirmation of a transaction, expressed in seconds.
- *Transaction Maximum Sustained Throughput (TrxMST)* – The maximum number of transactions that can be processed in a second, on average.

The benchmark was set by stress-testing, in a lab environment, actual Blockchain platforms. These were selected after a preliminary analysis of the permissioned Blockchains available from open source communities, using criteria like code maturity and, most importantly, finality of transactions. The only two platforms that passed the selection were Hyperledger Fabric (HLF) and NEO. The stress test was then conducted using BlockBench, a specialized testing framework [6], and a simple configuration of eight nodes on commodity hardware.

HLF emerged from tests as the only viable platform for CPS applications, given that NEO is penalized by a significant latency (~7s.), which is independent of workload (the expected result for a "classical" Blockchain architecture that aggregates transactions into blocks and defines a fixed delay for processing each block). On the contrary, HLF was able to accept a workload of up to 160 transactions per second with relatively low latency (0.1–1 s.). On heavier workloads, up to 1000 transactions per second, NEO is instead the clear winner, thanks to its constant latency, while HLF's performance progressively degrades (>50 s.). This workload profile however, while appealing for high-throughput scenarios (e.g., B2C payment networks), is not compatible with basic CPS requirements. Consequently, the Blockchain performance benchmark was set as follows:

- $<=$ TrxAL $<=$ 1.0
- 0 $<=$ TrxMST $<=$ 160

This is also considered the performance envelope of the FAR-EDGE Ledger Tier, as the HLF platform has been adopted as its baseline Blockchain implementation.

3.5.4 Validation use Cases

Having marked some boundaries, the FAR-EDGE project then proceeded with the identification of some pilot applications for the validation phase. The starting point was a set of candidate use cases proposed by our potential users, who were eager to tackle some concrete problems and experiment with some new ideas. The general framework of this exercise is described here.

As explained, the main objective in FAR-EDGE is to achieve flexibility in the factory through the decentralization of production systems. The catalyst of this transformation is the Blockchain, which – if used as a computing platform rather than a distributed ledger – allows the virtualization of the automation pyramid. The Blockchain provides a common *virtual space* where data can be securely shared and business logic can be consistently run. That said, users can leverage this opportunity in two ways: one easier but somewhat limited approach, and the other more difficult and more ambitious approach.

The easiest approach is of the brown-field type: just migrate (some of) the factory's centralized monitoring and control functionality to Ledger Services on the Ledger Tier. Thanks to the Gateway Tier, legacy centralized services can be "impersonated" on a local scale by Edge Gateways: the shopfloor – that hardest environment to tamper with in a production facility – is left untouched. The main advantages of this configuration are the mitigation of performance bottlenecks (heavy network traffic is confined locally, workload is spread across multiple computing nodes) and added resiliency (segments of the shopfloor can still be functional when temporarily disconnected from the main network). Flexibility is also enhanced, but on a coarse-grained scale, modularity is achieved by grouping a number of shopfloor Edge Nodes under the umbrella of one Edge Gateway, so that they all together become a single "module" with some degree of self-contained intelligence and autonomy. Advanced Industry 4.0 scenarios like plug-and-produce are out of reach.

The more ambitious approach is also a much more difficult and risky endeavour in real-world business, being of the green-field type. It is about delegating responsibility to Smart Objects on the shopfloor, which communicate with each other through the mediation of the Ledger Tier. The business logic in Ledger Services is of higher level with respect to the previous scenario: more about governance and orchestration than direct control. The Gateway Tier has a marginal role, mostly confined to Big Data analytics. In this configuration, central bottlenecks are totally removed and the degree of flexibility is extreme. The price to pay is that a complete overhaul of the shopfloor of existing factories is required, replacing PLC-based automation with intelligent machines.

In FAR-EDGE, both paths are explored with different use cases combining on automation, analytics and simulation. We here give one full example of each type.

The first use case follows the brown-field approach. The legacy environment is an assembly facility for industrial vehicles. The pilot is called *mass-customization*: the name refers to capability of the factory assembly line to handle individually customized products having a high level of variety. If implemented successfully, mass-customization can give a strategic advantage to target niche markets and meet diverse customer needs in a timely fashion. In particular, the pilot factory produces highly customized trucks. The product specification is defined by up to 800 unique variants, and the final assembly includes approximately 7000 manufacturing operations and handles a very high degree of geometrical variety (axle configurations, fuel tank positions etc.). Despite the high level of variety in the standard product, at some production sites, 60% of the produced trucks have unique customer adaption.

In the pilot factory, the main assembly line is sequential but feeds a number of finishing lines that work in parallel. In particular, the wheel alignment verification is done on the finishing assembly line and is one of the last active checks done on trucks before they leave the plant. This opens up an opportunity to optimize the workload. In the as-is scenario, wheel alignment stations are statically configured to accommodate specific truck model ranges: products must be routed to a matching station on arrival, creating a potential bottleneck if model variety is not optimal. As part of the configuration, a handheld nut runner tool needs to be instructed as to the torque force to apply.

In the to-be solution, according to the FAR-EDGE architectural blueprint, each wheel alignment station is represented at the Gateway Tier level by a dedicated Edge Gateway box. The EG runs some simple ad-hoc automation software that integrates the Field systems attached to the station (e.g., a barcode reader, the smart nut runner) using standard IoT protocols like MQTT. The EG also runs a peer node that is a member of the logical Ledger Tier. A custom Ledger Service deployed on the Ledger Tier implements the business logic of the use case. The instruction set for the products to be processed is sent in JSON format to the Ledger Service, once per day, by the central ERP-MES systems: from that point and until a new production plan is published, the Ledger and Gateway Tiers are autonomous.

When a new truck reaches the end of the main line, it is dispatched to the first finishing line available, achieving the desired result of product flow optimization. Then, when it reaches the *wheel alignment station*, the chassis ID is scanned by a barcode reader and a request for instructions is sent, through the automation layer on the EG, to the Ledger Service. The Ledger Service will retrieve the instruction set from the production

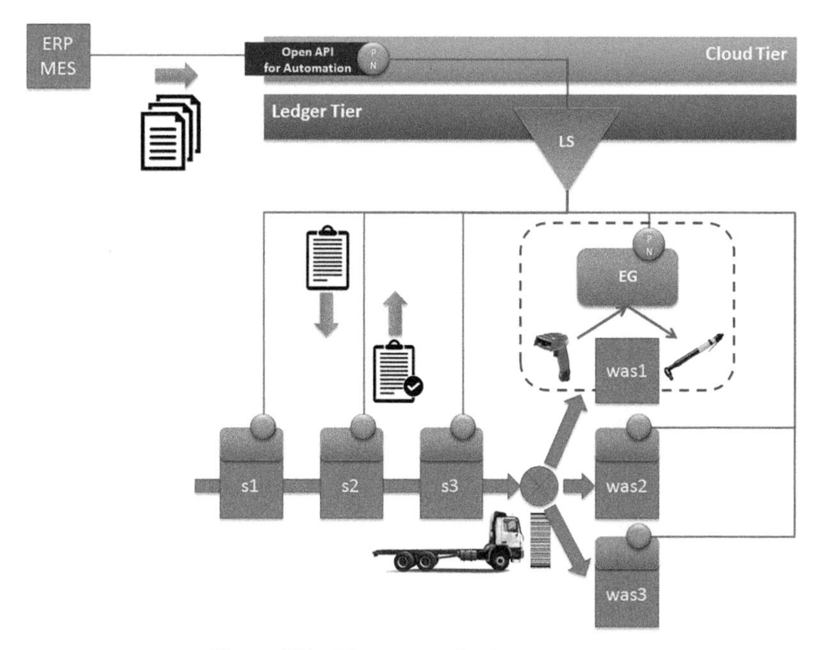

Figure 3.7 Mass-customization use case.

plan – which is saved on the Ledger itself – by matching the chassis ID. When the automation layer receives the instructions set, it parses the specific configuration parameters of interest and sends them to the nut runner, which adjusts itself. The wheel alignment operations will then proceed as usual. A record of the actual operations performed, which may differ from those in the instruction set, is finally set back to the Ledger and used to update the production plan. An overall view of the use case is given in Figure 3.7.

While the product flow optimization mentioned above is the immediate result of the pilot, there are some additional benefits to be gained either as a by-product or as planned extensions.

First, the wheel alignment station, together with its EG box, becomes an autonomous module that can be easily added/removed and even relocated in a different environment. This scenario is not as far-fetched as it may seem, because it actually comes from a business requirement: the company has a number of production sites in different locations all over the world, each with their own unique MES maps. The deployment of a new module with different MES maps is currently a difficult and costly process.

Second, in the future, the truck itself may become a Smart Object that communicates directly with the Ledger Tier. Truck–Ledger interactions will happen throughout the entire life cycle of the truck – from manufacturing to operation and until decommissioning – with the Ledger maintaining a digital twin of the truck.

The second use case follows instead the *heavyweight* green-field approach. The pilot belongs to a white goods (i.e., domestic appliances) factory. The objective of the pilot is "reshoring", which in the FAR-EDGE context means enabling the company to move production back from off-shore locations, thanks to a better support for the rapid deployment of new technologies (i.e., shopfloor Smart Objects) offered by the more advanced domestic plants. In this particular plant, a 1 km long conveyor belt moves pallets of finished products from the factory to a warehouse, where they are either stocked or forwarded for immediate delivery. The factory/warehouse conveyor is not only a physical boundary, but also an administrative one, as the two facilities are under the responsibility of two different business units. Moreover, once the pallet is loaded on a delivery vehicle, it comes under the responsibility of a third party who operates the delivery business.

In the as-is scenario, the conveyor feeds 19 shipping bays, or "lanes", in the warehouse. Each lane is simply a dead-end conveyor segment, where pallets are dropped in by the conveyor and retrieved by a manually operated forklift (basically, an FIFO queue). Simple mechanical actuators do the physical routing of the pallets, controlled by logic that runs on a central "sorter" PLC. The sorting logic is very simple: it is based on a production schedule that is defined once per day and on static mappings of the lanes to product types and/or final destinations. This approach has one big problem: production cannot be dynamically tuned to match business changes, or at least only to a very limited extent, because the fixed dispatching scheme downstream cannot sync with it. The problem is not only in software: the physical layout of the system is fixed.

In the to-be solution, the shipping bays become Smart Objects that can be plugged in and out at need (see Figure 3.8). They embed simple sensors that detect the number of pallets currently in their local queue, and a controller board that runs some custom automation logic and connects directly to the Ledger Tier (i.e., without the mediation of an Edge Gateway). A custom Ledger Service acts as a coordination hub: it is responsible for authorizing a new "smart bay" that advertise itself to join the system (plug-and-produce) and, once accepted, to apply the sorting logic. This is based on the current state of the main conveyor belt, where incoming and outgoing pallets are individually identified by an RFID tag, and on "capability update" messages that are sent by smart bays each time they undergo an internal state change

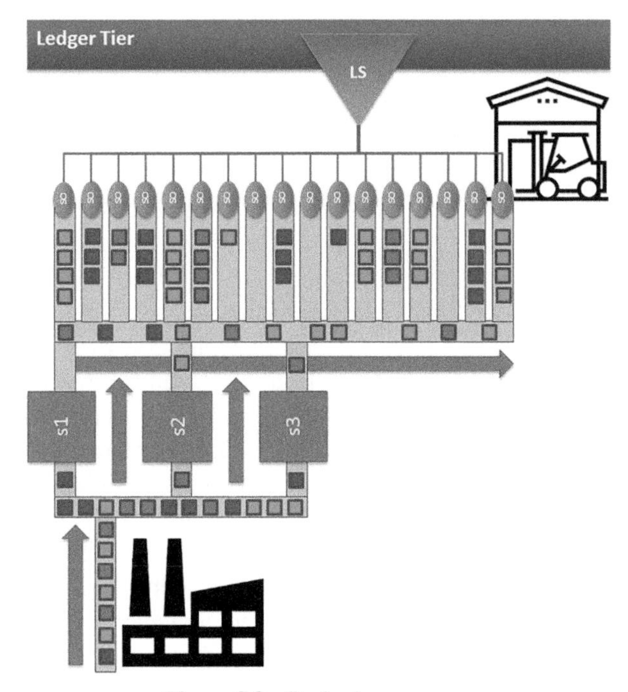

Figure 3.8 Reshoring use case.

(e.g., number of free slots in the local queue, preference for a product type). The production schedule is not required at all, because sorting is only calculated on the actual state.

3.6 Conclusions

FAR-EDGE is one of the few ongoing initiatives that focus on edge computing for factory automation, similarly to the IIC's edge intelligence testbed and EdgeX Foundry. However, the FAR-EDGE RA introduces some unique concepts. In particular, the notion of a special logical layer, the Ledger Tier, that is responsible for sharing process state and enforcing business rules across the computing nodes of a distributed system, thus permitting *virtual* automation and analytics processes that span multiple nodes – or, from a bottom-up perspective, autonomous nodes that cooperate to a common goal. This new kind of architectural layer stems from the availability of Blockchain technology, which, while being well understood and thoroughly tested in mission-critical areas like digital currencies, have never been applied before to industrial systems. FAR-EDGE aims at demonstrating how a pool of specific Ledger Services can enable decentralized factory automation in an effective, reliable, scalable and secure way. In this chapter, we also presented the general framework of the industrial pilot applications that are going to be run during the validation phase of the project.

References

[1] Karsten Schweichhart: Reference Architectural Model Industrie 4.0 – An Introduction, April 2016, Deutsche Telekom, online resource: https://ec.europa.eu/futurium/en/system/files/ged/a2-schweichhart-reference_architectural_model_industrie_4.0_rami_4.0.pdf

[2] Dagmar Dirzus, Gunther Koschnick: Reference Architectural Model Industrie 4.0 – Status Report, July 2015, VDI/ZVEI, online resource: https://www.zvei.org/fileadmin/user_upload/Themen/Industrie_4.0/Das_Referenzarchitekturmodell_RAMI_4.0_und_die_Industrie_4.0-Komponente/pdf/5305_Publikation_GMA_Status_Report_ZVEI_Reference_Architecture_Model.pdf

[3] H. Halpin, M. Piekarska, "Introduction to Security and Privacy on the Blockchain", IEEE European Symposium on Security and Privacy Workshops (EuroS & PW), Paris, 2017, pp. 1–3.

[4] L. Lamport, R. Shostak, M. Pease, "The Byzantine Generals problem", ACM Transactions on Programming Languages and Systems, volume 4 no. 3, p. 382–401, 1982.

[5] Z. Zheng, S. Xie, H. Dai, X. Chen, H. Wang, "An overview of Blockchain technology: architecture, consensus, and future trends", proceedings of IEEE 6th International Congress on Big Data, 2017.

[6] T. Dinh, J. Wang, G. Chen, R. Liu, C. Ooi, K. L. Tan, "BLOCKBENCH: a framework for analyzing private Blockchains", unpublished, 2017. Retrieved from: https://arxiv.org/pdf/1703.04057.pdf

4

IEC-61499 Distributed Automation for the Next Generation of Manufacturing Systems

Franco A. Cavadini[1], Giuseppe Montalbano[1], Gernot Kollegger[2], Horst Mayer[2] and Valeriy Vytakin[3]

[1]Synesis, SCARL, Via Cavour 2, 22074 Lomazzo, Italy
[2]nxtControl, GmbH, Aumühlweg 3/B14, A-2544 Leobersdorf, Austria
[3]Department of Computer Science, Electrical and Space Engineering, Luleå tekniska universitet, A3314 Luleå, Sweden
E-mail: franco.cavadini@synesis-consortium.eu;
giuseppe.montalbano@synesis-consortium.eu;
gernot.kollegger@nxtcontrol.com;
horst.mayer@nxtcontrol.com;
Valeriy.Vyatkin@ltu.se

Global competition in the manufacturing sector is becoming fiercer and fiercer, with fast evolving requirements that must now take much more into account: rising product variety; product individualization; volatile markets; increasing relevance of value networks; shortening product life cycles. To fulfil these increasingly complex requirements, companies have to invest on new technological solutions and to focus the efforts on the conception of new automation platforms that could grant to the shopfloor systems the flexibility and re-configurability required to optimize their manufacturing processes, whether they are continuous, discrete or a combination of both.

Daedalus is conceived to enable the full exploitation of the CPS' virtualized intelligence concept, through the adoption of a completely distributed automation platform based on IEC-61499 standard, fostering the creation of a digital ecosystem that could go beyond the current limits of manufacturing control systems and propose an ever-growing market of innovative solutions for the design, engineering, production and maintenance of plants' automation.

4.1 Introduction

European leadership and excellence in manufacturing are being significantly threatened by the huge economic crisis that hit the Western countries over the last years. More sustainable and efficient production systems, able to keep pace with the market evolution, are fundamental in the recovery plan aimed at innovating the European competitive landscape. An essential ingredient for a winning innovation path is a more aware and widespread use of ICT in manufacturing-related processes.

In fact, the rapid advances in ubiquitous computational power, coupled with the opportunities of de-localizing into the Cloud parts of an ICT framework, have the potential to give rise to a new generation of service-based industrial automation systems, whose local intelligence (for real-time management and orchestration of manufacturing tasks) can be dynamically linked to runtime functionalities residing in-Cloud (an ecosystem where those functionalities can be developed and sold). Improving the already existing and implemented IEC-61499 standard, these new "Cyber Physical Systems" will adopt an open and fully interoperable automation language (dissipating the borders between the physical shop floors and the cyber-world), to enable their seamless interaction and orchestration, while still allowing proprietary development for their embedded mechanisms.

These CPS based on real-time distributed intelligence, enhanced by functional extensions into the Cloud, will lead to a new information-driven automation infrastructure, where the traditional hierarchical view of a factory functional architecture is complemented by a direct access to the on-board services (non-real-time) exposed by the Cyber-Physical manufacturing system, composed in complex orchestrated behaviours. As a consequence, the current classical approach to the Automation Pyramid (Figure 4.1) has been recently addressed several times (Manufuture, ICT 2013 and ICT 2014 conference, etc.) and deemed by RTD experts and industrial key players to be inadequate to cope with current manufacturing trends and in need to evolve.

In the European initiative Daedalus, financed under the Horizon 2020 research programme, it has been acknowledged deeply that CPS intrinsic existence defies the concept of rigid hierarchical levels, being each CPS capable of complex functions across all layers. An updated version of the pyramid representation is therefore adopted (Figure 4.2), where CPS are hierarchically orchestrated in real time (within the shop floor) through the IEC-61499 automation language, to achieve complex and optimized behaviours (impossible to other current technologies), while still being singularly and directly

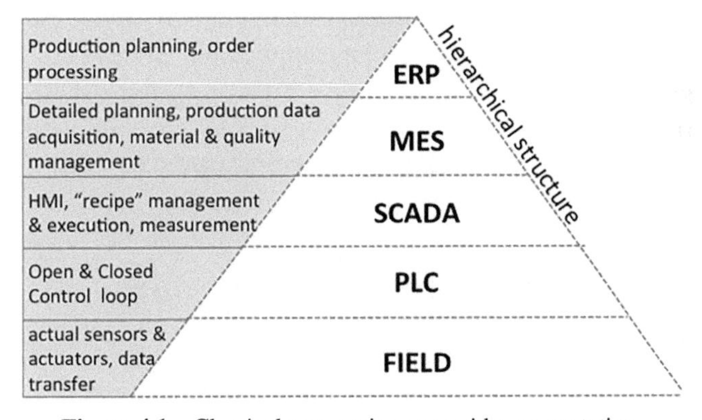

Figure 4.1 Classical automation pyramid representation.

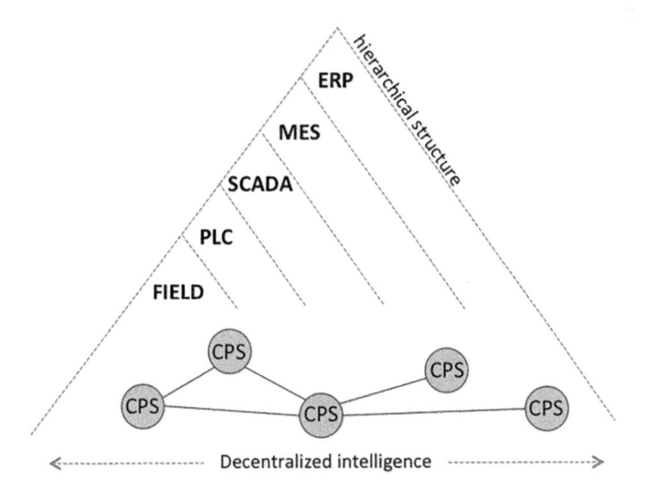

Figure 4.2 Daedalus fully accepts the concept of vertically integrated automation pyramid introduced by the PATHFINDER [1] road-mapping initiative and further developed with the Horizon 2020 Maya project [6].

accessible, at runtime, by whatever elements of the Factory ICT infrastructure that wants to leverage on their internal functionalities (and have the privileges to do that).

This innovative approach to the way of conceiving automated intelligence within a Factory – across the boundaries of its physically separated functional areas thanks to the constant and bidirectional connection to the cyber world – will be the enabler for a revolutionary paradigm shift within the market of industrial automation. The technological platform of Daedalus will become in fact also Economic Platform of a completely new multi-sided ecosystem, where the creation of added-value products and services by device producers, machine builders, systems integrators and application developers will go beyond the current limits of manufacturing control systems and propose an ever-growing market of innovative solutions for the design, engineering, production and maintenance of plants' automation (see also Chapter 13 of this book).

4.2 Transition towards the Digital Manufacturing Paradigm: A Need of the Market

Current worldwide landscape is seeing continuously growing value creation from digitization, with digital technologies increasingly playing the central role in value creation for the entire economy. More and more types of product are seeing a progressive transition to the "digital inside" model, where innovation is mostly related to the extension of the product-model to the service-model, through a deeper integration of digital representations. This means, in concrete terms, that even in very "classical" domains, the dissipation of borders between what is a product and what are the services that it enables is fostering a widespread "Business Model Innovation" need.

Looking at how global competition in the manufacturing sector is becoming fiercer and fiercer, with fast evolving requirements that must now take into account several concurrent factors, it is clear that European Manufacturing Companies have to focus the efforts on new automation solutions that could grant to the shop floor systems the flexibility and reconfigurability required to optimize their manufacturing processes (Figure 4.3).

To realize such a vision, current technological constraints must be surpassed through research and development activities focusing on the following topics:

- interoperability of data/information (versus compatibility) and robustness;

Evolving requirements of the manufacturing sectors:

- Rising product variety
- Increasing relevance of value networks
- Shortening product life cycles
- Quick variation of demand
- Request for high quality and customized products

Needs of new automation solutions to enable:

- *Flexibility and re-configurability*
- *Increased production performance*
- *Reduced energy consumption*
- *Better environmental footprint*

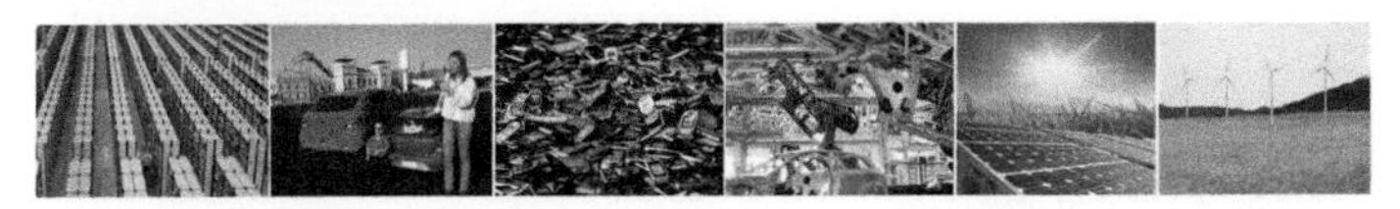

Figure 4.3 The industrial "needs" for a transition towards a digital manufacturing paradigm.

- integration of different temporal-decision scale data (real-time, near-time, anytime) and multiple data sources;
- integration of the real and the virtual data-information towards a predictive model for manufacturing; real-time data collection, analysis, decision, enforcement;
- optimization in complex system of systems infrastructures;
- seamless data integration across the process value chain;
- standardization and interoperability of manufacturing assets components, subsystems and services.

Within this context, the future of Europe's industry must be digital, as clearly highlighted by Commissioner Oettinger EU-wide strategy [2] to "ensure that all industrial sectors make the best use of new technologies and manage their transition towards higher value digitised products and processes" through the "Leadership in next generation open and interoperable digital platforms", opening incredible opportunities for high growth of vertical markets, especially for currently "non-digital" industries (Figure 4.4).

The core motivation for Daedalus was therefore born by the awareness that purely technological advancements in themselves are not enough to satisfy the need of innovation of the industrial automation market. New methodologies for the sector's main stakeholders to solve the new manufacturing needs of end-users must be conceived and supported by the creation of a technological and economic ecosystem built on top of a multi-sided platform.

In developing this concept, Daedalus takes into account a certain number of fundamental "non-functional" requirements:

- CPS-like interoperable devices must be "released" on the market together with their digital counterpart, both in terms of behavioural

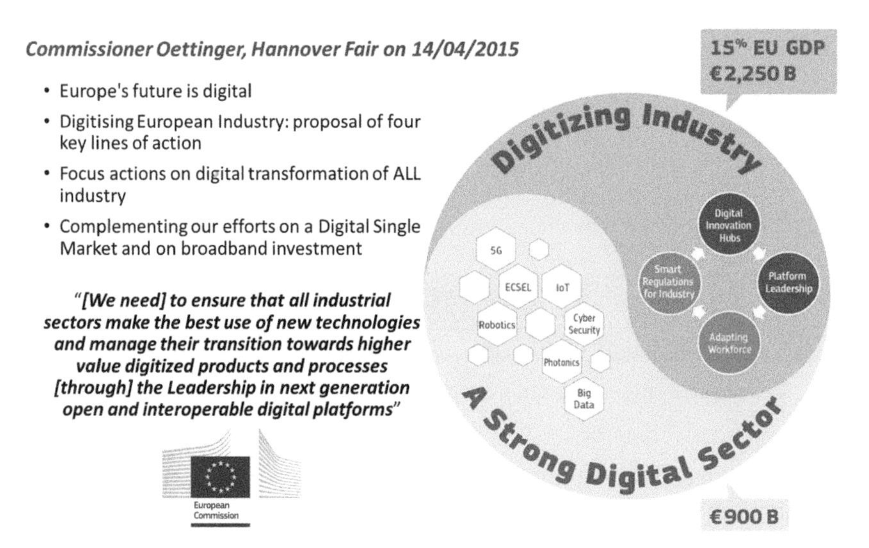

Figure 4.4 Commissioner Oettinger agenda for digitalizing manufacturing in Europe.

models and with the software "apps" that allows their simple integration and orchestration in complex system-of-systems architectures;

- The development of coordination (orchestration) intelligence by system integrators, machine builders or plant integrators (more in general, by aggregators of CPS) should rely on existing libraries of basic functions, developed and provided in an easy-to-access way by experts of specific algorithmic domains;
- Systemic performance improvement at automation level should rely on well-maintained SDKs that mask the complexity of behind-the-scenes optimization approaches;
- Large-scale adoption of simulation as a tool to accelerate development and deployment of complex automation solutions should be obtained by shifting the implementation effort of models to device/system producers;

This translates into an explicit involvement of all main stakeholders of the automation development domain, brought together in a multi-sided market. Such "Automation Ecosystem" must rely on a technological platform that, leveraging on standardization and interoperability, can mask the complexity of interconnecting these Complementors.

4.3 Reasons for a New Engineering Paradigm in Automation

The core conceptual idea launched at European level by the German "Industrie 4.0" initiative is that embedding intelligence into computational systems distributed throughout the factory should enable vertical networking with business process at management level, and horizontal connection among dispersed value networks.

The RAMI 4.0 framework has been therefore developed to highlight this new degree of integration between different aspects of the manufacturing domain, which does not exist only within the usual hierarchy of automation (functional layers) but also across life cycle and aggregation levels (Figure 4.5). The core issue (tackled by Daedalus), which is not apparent enough in this framework, is that the evolution of the Hierarchy Levels, those that characterize the progressive aggregation of physical systems into more complex one, is currently limited by a technological gap between the shop floor and the office floor automation.

In fact, two specific limits hinder the transition towards the next step of the shop floor automation:

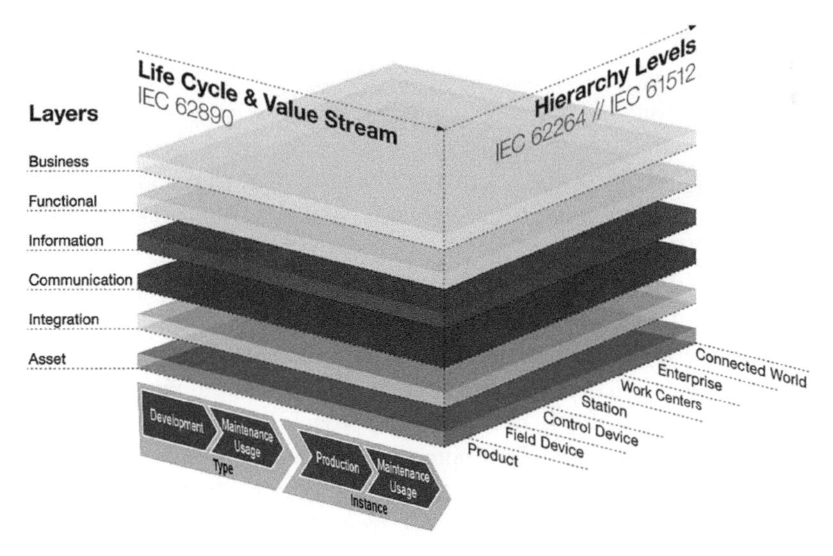

Figure 4.5 RAMI 4.0 framework to support vertical and horizontal integration between different functional elements of the factory of the future.

- Current PLC technology, which dominates the deployment of industrial automation applications, is a legacy of the 1980s, unsuited for sustaining complex "system of intelligent systems" functional architectures;
- Automation languages of the IEC-61131 standard, basis of the aforementioned PLCs, are antiquated from a software engineering point of view; additionally, they have been implemented by each vendor through specific "dialects" that prevent real interoperability.

In technological terms, this has a very specific impact: while products of the automation domain are still completely based on an engineering approach built over the concept of a time cycle (and, consequently, its programming languages), the ICT domain has been working for decades through object orientation and, most importantly, event-based programming. Trying to bring together these two worlds, to guarantee the new levels of integration envisioned by Industry 4.0, is going to be practically impossible if nothing changes in the way industrial automation is conceived and then deployed.

During the last 20 years, the standardization and research efforts related to control software for industrial automation was focused on improving quality and reliability while reducing development time. As explained previously, distributed automation is considered the needed innovation step; however, the current automation paradigm, based on the use of programmable logic controllers (PLC), according to the IEC 61131-3 standard, is not suitable for distributed systems, as it was conceived for centralized ones. This device-centric and monolithic engineering approach is not well apt for regular changes of the executed control applications, while the multiple engineering tools required for adapting them greatly increases the engineering time, because the majority of vendors implements specific extensions or only partial support of IEC 61131-3.

The IEC took this into account for the development of the IEC 61499 architecture in order to support such new features of next-generation industrial automation systems as distribution and reconfiguration [3], offering modern platform-independent approach to system design, similar to the Model-Driven Architecture [4]. The MDA approach has greatly improved flexibility and efficiency of the development process for embedded systems [5] on account of re-using elements of the solutions, described in high-level languages. We can expect similar benefits from IEC 61499 for industrial automation that MDA brought to software engineering and embedded system development.

The solution is therefore to propose a technological foundation to CPS that could be used to overstep these constraints and consequently enable

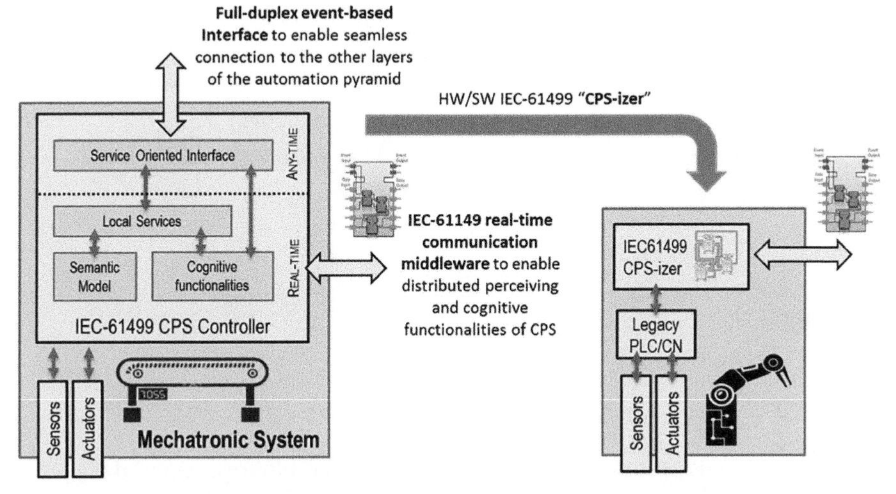

Figure 4.6 Qualitative representation of the functional model for an automation CPS based on IEC-61499 technologies; concept of "CPS-izer".

the additional functionalities needed by the Automation Digital Platform envisioned by the project. By exploiting the already existing features of the IEC-61499 international standard for distributed automation, the idea is to propose a functional model for CPS that blends coherently real-time coordination of its automation tasks with the "anytime" provision of services to other elements of the automation pyramid (Figure 4.6).

This extension of the IEC-61499 functionalities adopts the openness and interoperability of implementation that the standard proposes, guaranteeing that CPS developed independently will be able to communicate and be orchestrated. But it is not just a matter of interoperable communication between CPS at shop floor level; transition towards an effective digitalization requires other composing elements:

- The real-time automation logic of a CPS must be programmed under an object-oriented paradigm and taking into account the transition from the time-based approach of the low-level control and the event-based needs of a service-oriented paradigm;
- The controller of a CPS must also contain a high-level semantic description of the behavioural models of the system it governs, mapping the automation tasks on top of it; this is needed to allow external modules (in the digital domain) to be capable of reading the raw data generated at shop floor level with the appropriate level of semantic context;

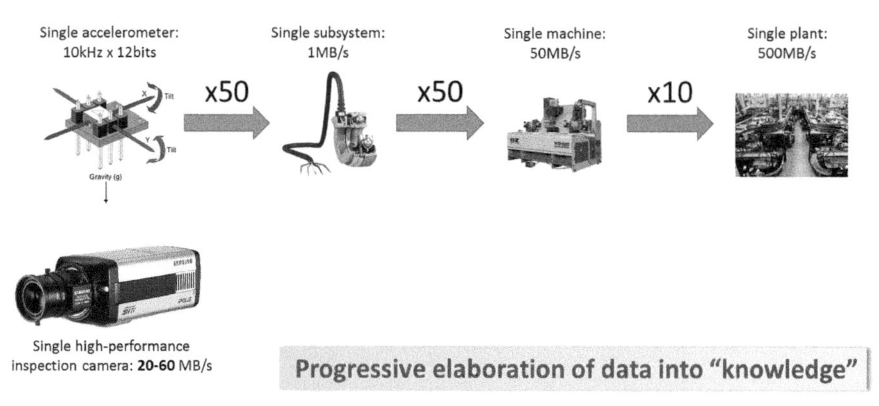

Figure 4.7 The need for local cognitive functionalities is due to the requirements of Big Data elaboration.

- A certain degree of cognitive functionalities must be programmed directly within the CPS, to guarantee that elaboration and modelling of data is done near to the sources of such data (Figure 4.7);

Finally, the "exposition" of services to the digital domain must be conceived by the automation engineer coherently and concurrently to the design of the internal automation tasks, enabling a secure interaction between internal (real-time) automation tasks and "external" requests for asynchronous functionalities.

This notwithstanding, the project understands and accepts the need of CPS vendors (developers) to protect their IP and/or continue using proprietary engineering technologies: the proposed approach supports different levels of "protection" to the inner working mechanism of a system, from a fully IEC61499-compliant but closed (= not accessible by users) implementation, to the "wrapping" of legacy PLCs.

Figure 4.8 therefore shows how the concept of an IEC-61499 CPS (networked in real time with similar systems, compliant with the standard) is only an enabler for a much more complex shopfloor automation, where horizontal integration with other platforms (eventually still in real time) is guaranteed by support to an extensive set of communication protocols and middleware (such as OPC-UA and DDS), while vertical integration through a service-oriented

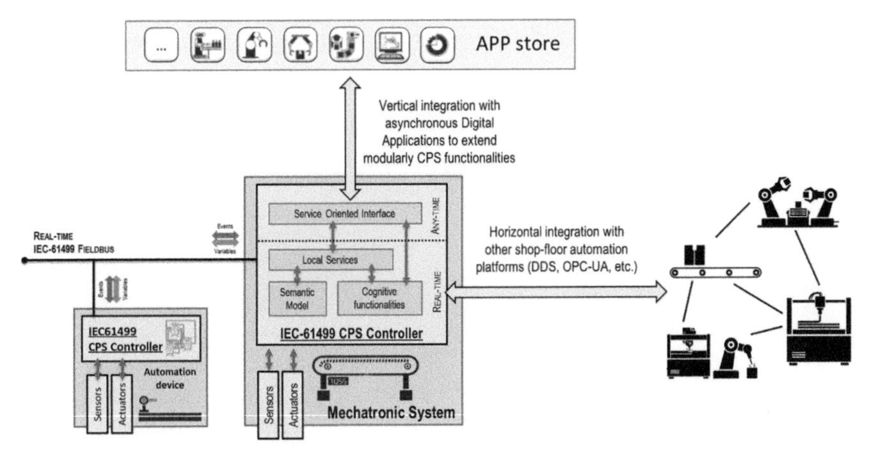

Figure 4.8 Framing of an IEC-61499 CPS within a complex shopfloor.

approach enables the extension of automation functionalities into the digital domain, where the concept of an APP store can greatly facilitate diffusion at market level of this approach.

4.3.1 Distribution of Intelligence is Useless without Appropriate Orchestration Mechanisms

Providing automation devices as IEC61499-compliant CPS is just the enabler for the cornerstone of the project. In fact, the real complexity of future shop floors (and, thus, the opportunities for new manufacturing paradigms) resides in the possibility to develop easily the multi-level orchestration intelligence needed to coordinate the behaviour of all the CPS composing a shop floor.

In fact, the paradigm of decentralization of computing power into smaller devices cannot be deployed only by solving issues about communication about them. Previous attempts to bring the concepts of service orientation into the automation domain has failed when facing the "servers-only issue": even if an intelligent systems is programmed to "expose" its functionalities as services to be invoked (a "server", using the vocabulary of SoA), the moment we have several of these servers, the problem that remains is who is going to coordinate those services in an orchestrated way (the "client") and, most importantly, in which programming language should such a client be designed.

The adoption of IEC-61499 presents automatically the solution to this issue, with an industry-ready approach (validated in several production

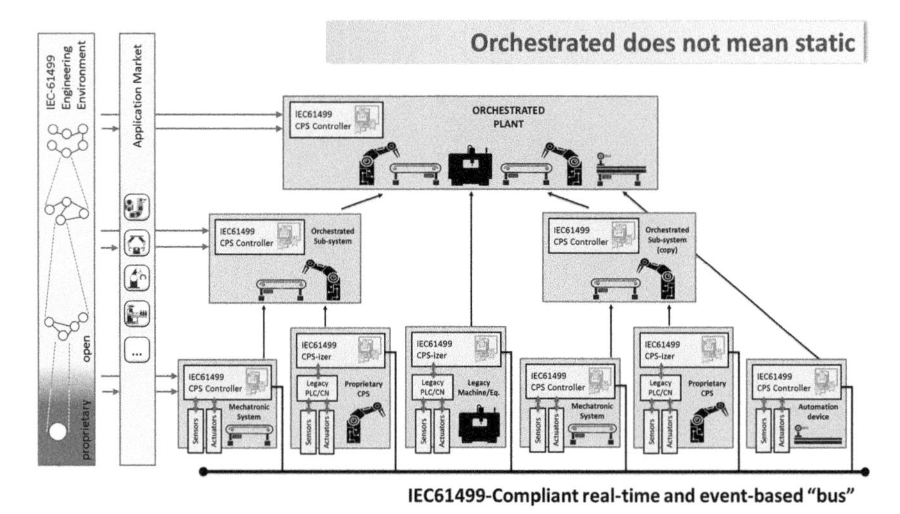

Figure 4.9 Hierarchical aggregation of CPS orchestrated to behave coherently.

environments) that already satisfies the major needs for engineering complex orchestrating applications: interoperability between devices, real-time communication between distributed systems, hardware abstraction, automatic management of low-level variable binding between CPS, a modern development language (and environment), etc. This set of functionalities just needs to be "completed" with additional ones that will make it the undisputed standard at European level.

Figure 4.9 therefore shows how a real "hierarchy" of CPS can be imagined in the shop floor of future factories, where the physical aggregation of equipment and devices to generate more complex systems (typical of the mechatronic approach) must be equally supported by a progressive orchestration of their behaviour, accepting the so-called "Automation Object Orientation" (A-OO, see also Section 4.5 for details) and taking into account that each subsystem may exist with its own controller and internally developed control logics.

The strength of this approach, that is already supported in all its basic and fundamental functionalities by the IEC-61499 standard and programming language, is highlighted in Figure 4.10.

A single CPS, independently from being a basic one or obtained through aggregation of others, can be seen internally (from the perspective of the developer of that CPS) as an intelligent system, which must be programmed

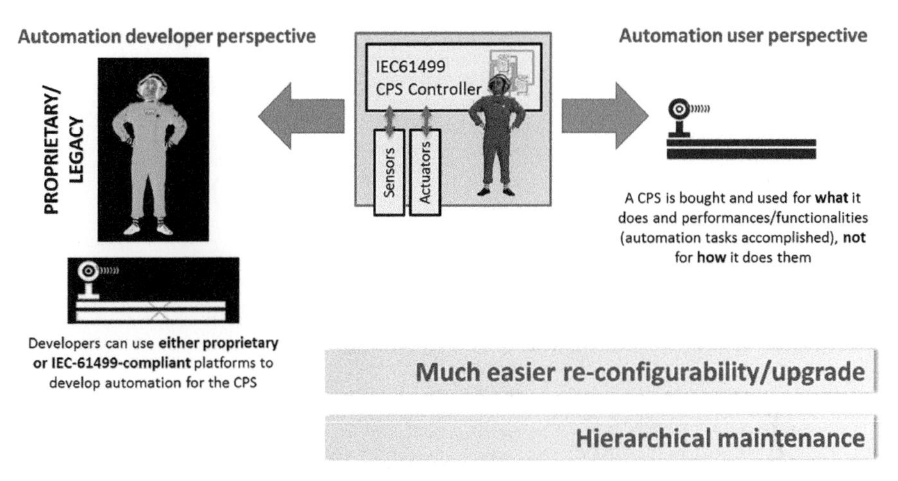

Figure 4.10 Progressive encapsulation of behaviour in common interfaces.

(eventually in proprietary technologies) to exhibit a certain behaviour and expose it over an IEC-61499 interface. On the other hand, seen from outside, the CPS will be a "black box" guaranteeing certain functionalities. This simplifies greatly both the activities of re-configurability and upgrade, and the progressive hiding of maintenance-related details.

Thanks to this unique and innovative approach, new automation systems will be capable of providing simple-to-deploy aggregation of already existing CPS, each one with its own on-board intelligence, to compose articulated "Systems of Cyber-Physical Systems" that, for the final user, will be nothing more than "bigger" CPS, exhibiting concerted behaviours that will mask their internal working mechanisms based on the design decision of the CPS provider.

The adoption of IEC-61499 provides also another opportunity, which is enabled by its natural object orientation (not only at software level but also in dealing with hardware topology through an appropriate abstraction layer): highly increase re-usability of code and applications.

Figure 4.11 shows how the development and IP generation value chain would be applied in the case of high code re-usability enabled by the usage of IEC-61499, where software components of increasing complexity (and aggregation of functionalities) would be progressively employed by different users of the automation domain (further explored in Chapter 13 in its large-scale consequences on the market).

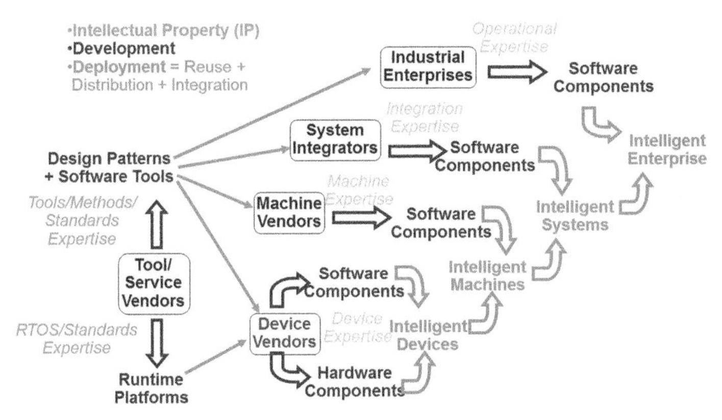

Figure 4.11 IEC-61499 CPS Development, the IP Value-Add Chain.

4.3.2 Defiance of Rigid Hierarchical Levels towards the Full Virtualization of the Automation Pyramid

While the design of orchestrating intelligence supported by IEC-61499 allows the conception of complex aggregated system of systems with advanced behaviour, the CPS functional model (at multiple levels) and the corresponding direct access to non-real-time "services" enables the complete restructuring of the concepts of a factory automation pyramid.

New levels of vertical and horizontal integration can be envisioned thanks to the peculiar service-oriented approach proposed by Daedalus. In fact, current MES and ERP can extend their scopes of application towards the shop-floor by being capable of directly accessing the information flows and elaboration functionalities of the automation CPS; moreover, non-real-time and bidirectional exchange of information can exist between devices even if they are not explicitly orchestrated, such as among products and manufacturing equipment, or between systems of different departments (across the production value chain).

Figure 4.12 proposes a different vision of the factory, extending the point of view outside of the shopfloor and into the so-called "digital" domain, where all the ICT tools of a company exists, from the MES up to the ERP. Hiding temporarily the hierarchy of CPS at shopfloor level shown before (for ease of readability), the picture shows how each IEC-61499 CPS of Daedalus, based on the functional model of Figure 4.6, can connect directly and independently from the other to any "digital module" allowed to do that from a security perspective. This means in practice that:

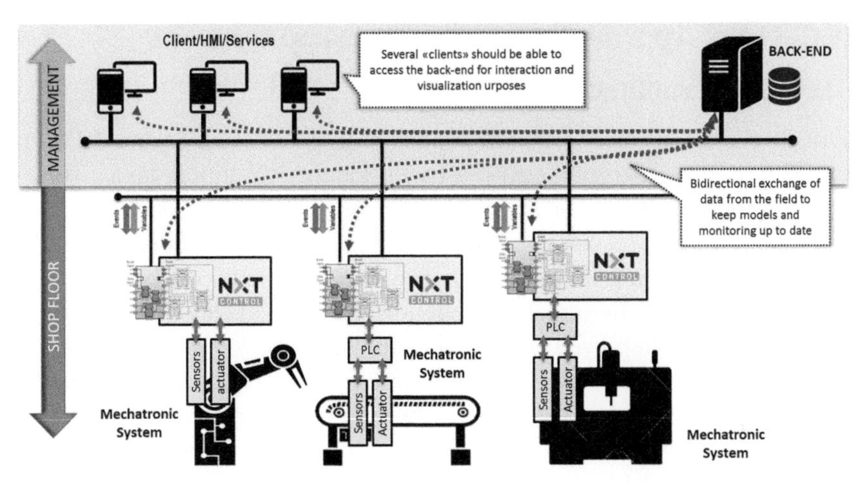

Figure 4.12 Direct and distributed connection between the digital domain and the Daedalus' CPS.

- Asynchronous connections can be established and maintained between a specific CPS and whatever ICT module has the privileges to do so, for instance, for extensive data gathering with semantic description attached; the level of access (within the shopfloor hierarchy of aggregation) is limited only by the granularity enabled by automation developers;
- Each CPS can be programmed to "expose" only the connections and functionalities that its automation developer deems appropriate, increasing at design level the security of the overall connection (apart from specific cyber-security mechanisms);
- Real-time automation functionalities governing the behaviour of the system can be "augmented" by asynchronous access to digital modules conceived to offer specific tools to the automation developer, exploiting, for instance, the higher computational power of a local or cloud server.

As an explicit consequence, the "Industrie 4.0"-envisioned bridging between the execution of the lowest-level manufacturing operations on the shop floor and the highest-level decision making of the top management of a factory is automatically obtained.

4.4 IEC-61499 Approach to Cyber-Physical Systems

4.4.1 IEC-61499 runtime

Based on an overall vision of CPS introduced in Daedalus (see Figure 4.13), an IEC61499 runtime enables the 61499-execution model running on a given OS and hardware platform, for example, the Linux Debian OS running on an ARM cortex platform. The runtime includes an event scheduler module responsible for scheduling the execution of algorithms; a resource management module to handle the creation, deletion, and life cycles of managed function blocks in a deployed application and modules to provide timer, memory, logging, IO access and communication services. The combination of hardware, OS services and the IEC 61499 runtime are collectively known as a device in the 61499 context, and a generic architecture for such a device is illustrated in Figure 4.14.

A control application is developed using an IEC-61499 compliant Engineering tool and then deployed to the device where, when necessary, it utilizes different communication protocols and OS services to interact with other CPS and the physical world (e.g., IO access). The IEC 61499 runtime can be extended to support different communication stacks, field buses and OS service and they are to be encapsulated as SIFB function blocks where

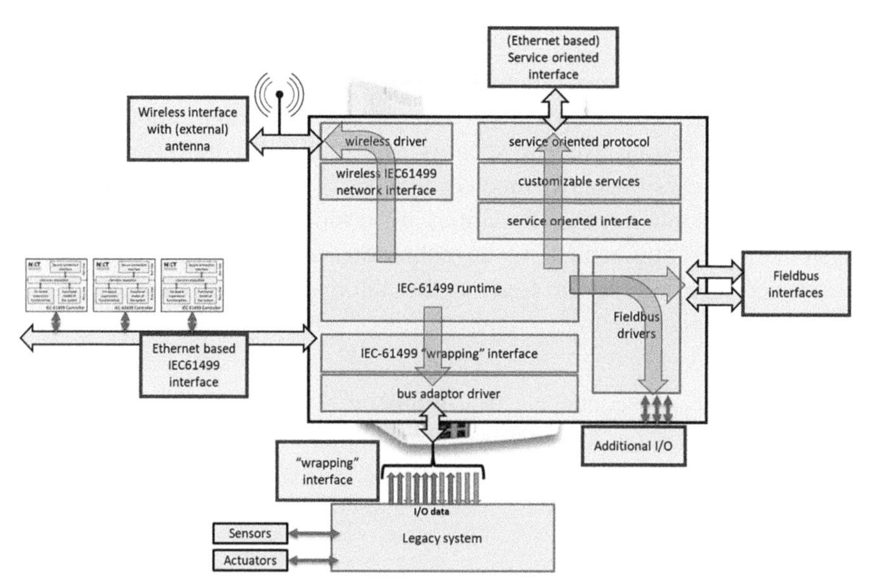

Figure 4.13 Qualitative functional model of an automation CPS based on IEC-61499.

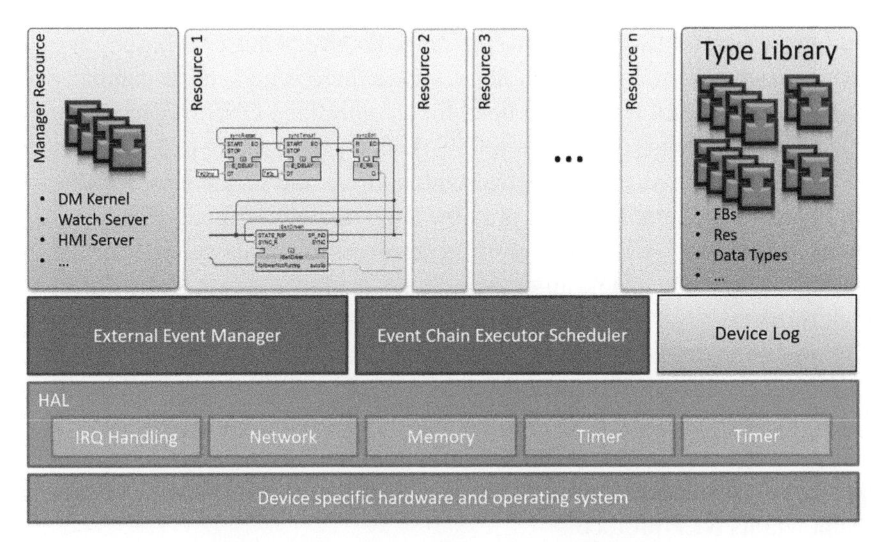

Figure 4.14 IEC 61499 runtime architecture.

the control application can access their services by making event and data connections to them. In this way, the application designer does not require any knowledge about the technical details how the communication will be established. For the platform to be widely applicable, it also needs the ability to communicate with other wireless CPS devices (see Section 4.5).

To enable faster, easier and less error-prone configuration of a network of CPSs in a dynamic changeable network topology, in Daedalus, auto-discovery and self-declaration have been added to the IEC61499 Runtime. To allow this, each device must be capable of creating semantic description of its own interface and functional automation capabilities, making its existence on the network (presence) known to other devices by advertising its entrance and leaving of the network and make necessary exchange of information in standardized, unambiguous syntax and semantics.

The first step is to develop a semantic meta-model for describing the functionalities provided by the CPS. The model must describe the physical interface of the device (parameters) and logical interface to access the automation capabilities it provides. Once the model has been automatically created, it can be exchanged with other CPS in predefined, extensible .xml format.

For the CPS to easily adapt to the dynamic network topology (imagine wireless CPS devices on a mobile platform), where CPS or SoA entities may join and leave local network at will, the auto-discovery must be based on a zero-configuration (zeroconf) technology, where there is no need to manually reconfigure the network layout or a need for a centralized DNS server, where it becomes a single point of failure. A CPS device participating in a zeroconf network will be automatically assigned with address and hostnames, making low-level network communication possible immediately after a device joins a network. Multicast DNS, a subset of the zeroconf technology, will further allow CPS to subscribe and be automatically notified of changes to the layout of the network.

To support the exchange of semantic information used for identification of other CPS's capabilities in the network, a new communication protocol based on XMPP has been chosen to be included in the IEC 61499 runtime. XMPP is chosen to leverage on mature standards that will encourage a broader acceptance of the solution implemented as well as its intrinsic nature of being extensible via its XEP protocol.

4.4.2 Functional Interfaces

4.4.2.1 IEC-61499 interface

The IEC-61499 interface enables the CPS to connect to a network of IEC-61499-based controllers leveraging a communication profile compliant with the IEC-61499 standard and enabling, as a consequence, a unified and globally recognized communication means with a network of automation devices.

This interface is mainly dedicated to the exchange of real-time data among the CPSs participating to the same IEC-61499 distributed control application, but it is also exploited by other systems to interact with a CPS to accomplish to specific tasks, as for example:

- to configure the IEC-61499 runtime;
- to deploy the IEC-61499 code in the CPS;
- to monitor and debug the IEC-61499 control application.

The IEC-61499 interface is also the interface that is going to host a strong real-time synchronization mechanisms.

From a hardware perspective, the IEC-61499 interface can be implemented both as a wired Ethernet interface, allowing wired and strong reliable connections, and as a wireless interface, providing flexibility in the implementation of a communication network. It is relevant to highlight, however, that the wireless connectivity will pose some limits in the performance that can be expected for the coordination of the distributed CPS in that network.

4.4.2.2 Wireless interface

The Wireless interface of DAEDALUS' CPS is mainly dedicated to the interfacing with remote devices based on dedicated communication protocols, for application-specific tasks. When the considered task follows in the context of connectivity among IEC-61499 nodes, this interface can partially overlap in terms of functionalities the IEC-61499 interface (in the wireless version). However, while the IEC-61499 interface is designed to be a general communication interface for cooperation of distributed control devices over an IP-based network, the Wireless interface is specialized to support specific communication links. Some examples of specific communication channels for which the Wireless interface would be appropriate are:

- Point-to-point bus communication over a specific wireless technology (different from 802.11a,b,g,n) between two CPSs to support IEC-61499 connectivity;
- Connection to remote device for mono-/bi-directional exchange of data, for example:
    - to a remote I/O module;
    - to a DAEDALUS CPS behaving as a supervisor node;
    - to a third-party technology gateway.

To enable an effective approach, which can make easier to extend in future this to support additional wireless communication technologies, this interface is structured on a dual layer:

- a hardware abstraction layer, which provides the mechanisms to leverage the Wireless interface within an IEC-61499 application, and that hides the details of the communication technology adopted underneath;
- a technology-specific driver, which is leveraged by the abstraction layer to map the expected functionalities over the specific features offered by the selected communication protocol.

4.4.2.3 Wrapping interface

The Wrapping interface constitutes the enabler for an IEC-61499-based controller to operate as a CPS-izer. This interface has to enable the communication with a "legacy" controller through a communication channel not based on the IEC-61499 protocol.

While from the communication protocol perspective, we can foresee different implementations of this interface based on the specific protocol adopted by the CPS to connect to a non-IEC61499-based controller. The main characteristic of this interface is to present a well-defined mechanism to enable interaction with third-party control applications.

Through this interface, it will be possible to enable the cooperation between the event-based approach of a DAEDALUS' CPS with the scan-based mechanism adopted by classic controllers. This enables us to consider the CPS as a wrapper that extends the capabilities of the legacy controller with the IEC-61499 features.

4.4.2.4 Service-oriented interface

The service-oriented interface of a DAEDALUS CPS is fully integrated in the IEC 61499 runtime platform and conceived to enable a dynamic interaction among the CPSs and between the CPSs and the higher automation layers. By means of that interface, a CPS will be able to connect to other systems at the shop floor or at the supervisory/management levels for acquisition of data reflecting the current state of the manufacturing process and therefore extending its perceiving capabilities over the limits of its directly connected sensors.

The service-oriented interface enables a unified methodology of interaction among the intelligent units of the manufacturing plant and, at the same time, the possibility for an orchestrating unit at supervisory/management level to interact directly with the network of cyber-devices and coordinate their action, without requiring compliancy with the IEC 61499.

A CPS exposes through its service-oriented interface a set of functionalities that are exploited by an orchestrating intelligence to reconstruct a better understanding of the actual condition of the manufacturing process and of the CPS' behaviour and to elaborate more accurate and effective coordination plans, which are then used to instruct appropriately the single automation units.

The service-oriented interface provides a flexible communication mechanism that does not require the specification of all the nodes involved in the communication at design stage, hence making the application easy to scale.

The specification of the service-oriented interface defines (among other aspects):

- The architectural mechanism to integrate the service-oriented interface within an IEC-61499 runtime;

- The protocols supported by the initial implementation of the DAEDALUS platform;
- The set of services implemented as a first prototype of the interface.

4.4.2.5 Fieldbus interface(s)

To enable a DAEDALUS CPS to be applicable to different application scenarios, the CPS should support connectivity toward other automation devices through common fieldbus technologies.

The fieldbus interface(s) can be of different types and the specific implementation will depend on the types of technologies, for which the appropriate driver will be available/implemented, and the application requirements.

The general goal of this interface is to provide I/O communication with other automation devices. Some of the common fieldbus technologies that are planned to be supported are EtherCAT and Modbus TCP/IP.

4.4.2.6 Local I/O interface

The Local I/O interface represents a specific interfacing mechanism toward the I/O modules locally installed in the same HW platform of the CPS.

From a functional point of view, this interface is similar to the Fieldbus interface, but it is specialized to enable the exploitation of the resources characterizing a specific implementation of DAEDALUS CPS: those resources can leverage custom/proprietary communication mechanisms, instead of common standards.

4.5 The "CPS-izer", a Transitional Path towards Full Adoption of IEC-61499

The technological concept is that of a CPS-izer: a small-footprint (and costs) controller, based on the IEC-61499 technologies of Daedalus, is also capable of interfacing with usual PLCs through standard communication buses (Figure 4.15). This could provide a path for transition towards digital automation to two major families of users:

- End-users will have access to a product that can be easily installed on existing machines and manufacturing systems and, with a limited engineering effort, used to upgrade their plants prolonging functional life;
- Other developers of automation platforms compliant with IEC-61131 will have a temporary solution to make their systems at least partially coherent with the new IEC-61499 standard.

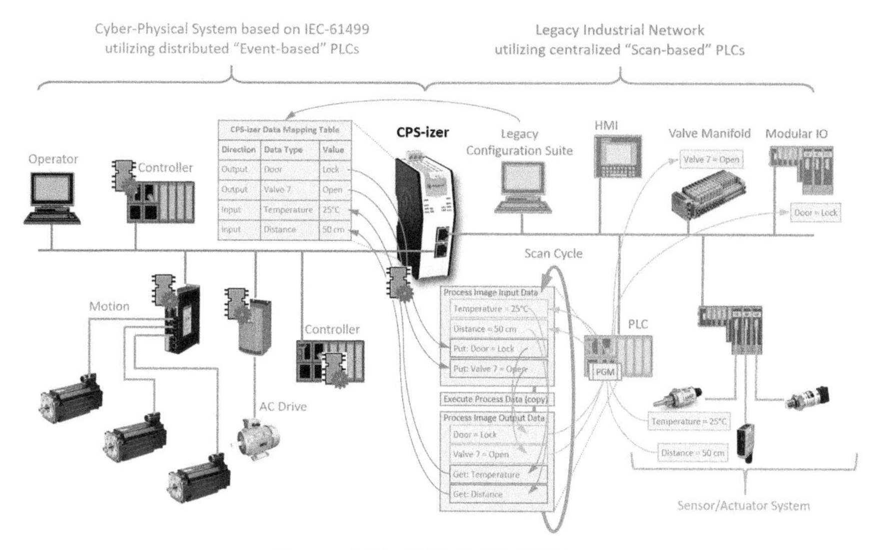

Figure 4.15 IEC 61499 CPS-izer.

Finally, the object-oriented approach that the standard adopts will not be limited a priori to automation algorithms only, but it can be extended to further "dimensions of existence" of the system, guaranteeing two important added values. Behavioural models of CPS (needed for several purposes, such as simulation) will become explicit elements of the device virtual representation (avatar), enabling seamless (= transparent to the end-user) connectivity between the device deployed on field and its models memorized in-Cloud. In addition, synchronization and co-simulation in near-real-time will be automatically achieved as already part of the functional IEC-61499 architecture, with the event-based nature of the standard perfectly suited to deal with the management of Big Data coming from the field.

The CPS-izer follows the same common requirements like for an IEC-61499 Controller device, but deviations of the implementation of the common requirements for the CPS-izer in comparison to an IEC-61499 Controller device are possible. Besides these common requirements, there are other requirements and constraints defined for the CPS-izer. First of all, the CPS-izer needs to provide support for legacy industrial networks.

Legacy industrial networks are characterized by means of their physical and data link layers (e.g., Serial, CAN, Ethernet) and the transport layers up to the application layers depending on the implemented technologies (e.g., Modbus/RTU, PROFIBUS, CANopen, PROFINET).

The CPS-izer supports these legacy industrial networks by means of the adaptation of an interface, which could be implemented by hardware, software or IP-core. The preferred solution for the CPS-izer in Daedalus is the HMS Anybus® (see https://www.anybus.com/products/embedded-index) embedded product family of various industrial interfaces. The CPS-izer only implements a slave/server/device in terms of the applied industrial networking technology. As a restriction, here the CPS-izer cannot be used as a master/client/controller in any industrial network.

One constraint of this device is that it will only support connectivity to legacy industrial networks but no IO data as signals, neither discrete nor analogue. As for an example, other IO modules connected in the legacy industrial network controlled by a PLC in that system can be used if additional IO signals are needed. Also, no support for IO data like in industrial sensor/actuator system (e.g., AS-Interface, IO-Link) will be provided. Those systems would require a master to be implemented, which is out of scope for the realization of the CPS-izer. If IO data from such systems need to be exchanged with CPS, this shall be realized by using an appropriate master in the legacy industrial network controlled by a PLC in that system. The CPS-izer may have limited resources for IEC 61499 functionalities compared to the IEC 61499 Controller when it comes to the implementation of the runtime system. It must of course implement function block(s) and driver(s)/interface(s) to handle the data transfer to the legacy industrial network connected.

The CPS-izer will map input data and output data between CPS and legacy industrial networks. For this, the CPS-izer will implement some kind of a shared memory (in either physical or logical way) to exchange data. The data mapped to this area will be consistent in common for all inputs and outputs mapped with the legacy industrial network. It may be consistent to a finer granularity depending on the types of devices connected.

Since all legacy industrial networks share the same implementation approach of mapping data, this will be the lowest common denominator of all such systems. So, the CPS-izer will follow this philosophy. Some – but not all – legacy industrial networks provide events like alarms or diagnostic messages. That implementation is always specific to the industrial network, but no generic solution will be available for this. So, the CPS-izer will not support events of the legacy industrial networks.

The configuration of the available input and output data in CPS-izer will be specific to the legacy industrial network it is connected to. The tools and methods typically for such networks are applied. The PLC in that system is responsible to get the input data from CPS-izer and write them to the outputs of the devices. On the other hand, it will collect the inputs from the devices and put them into the output data of the CPS-izer. The processing of output and input data in the PLC will follow the common approach for a scan cycle as it is implemented in automation industries since decades: read inputs – execute process data – write outputs.

In terms of such PLC systems, the CPS-izer will put output data from the legacy industrial network to the CPS, which is seen there as input. It will get output data from the CPS, which is seen as input in the legacy industrial network. For the CPS-izer, the execution of the process data in the PLC is just a copy function to copy data from the process image input to the process image output.

Some legacy industrial networks like EtherCAT or PROFINET provide real-time capabilities to transport IO data in the ms or even s range of cycle times. This real-time behaviour will not be made transparent to the CPS. The CPS-izer will only guarantee data consistency between CPS and legacy industrial network related to the cycle time running in that network, but it cannot guarantee real-time transport between both systems.

The CPS-izer should be realized in a small industrial-approved plastic housing, which could be easily mounted at a machine or in a cabinet using DIN-rail mechanics. It should require a single 24 V power supply as used in standard industrial automation systems. Furthermore, it should realize a common way to connect to legacy industrial networks by means of front plugs/connectors and indicators.

The CPS-izer should follow requirements for industrial grading like temperature range, shock and vibration, EMC and others for common cabinet mounting. It must adhere to CE compliance.

Harsh industrial requirements like IP67, sealed connectors and housing and higher temperature range are not in the focus of the realization of the CPS-izer.

4.6 Conclusions

This chapter has explored a new generation of functional architecture for industrial automation, centred around the concepts, methodologies and technologies of the IEC-61499 standard but exploiting and extending them for a concrete implementation of what are called "Cyber-Physical Systems".

The transition to this type of model is not just a matter of installing new devices into a shopfloor, but it requires a real paradigm shift in the way real-time control and automation in manufacturing are engineering, introducing new concepts of design and the corresponding skills.

Daedalus project is developing all the tools to enable such a transition, considering both green-field and brown-field scenarios, accepting that Industry 4.0 full implementation will need a radical change in the way existing PLCs work.

Acknowledgements

This work was achieved within the EU-H2020 project DAEDALUS, which received funding from the European Union's Horizon 2020 research and innovation programme, under grant agreement No 723248.

References

[1] http://www.pathfinderproject.eu
[2] https://ec.europa.eu/digital-agenda/en/digitising-european-industry
[3] Zoitl, Alois, and Valeriy Vyatkin. "IEC 61499 Architecture for Distributed Automation: the 'Glass Half Full' View", IEEE Industrial Electronics Magazine 3.4: 7–23, 2009.
[4] Object Management Group, "Model Driven Architecture", Online Available: http://www.omg.org/mda/faq_mda.htm, Jun. 2009.
[5] B. Huber, R. Obermaisser, and P. Peti, "MDA-based development in the DECOS integrated architecture–modeling the hardware platform", in Object and Component-Oriented Real-Time Distributed Computing, 2006. ISORC 2006. Ninth IEEE International Symposium in April 2006, p. 10.
[6] http://www.maya-euproject.com

5

Communication and Data Management in Industry 4.0

Maria del Carmen Lucas-Estañ[1], Theofanis P. Raptis[2], Miguel Sepulcre[1], Andrea Passarella[2], Javier Gozalvez[1] and Marco Conti[2]

[1]UWICORE Laboratory, Universidad Miguel Hernández de Elche (UMH), Elche, Spain
[2]Institute of Informatics and Telematics, National Research Council (CNR), Pisa, Italy
E-mail: m.lucas@umh.es; theofanis.raptis@iit.cnr.it; msepulcre@umh.es; andrea.passarella@iit.cnr.it; j.gozalvez@umh.es; marco.conti@iit.cnr.it

The Industry 4.0 paradigm alludes to a new industrial revolution where factories evolve towards digitalized and networked structures where intelligence is spread among the different elements of the production systems. Two key technological enablers to achieve the flexibility and efficiency sought for factories of the future are the communication networks and the data management schemes that will support connectivity and data distribution in Cyber-Physical Production Systems. Communications and data management must be built upon a flexible and reliable architecture to be able to efficiently meet the stringent and varying requirements in terms of latency, reliability and data rates demanded by industrial applications. To this aim, this chapter presents a hierarchical communications and data management architecture, where decentralized and local management decisions are coordinated by a central orchestrator that ensures the efficient global operation of the system. The defined architecture considers a multi-tier organization, where different management strategies can be applied to satisfy the different requirements in terms of latency and reliability of different industrial applications. The use of virtualization and softwarization technologies as RAN Slicing and

Cloud RAN will allow to achieve the flexibility, scalability and adaptation capabilities required to support the high-demanding and diverse industrial environment.

5.1 Introduction

In future industrial applications, the Internet of Things (IoT) with its communications and data management functions will help shape the operational efficiency and safety of industrial processes through integrating sensors, data management, advanced analytics, and automation into a mega-unit [1]. The future and significant participation of intelligent robots will enable effective and cost-efficient production, achieving sustainable revenue growth. Industrial automation systems, emerging from the Industry 4.0 paradigm, count on sensors' information and the analysis of such information [2]. As such, connectivity is a crucial factor for the success of industrial Cyber-Physical-Systems (CPS), where machines and components can talk to one another. Moreover, in the context of Industry 4.0 and to match the increased market demand for highly customized products, traditional pilot lines designed for mass production are now evolving towards more flexible "plug & produce" modular manufacturing strategies based on autonomous assembly stations [3], which will make increased use of massive volumes of Big Data streams to support self-learning capabilities and will demand real-time reactions of increasingly connected mobile and autonomous robots and vehicles. While conventional cloud solutions will be definitely part of the picture, they will not be enough. The concept of centrally organized enterprises at which large amounts of data are sent to a remote data center do not deliver the expected performance for Industry 4.0 scenarios and applications. Recently, moving service supply from the cloud to the edge has enabled the possibility of meeting application delay requirements, improves scalability and energy efficiency, and mitigates the network traffic burden. With these advantages, decentralized industrial operations can become a promising solution and can provide more scalable services for delay-tolerant applications [4].

Two technological enablers of Industry 4.0 are: (i) the communication infrastructure that will support the ubiquitous connectivity of Cyber-Physical Production Systems (CPPS) and (ii) the data management schemes built upon the communication infrastructure that will enable efficient data distribution within the Factories of the Future [5]. In the industrial environment, a wide set of applications and services with very different communication requirements will coexist, being one of the most demanding verticals with respect to the number of connected nodes, ultra-low latencies, ultra-high reliability, energy efficiency, and ultra-low communication costs [6]. The varying and stringent communication and data availability requirements of the industrial applications pose an important challenge for the design of the communication network and of the data management systems. The communication network and the data management strategy must be built upon a flexible architecture capable of meeting the communication requirements of the industrial applications, with particular attention on time-critical automation.

The architecture reviewed in this chapter is the reference communications and data management architecture of the H2020 AUTOWARE project [7]. The main objective of AUTOWARE is to build an open consolidated ecosystem that lowers the barriers of small, medium- & micro-sized enterprises (SMMEs) for cognitive automation application development and application of autonomous manufacturing processes. Communications and data management are two technological enablers within the AUTOWARE Framework (Figure 5.1 and presented in detail in Chapter 2). Within the AUTOWARE framework, the AUTOWARE Reference Architecture establishes four layers: Enterprise, Factory, Workcell/Production Line, and Field Devices. In addition, the AUTOWARE Reference Architecture also includes two transversal layers: (i) the Fog/Cloud layer, since applications or services

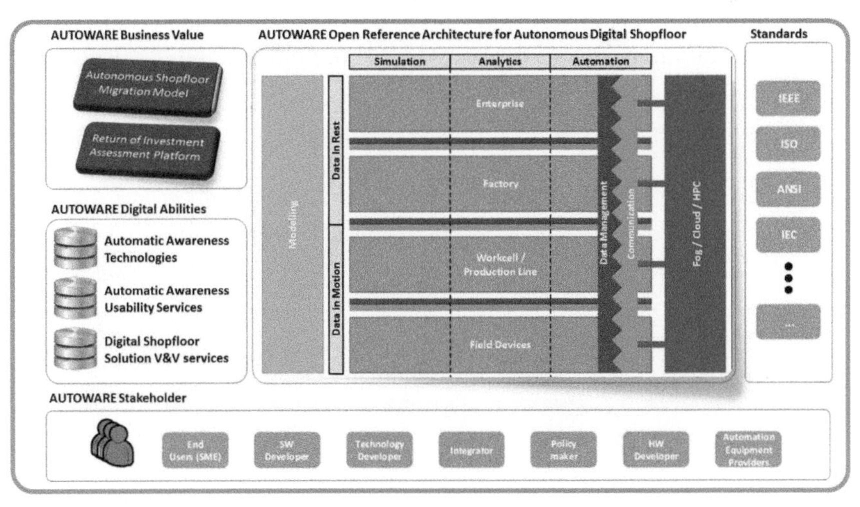

Figure 5.1 The AUTOWARE framework.

in all the layers can be included or implemented in the Fog/Cloud, and (ii) the Modelling layer, since different technical components inside the different layers can be modelled, and it could be possible to have modeling approaches that take the different layers into account. The communications and data management architecture proposed in AUTOWARE supports the communication network and the data management system and enables the data exchange between the different AUTOWARE components, exploiting the Fog and/or Cloud concepts. It provides communication links between devices, entities, and applications implemented in different layers, and also within the same layer. Within the AUTOWARE Reference Architecture (defined in the H2020 AUTOWARE Project), the communication network and data management system can be represented as a transversal layer that interconnects all the functional layers of the AUTOWARE Reference Architecture (see Figure 5.2). The communications and data management architecture presented in this chapter provides the communication and data distribution capabilities required by the different systems or platforms developed within the AUTOWARE framework.

AUTOWARE proposes the use of a heterogeneous network that integrates different communication technologies covering the industrial environment. The objective is to exploit the abilities of different wired and wireless communication technologies to meet the broad range of communication requirements posed by Industry 4.0 in an efficient and reliable way. To this aim, inter-system interferences between different wireless technologies operating in the same unlicensed frequency band need to be monitored and controlled, as well as inter-cell interferences for wireless technologies using the licensed

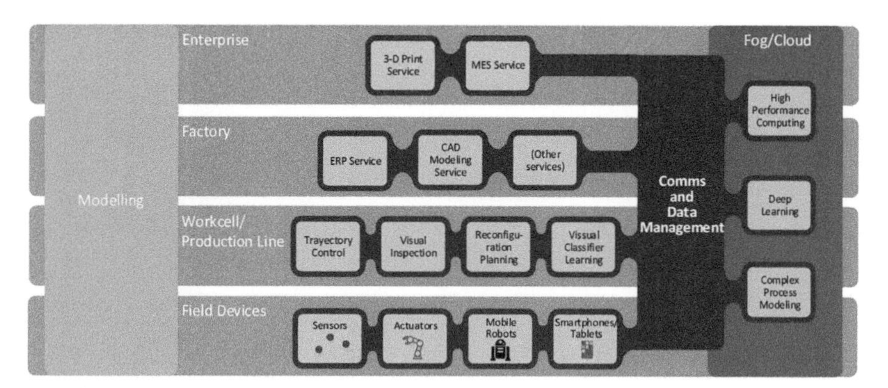

Figure 5.2 Communication network and data management system into the AUTOWARE Reference Architecture.

spectrum. From a data management standpoint, real-time data availability requirements, optimized utilization of IT resources (particularly for SMMEs), and data ownership constraints call for distributed data management schemes, whereby data are stored, replicated, and accessed from multiple locations in the network, depending on data generation and data access patterns, as well as the status of physical resources at the individual nodes.

To efficiently integrate the different communication technologies in a unique network and handle the data management process, we adopt a software-defined hierarchical approach where a central entity guarantees the coordination of local and distributed managers resulting in a mix of centralized management (orchestration) and decentralized operation of the communication and data management functions. Communication links are organized in different virtual tiers based on the performance requirements of the application they support. Different communications and data management strategies can then be applied at each tier to meet the specific communication and data availability requirements of each application. To implement the proposed hierarchical and multi-tier management architecture, we consider the use of RAN (Radio Access Network) Slicing and Cloud RAN as technological enablers to achieve the flexibility, scalability, and adaptation architectural capabilities needed to guarantee the stringent and varying communication and data distribution requirements of industrial applications.

This chapter is organized as follows. Section 5.2 presents the requirements imposed by Industry 4.0 to the communications and data management system. Section 5.3 reviews communication architectures proposed for Industrial Wireless Networks, and Section 5.4 presents traditional and current trends on the design of data management strategies in industrial environments. Section 5.5 presents the proposed communications and data management architecture, and the technological enablers considered to build up the architecture, RAN Slicing and Cloud RAN. Section 5.6 describes the possibilities offered by the proposed hierarchical architecture to implement hybrid management schemes to introduce flexibility in the management of wireless connections while maintaining a close coordination with a central network manager. Section 5.7 presents examples of early adoption of communication and data management concepts supported by the suggested architecture. How the reference communications and data management architecture fits into the overall AUTOWARE framework is presented in Section 5.8. Section 5.9 summarizes and concludes the chapter.

5.2 Industry 4.0 Communication and Data Requirements

Industry 4.0 poses a complex communication environment because of the wide set of different industrial applications and services that will coexist, all of them demanding very different and stringent communication requirements. The 5GPPP classifies industrial use cases in five families, each of them representing a different subset of communication requirements in terms of latency, reliability, availability, throughput, etc. [6]. Instant process optimization based on real-time monitoring of the manufacturing performance and the quality of produced goods is one of the most demanding use case families in terms of latency and reliability. Some of the sensors may communicate at low bitrates but with ultra-low latency and ultra-high reliability, whereas vision-controlled robot arms or mobile robots may require reliable high-bandwidth communication. Inside the factory, there are also applications or services without time-critical requirements, such as the localization of assets and goods and logistic processes, non-time critical quality control, or data capturing for later usage in virtual design contexts. The challenge in this second use case family is to ensure high availability of the wireless networks, given the harsh industrial environment. Remotely controlling digital factories requires end-to-end communications between remote workers and the factory. This use case family could simply involve the use of tablets or smartphones, or more complex scenarios with augmented reality devices that facilitate the creation of virtual back office teams that exploit the collected data for preventives analytics. In this use case family, there is a less stringent need for low latency, but high availability is key to ensure that emergency maintenance actions can take place immediately. The fourth use case family identified involves the connectivity between different production sites as well as with further actors in the value chain (e.g. suppliers, logistics) seamlessly. A high level of network and service availability and reliability including wireless link is one of the key requirements. The last use case family identified by the 5G-PPP considers that factories will play an important role in the provisioning of the connected goods that are produced, for which autonomy is a key requirement. Table 5.1 summarizes the communication requirements for each of the five use case families identified by the 5G-PPP.

The International Society of Automation (ISA) and ETSI also highlight the diverse communication requirements of industrial applications. For example, ISA establishes safety, control, and monitoring applications in six different classes based on the importance of message timeliness [9]. ETSI has also investigated the communication requirements of industrial automation

Table 5.1 5G-PPP use case families for manufacturing [6]

Use Case Family	Representative Scenarios	Latency	Reliability	Bandwidth
1. Time-critical process optimization inside factory	• Real-time closed-loop communication between machines to increase efficiency and flexibility • 3D augmented reality applications for training and maintenance • 3D video-driven interaction between collaborative robots and humans	Ultra-low	Ultra-high	Low to high
2. Non-time-critical in-factory communication	• Identification/tracing of objects/goods inside the factory • Non-real-time sensor data capturing for process optimization • Data capturing for design, simulation, and forecasting of new products and production processes	Less critical	High	Low to high
3. Remote control	• Remote quality inspection/diagnostics • Remote virtual back office	Less critical	High	Low to high
4. Intra-/Inter-enterprise communication	• Identification/tracking of goods in the end-to-end value chain • Reliable and secure interconnection of premises (intra-/inter-enterprise) • Exchanging data for simulation/design purposes	Ultra-low to less critical	High	Low to high
5. Connected goods	• Connecting goods during product lifetime to monitor product characteristics, sensing its surrounding context and offering new data-driven services	Less critical	Low	Low

in [10] and differentiated two types of applications. The first type involves the use of sensors and actuators in industrial automation and its main requirement is the real-time behavior or determinism. The second type of applications involves the communication at higher levels of the automation hierarchy, e.g. at the control or enterprise level, where throughput, security, and reliability become more important. Automation systems are subdivided into three main classes (manufacturing cell, factory hall, and plant level) with different needs in terms of latency (from 5 to 20 ms). Their requirements in terms of latency, update time, and number of devices can notably differ between them (see Table 5.2). However, all three classes require a 10^{-9} packet loss rate and a 99.999% application availability.

The timing requirements depend on different factors. As presented by the 5GPPP in [6], process automation industries (such as oil and gas, chemicals, food and beverage, etc.) typically require cycle times of about 100 ms. In factory automation (e.g. automotive production, industrial machinery, and consumer products), typical cycle times are 10 ms. The highest demands

Table 5.2 Performance requirements for three classes of communication in industry established by ETSI [10]

	Manufacturing Cell	Factory Hall	Plant Level
Indoor/outdoor application	Indoor	Mostly indoor	Mostly outdoor
Spatial dimension L×W×H (m^3)	10×10×3	100×100×10	1000×1000×50
Number of devices (typically)	30	100	1000
Number of parallel networks (clusters)	10	5	5
Number of such clusters per plant	50	10	1
Min. number of locally parallel devices	300	500	250
Network type	Star	Star/Mesh	Mesh
Packet size (on air, byte)	16	200	105
Max. allowable latency (end-to-end) incl. jitter/retransmits (ms)	5 ± 10%	20 ± 10%	20 ± 10%
Max. on-air duty cycle related to media utilization	20%	20%	20%
Update time (ms)	50 ± 10%	200 ± 10%	500 ± 10%
Packet loss rate (outside latency)	10^{-9}	10^{-9}	10^{-9}
Spectral efficiency (typically) (bis/s/Hz)	1	1.18	0.13
Bandwidth requirements (MHZ)	8	34	34
Application availability		Exceeds 99.999%	

Table 5.3 Timing requirements for motion control systems [6]

Requirement	Value
Cycle time	1 ms (250 μs . . . 31.25 μs)
Response time/update time	. . . 100 μs
Jitter	<1 μs . . . 30 ns
Switch latency time	. . . 40 ns
Redundancy switchover time	<15 μs
Time synchronization accuracy	. . . 100 ns

Table 5.4 Communication requirements for some industrial applications [5]

	Motion Control	Condition Monitoring	Augmented Reality
Latency/cycle time	250 μs–1 ms	100 ms	10 ms
Reliability (PER)	1e-8	1e-5	1e-5
Data rate	kbit/s–Mbit/s	kbit/s	Mbit/s–Gbit/s

are set by motion control applications (printing machines, textiles, paper mills, etc.) requiring cycle times of less than 1 ms with a jitter of less than 1 μs. For motion control, current requirements are shown in Table 5.3. Table 5.4 also shows the communication requirements of three relevant application examples (extracted from [5]) that illustrate the range of diverging and stringent communications requirements imposed by Industry 4.0.

These requirements have been confirmed within AUTOWARE. The communication requirements of several industrial use cases that are being developed within AUTOWARE have been analyzed. For example, in the PWR Pack AUTOWARE use case presented in [11], a stringent latency bound of 1 ms with a data rate lower than 100 kb/s is imposed to transmit commands from a Programmable Logic Controller (PLC) to a robot to control the servomotors and the movement of the robot, while 1–100 Mb have to be transmitted per image from a camera to a 3D visualization system tolerating a maximum 5 ms latency. On the other hand, the communication between a fixed robot and a component supplier mobile robotic platform within the neutral experimentation facility for collaborative robotics that is being developed by IK4-Tekniker [12] requires robust, flexible, and highly reliable wireless communication with latency bounded to some hundreds of milliseconds to guarantee the coordination and interoperation of both robots.

Due to the fact that the application functions should be applicable to different types of network nodes, they cannot rely only on specific communication functions, but include additional functions like smart data distribution and management. It should be worth noting that the ultimate

Table 5.5 Additional requirements for different application scenarios [13, 14]

	Desired Value	Application Scenario
Connectivity	300.000 devices per AP	Massive M2M connectivity
Battery life	>10 years	Hard-to-reach deployments
Reliability	99.999%	Protection and control
Seamless and quick connectivity	–	Mobile devices

Industry 4.0 application performance is the result of the concurrent operation and synergies across communication architectures and data distribution strategies. Table 5.5 shows some additional requirements for different application scenarios that impose additional constraints to manage the communications network and impose specific constraints to data management schemes [13, 14]. A massive M2M (machine to machine) connectivity will require an Access Point (AP) to support hundreds of thousands of field devices, with obvious limitations on the data rates each can support, and thus on rates at which they are enquired for (new) data. Maintenance for such large connectivity should be very low; thus, a very long battery period for such devices will be a necessity. A battery life for wireless devices greater than 10 years will mean that many hard-to-reach sensors and actuators could only sustain very low data rates. Reliability will play a critical role in industrial requirements with safety protection and control applications, calling for resilient data management schemes. In addition to all these requirements, a network should also be able to provide pervasive connectivity experience for the devices that may transition from outdoors to indoors location in a mobile scenario. Finally, data availability issues impose other specific requirements. For example, depending on applications, data might not be replicated outside of a set of devices or a geographical area for ownership reasons. Data might have to be replicated, instead, on other groups of nodes for data availability. Conversions across data formats might be needed, to guarantee interoperability across different factory or enterprise systems. All these issues belong to the broader concept of data sovereignty that is the main focus of the Industrial Data Space (IDS) initiative [15].

5.3 Industrial Wireless Network Architectures

Traditionally, communication networks in industrial systems have been based on wired fieldbuses and Ethernet-based technologies, and often on proprietary standards such as HART, PROFIBUS, Foundation Fieldbus H1, etc. While

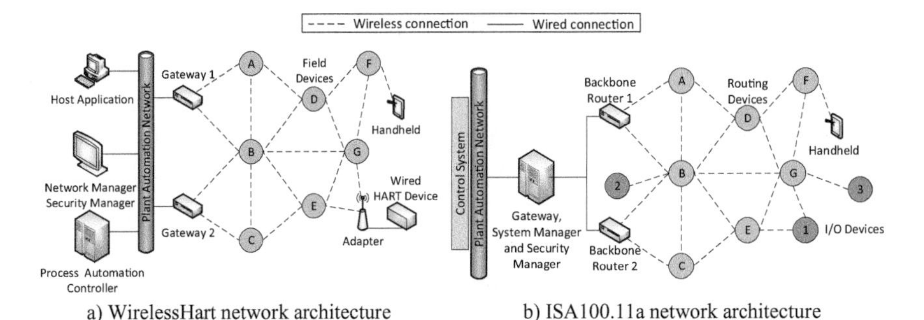

a) WirelessHart network architecture b) ISA100.11a network architecture

Figure 5.3 Examples of centralized management architectures.

wired technologies can provide high communications reliability, they are not able to fully meet the required flexibility and adaptation of future manufacturing processes for Industry 4.0. Wireless communication technologies present key advantages for industrial monitoring and control systems. They can provide connectivity to moving parts or mobile objects (robots, machinery, or workers) and offer the desired deployment flexibility by minimizing and significantly simplifying the need of cable installation. Operating in unlicensed frequency bands, WirelessHART, ISA100.11a, and IEEE 802.15.4e, are some of the wireless technologies developed to support industrial automation and control applications. These technologies are based on the IEEE 802.15.4 physical and MAC (Medium Access Control) layers, and share some fundamental technologies and mechanisms, e.g., a centralized network management and Time Division Multiple Access (TDMA) combined with Frequency Hopping (FH). Figure 5.3 shows the network architecture for WirelessHart and ISA100.11a. In both examples, there is a central network management entity referred to as Network Manager in a WirelessHart network and System Manager in the ISA100.11a network that is in charge of the configuration and management at the data link and network levels of the communications between the different devices (gateways, routers, and end devices).

The main objective of having a centralized network management is to achieve high communications reliability levels. However, the excessive overhead and reconfiguration time that results from collecting state information by the central manager (e.g. the Network Manager in a WirelessHart network or the System Manager in a ISA100.11a network) and distributing management decisions to end devices limits the reconfiguration and scalability capabilities of networks with centralized management, as highlighted in [16] and [17]. To overcome this drawback, the authors of [17–21] proposed to divide a large network into multiple subnetworks and considered a hierarchical management architecture. In this context, each subnetwork has its own manager that deals with the wireless dynamics within its subnetwork. A global entity is in charge of the management and coordination of the entire network with the subnetwork managers. Proposals in [19–21] rely on hierarchical architectures and also propose the integration of heterogeneous technologies to efficiently guarantee the wide range of different communication requirements of industrial applications; the need of using heterogeneous technologies in manufacturing processes was already highlighted by ETSI in [10]. For example, the approach proposed in [19], and shown in Figure 5.4(a), considers the deployment of several subnetworks in the lowest level of the industrial network architecture connecting sensors and actuators. The deployed devices collect data and send it to a central control and management system, which is located at the highest level of the network architecture. This IWN integrates and exploits various wireless technologies with different communication capacities at different levels of the architecture. Coordinators at each subnetwork act as sink nodes and collect data from different low-bandwidth sensors and transmit it to gateway nodes using higher-bandwidth wireless technologies. The gateway nodes are usually deployed so that they can collect and transmit data from various sink nodes to the central control and management system through high-bandwidth technologies. Another example is the network architecture proposed in the framework of the DEWI (Dependable Embedded Wireless Infrastructure) project [22]. The DEWI hierarchical architecture [20] is depicted in Figure 5.4(b). This architecture is based on the concept of DEWI Bubbles. A DEWI Bubble is defined as a high-level abstraction of a set of industrial wireless sensor networks (WSN) located in proximity with enhanced inter-operability, technology reusability, and cross-domain development. In ref. [20], standard interfaces are defined to allow WSNs that can implement different communication technologies to exchange information among them. Each WSN has its own Gateway that is in charge of the WSN management and protocol translation. The use of resources at different WSNs inside a Bubble is coordinated by a higher-level gateway that also provides protocol translation functionalities for the WSN under its support. Communication between different Bubbles is possible through their corresponding Bubble Gateways. Interfaces, services, and interoperability features of the different nodes and gateways are described in [20]. Ref. [20] is focused on IoT systems and provides connectivity to

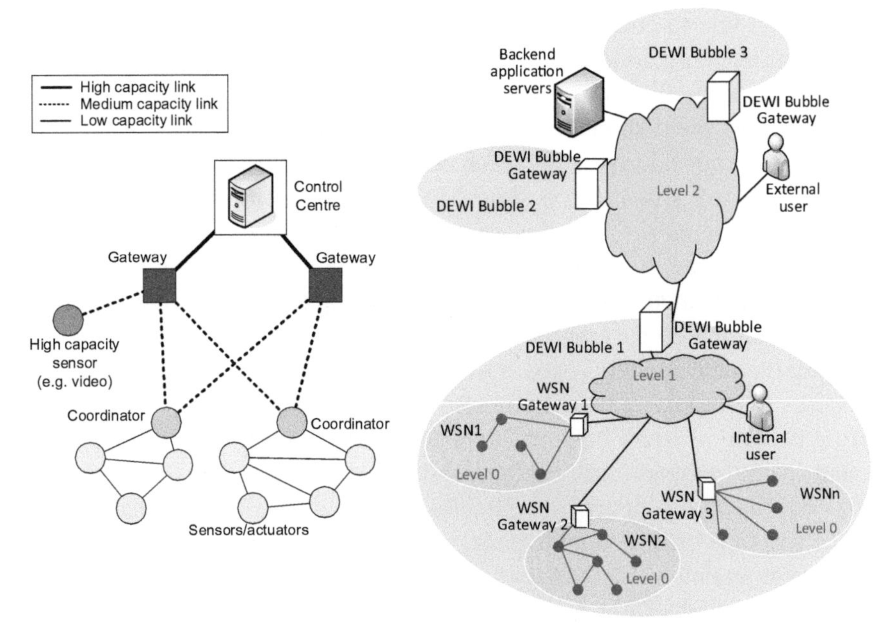

a) IWN architecture adapted from [19] b) IWN architecture adapted from [20]

Figure 5.4 Examples of hierarchical IWN architectures.

a large number of communication devices. However, it does not particularly consider applications with very stringent latency and reliability requirements.

Another interesting hierarchical management architecture that considers the use of heterogeneous wireless technologies is presented in [21], and has been developed in the framework of the KoI project [23]. The architecture presented in [21] proposes a two-tier management approach for radio resource coordination to support mission-critical wireless communications. To guarantee the capacity and scalability requirements of the industrial environment, ref. [21] considers the deployment of multiple small cells. Each of these small cells can implement a different wireless technology, and has a Local Radio Coordinator (LRC) that is in charge of the fine-grained management of radio resources for devices in its cell. On a higher level, there is a single Global Radio Coordinator (GRC) that carries out the radio resource management on a broader operational area and coordinates the use of radio resources by the different cells to avoid inter-system (for wireless technologies using unlicensed bands) and inter-cell (for those working on licensed bands) interference among them. In ref. [21], the control plane and the data plane are split following the Software-Defined Networking (SDN) principle. Control management is carried out in a centralized mode at LRCs and the GRC. For the data plane, centralized and assisted Device-to-Device (D2D) modes are considered within each cell.

5G networks are also being designed to support, among other verticals, Industrial IoT systems [24]. To this end, the use of Private 5G networks is proposed [25]. Private 5G networks will allow the implementation of local networks with dedicated radio equipment (independent of traffic fluctuation in the wide-area macro network) using shared and unlicensed spectrum, as well as locally dedicated licensed spectrum. The design of these Private 5G networks to support industrial wireless applications considers the implementation of several small cells to cover the whole industrial environment integrated in the network architecture as shown in Figure 5.5. Private 5G networks will have to support Ultra Reliable Low Latency Communications (URLLC) for time-critical applications, and Enhanced Mobile Broadband services for augmented/virtual reality services. In addition, the integration of 5G networks with Time Sensitive Networks (TSN)[1] is considered to guarantee deterministic end-to-end industrial communications, as presented in [24]. Figure 5.6 summarizes these key capabilities of Private 5G networks for Industrial IoT systems.

The reference communication and data management architecture designed in AUTOWARE is very aligned with the concepts that are being studied for Industrial 5G networks. The support of very different communication requirements demanded for a wide set of industrial applications (from time-critical applications to ultra-high demanding throughput applications) and the integration of different communication technologies (wired and wireless) are key objectives of the designed AUTOWARE communication and data management architecture to meet the requirements of Industry 4.0. In fact, AUTOWARE focuses on the design of a communication architecture that is able to efficiently meet the varying and stringent communication

[1]TSN is a set of IEEE 802 Ethernet sub-standards that aim to achieve deterministic communication over Ethernet by using time synchronization and a schedule that is shared between all the components (i.e. end systems and switches) within the network. By defining various queues based on time, TSN ensures a bounded maximum latency for scheduled traffic through switched networks, thereby guaranteeing the latency of critical scheduled communication. Additionally, TSN supports the convergence of having critical and non-critical communication sharing the same network, without interfering with each other, resulting in a reduction of costs (reduction of required cabling).

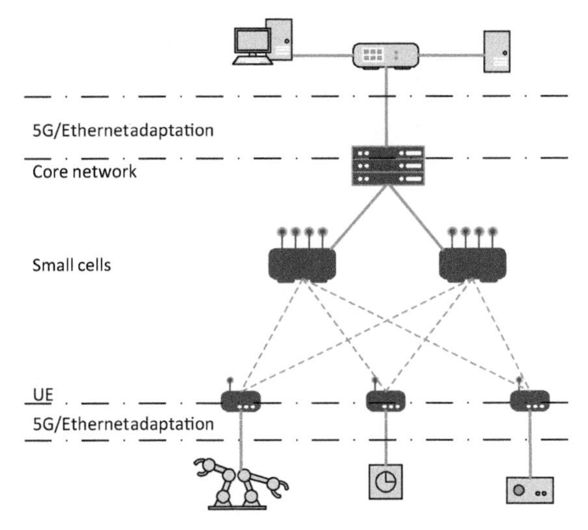

Figure 5.5 Private 5G Networks architecture for Industrial IoT systems [24].

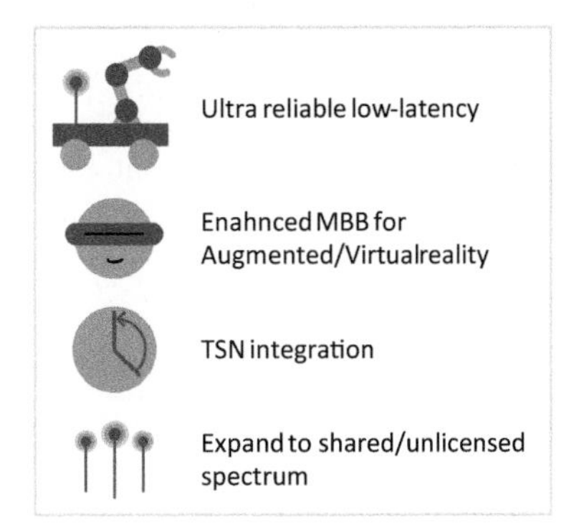

Figure 5.6 Key capabilities of Private 5G Networks for Industrial IoT systems [24].

requirements of the wide set of applications and services that will coexist within the factories of the future; in contrast to the architectures proposed in [20] and [21], which are mainly designed to guarantee communication requirements of a given type of service (to provide connectivity to a large number of communication devices in [20], and mission-critical wireless communications in [21]). In addition, this work goes a step further and analyzes the requirements of the communication architecture from the point of view of the data management and distribution.

5.4 Data Management in Industrial Environments

Traditionally, industrial application systems tend to be entirely centralized. For this reason, distributed data management has not been studied extensively in the past, and the emphasis has been put on the efficient wireless and wired communication within the industrial environment. The reader can find state-of-the-art approaches on relevant typical networks in [19, 26–28].

However, there have been some interesting works on various aspects of the data management process, e.g., end-to-end latency provisioning. In [29], the authors present a centralized routing method, and, consequently, they do not use proxies, data handling special nodes, or hierarchical data management. In [30], the authors address different optimization objectives, focusing on minimizing the maximum hop distance, rather than guaranteeing it as a hard constraint. Also, they assume a bounded number of proxies and they examine only on the worst-case number of hops. In [31], the authors present a cross-layer approach, which combines MAC-layer and cache management techniques for adaptive cache invalidation, cache replacement, and cache prefetching. In [32], the authors consider a different data management objective: replacement of locally cached data items with new ones. As the authors claim, the significance of this functionality stems from the fact that data queried in real applications is not random but instead exhibits locality characteristics. Therefore, the design of efficient replacement policies, given an underlying caching mechanism, is addressed. In [33], although the authors consider delay aspects and a realistic industrial IoT model (based on WirelessHART), their main objective is to bound the worst-case delay in the network. Also, they do not exploit the potential presence of proxy nodes, and consequently, they stick to the traditional, centralized industrial IoT setting. In [34], the authors consider a multi-hop network organized in clusters and provide a routing algorithm and cluster partitioning. Our distributed data management concepts and algorithms can work on top of this approach (and of any clustering approach), for example, by allocating the role of proxies to cluster-heads. In fact, clustering and our solutions address two different problems.

5.5 Hierarchical Communication and Data Management Architecture for Industry 4.0

The network architecture presented in this chapter is designed to provide flexible and efficient connectivity and data management in Industry 4.0.

AUTOWARE proposes a hierarchical management architecture that supports the use of heterogeneous communication technologies. The proposed architecture also establishes multiple tiers where communication cells are functionally classified; different tiers establish different requirements in terms of reliability, latency, and data rates and impose different constraints on the management algorithms and the flexibility to implement them.

5.5.1 Heterogeneous Industrial Wireless Network

As presented in Section 5.2, industrial applications demand a wide range of different communication requirements that are difficult to be efficiently satisfied with a single communication technology. In this context, the proposed architecture exploits the different capabilities of the available communication technologies (wired and wireless) to meet the wide range of requirements of industrial applications. For example, unlicensed wireless technologies such as WirelessHART, ISA100.11a, or IEEE 802.15.4e must implement mechanisms to minimize the interference generated to other potential devices sharing the same band, as for example, listen-before-talk-based channel access schemes. Although these wireless technologies are suitable to efficiently meet the requirements of non-time-critical monitoring or production applications, they usually fail to meet the stringent latency and reliability requirements of time-critical automation and control applications. In addition, these technologies were designed for static and low-bandwidth deployments, and the digitalization of industries requires significantly higher bandwidth provisioning and the capacity to integrate moving robots and objects in the factory. On the other hand, cellular standards operating on licensed frequency bands introduced in Release 14 [35] mechanisms for latency reduction in order to support certain delay critical applications. Moreover, Factories of the Future represent one of the key verticals for 5G-PPP, and 5G technologies are being developed to support a large variety of applications scenarios, targeting URLLC with a latency of about 1 ms and reliability of $1–10^{-9}$ [36]. Also, Private LTE and Private 5G networks will be relevant technologies to be used in industrial environments [25]. As a complement of wireless technologies, the use of wired communication technologies, as for example TSN, can also be considered for communication links between static devices.

In this context, we propose that several subnetworks or cells (we will use the term cell throughout the rest of the document) implementing heterogeneous technologies cover the whole industrial plant (or several plants). We adopt and use the concept of cell to manage the communications and data management resources and improve the network scalability. Different cells can use different communication technologies. Cells using different communication technologies could overlap in space. Also, cells using the same technology but in a different channel could cover the same area (or partially). Each network node is connected to the cell that is able to most efficiently satisfy its communication needs. For example, WirelessHART can be used to monitor a liquid level and control a valve, while 5G communications can be employed for time-critical communications between a sensor and an actuator. TSN could be a good candidate to implement long-distance backhaul links between static devices. Figure 5.7 illustrates the concept of cells in the proposed heterogeneous architecture with five cells implementing two different technologies. Technology 1 and Technology 2 could represent WirelessHART and 5G technologies. Technology 3 is used to connect each cell through a local management entity, referred to as Local Manager (LM), to a central management entity represented as Orchestrator in Figure 5.7 (roles of LMs and the Orchestrator in the proposed reference communication and data management architecture are presented in the next section), and it could be implemented with TSN (the communication link between LMs and the Orchestrator could also be implemented by a multi-hop link using also heterogeneous technologies for improved flexibility and scalability (e.g., IEEE 802.11 and TSN)).

Cells implementing wireless communication technologies that operate in unlicensed spectrum bands can suffer from inter-system and intra-system interferences. Mechanisms to detect external interferences are needed, and cells need to be coordinated to guarantee interworking and coexistence between concurrently operating technologies. Cells implementing a communication technology using licensed spectrum, as for example, LTE or 5G networks, are also possible. Although the use of licensed spectrum bands guarantees communications free of external interference, planning and coordination among multiple cells is still needed to control inter-cell interference. Considering the highly dynamic and changing nature of industrial environments, coordination among cells need to be carried out dynamically in order to guarantee the stringent communication requirements of industrial automation processes.

5.5.2 Hierarchical Management

The proposed reference communication and data management architecture considers a hierarchical structure that combines local and decentralized

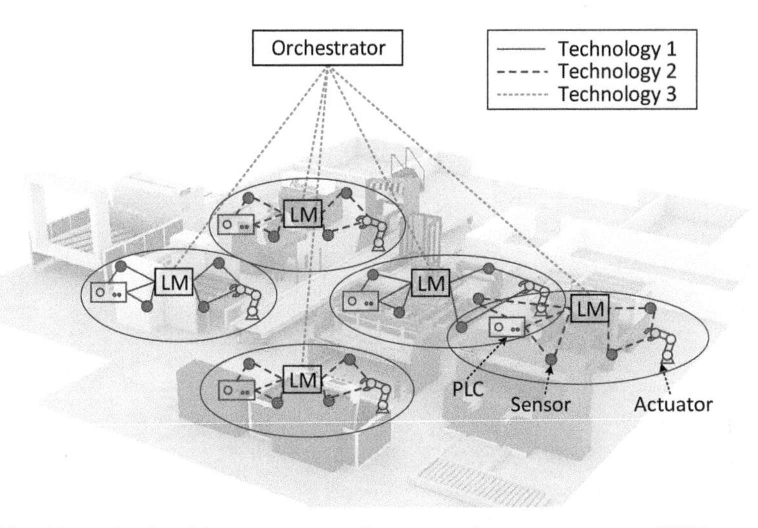

Figure 5.7 Hierarchical and heterogeneous reference architecture to support CPPS connectivity and data management.

management with centralized decisions to efficiently use the available communication resources and carry out the data management in the system. The management structure is depicted in Figure 5.7, and the functions of the two key components, the Orchestrator and the LMs, are next described.

5.5.2.1 Hierarchical communications

The Orchestrator is in charge of the global coordination of the radio resources assigned to the different cells. It establishes constraints to the radio resource utilization that each cell has to comply with in order to guarantee coordination and interworking of different cells, and finally guarantee the requirements of the industrial applications developed in the whole plant. For example, the Orchestrator must avoid inter-cell interferences between cells implementing the same licensed technology. It must also guarantee interworking among cells implementing wireless technologies using unlicensed spectrum bands in order to avoid inter-system interferences, as for example, dynamically allocating non-interfering channels to different cells based on the current demand. LMs are implemented at each cell. An LM is in charge of the local management of the radio resources within its cell and makes local decisions to ensure that communication requirements of nodes in its cell are satisfied.

As shown in Figure 5.8, LMs are in charge of management functions such as Radio Resource Allocation, Power Control, or Scheduling. These functions

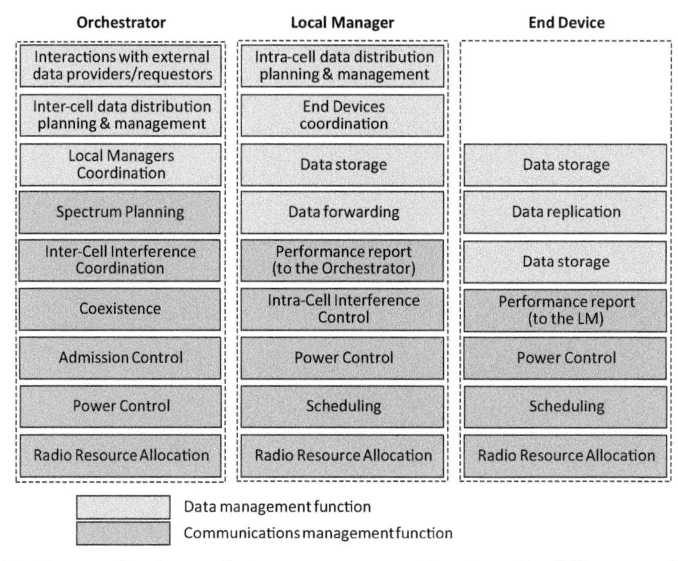

Figure 5.8 Communication and data management functions in different entities of the hierarchical architecture.

locally coordinate the use of radio resources among the devices attached to the same cell and require very short response times. Intra-Cell Interference Control needs to be carried out also by the LM if several transmissions are allowed to share radio resources within the same cell. LMs also report the performance levels experienced within its cell to the Orchestrator. Thanks to its global vision, the Orchestrator has the information required and the ability to adapt and (re-)configure the whole network. For example, under changes in the configuration of the industrial plant or in the production system, the Orchestrator can reallocate frequency bands to cells implementing licensed technologies based on the new load conditions or the new communication requirements. It could also establish new interworking policies to control interferences between different cells working in the unlicensed spectrum. The Orchestrator can also establish constraints about the maximum transmission power or the radio resources to allocate to some transmissions to guarantee the coordination between different cells. It is also in charge of the Admission Control. In this context, the Orchestrator also decides to which cell a new device is attached to consider the communication capabilities of the device, the communication requirements of the application, and the current operating conditions of each cell.

The described hierarchical communication and data management architecture corresponds to the control plane. We consider that control plane and user plane[2] are separated. Therefore, although a centralized management is adopted within a cell, nodes in proximity might communicate directly using D2D communications. In some cells, end-devices might also participate in management functions, for example, if distributed radio resource allocation algorithms are considered for D2D communications in 5G cells. End devices can also participate in other management functions such as Power Control or Scheduling (see Figure 5.8).

5.5.2.2 Data management

The Orchestrator plays an important role in facilitating the development of novel smart data distribution solutions that cooperate with cloud-based service provisioning and communication technologies. Smart proactive data storage/replication techniques can be designed, ensuring that data is located where it can be accessed by appropriate decision makers in a timely manner based on the performance of the underlying communication infrastructure. Consequently, the Orchestrator serves as a great opportunity to implement different types of data-oriented automation functions at reduced costs, like interactions with external data providers or requestors, inter-cell data distribution planning, and management and coordination of the LMs.

On the other hand, it is widely recognized that entirely centralized solutions to collect and manage data in industrial environments are not always suitable [38, 39] This is due to the fact that in order to assure quick reaction, process monitoring and automation control may span among multiple physical locations. Additionally, the adoption of IoT technologies with the associated massive amounts of generated data makes decentralized data management inevitable. A significant challenge is that, when data are managed across multiple physical locations, data distribution needs to be carefully designed, so as to ensure that industrial process control is not affected by the well-known issues related to communication delays and jitters [26, 40].

For data management, allocation of roles on the Orchestrator, LMs, and individual devices is less precisely defined in general, and can vary significantly on a per-application and per-scenario basis. In general, we expect

[2]The User Plane carries the network user traffic, i.e., the data that is generated and consumed by the AUTOWARE applications and services. The Control Plane carries signaling traffic, and is critical for the correct operation of the network. For example, signaling messages would be needed to properly configure a wired/wireless link to achieve the necessary latency and reliability levels to support an application. They would also be needed to intelligently control the data management process. The Control Plane therefore is needed to enable the user data exchange between the different AUTOWARE components.

that the Orchestrator would decide on which cells (controlled by one LM each) data need to be available and thus replicated. Also, it would decide out of which cells they must not be replicated due to ownership reasons. It would implement, in collaboration with cloud platforms, authentication of users across cells and, when needed, data transcoding functions. Thus, we expect the Orchestrator to be responsible for managing the heterogeneity issues related to managing data across a number of different cells, possibly owned and operated by different entities. LMs would manage individual cells. They would typically decide where, inside the cell, data need to be replicated, stored, and moved dynamically, based on the requirements of the specific applications, and the resources available at the individual nodes. Note that data will in general be replicated across the individual nodes, and not exclusively at the LMs, to guarantee low delays and jitters, which might be excessive if the LMs operate as unique centralized data managers. In some cases, end-devices can also participate in management functions, for example, by exploiting D2D communications to directly exchange data between them, implementing localized data replication or storage policies. In those cases, the data routing is not necessarily regulated centrally, but can be efficiently distributed, using appropriate cooperation schemes. In the architecture, therefore, the control of data management schemes can be performed centrally at the Orchestrator, locally at the LMs, or even at individual devices, as appropriate. Data management operations become distributed, and they exploit devices that lie between source and destination devices, like the use of proxies for data storage and access.

5.5.3 Multi-tier Organization

In the proposed reference communication and data management architecture, cells are organized in different tiers depending on the communication requirements of the industrial application they support. LMs of cells in different tiers consider the use of different management algorithms to efficiently meet the stringent requirements of the different industrial applications they support. For example, regarding scheduling, a semi-persistent scheduling algorithm could be applied in LTE cells to guarantee ultra-low latency communications; semi-persistent scheduling algorithms avoid delays associated to the exchange of signaling messages to request (from the device to the base station or eNB) and grant (from the base station or eNB to the device) access to the radio resources. However, semi-persistent scheduling algorithms might not be adequate for less demanding latency requirements due to the potential

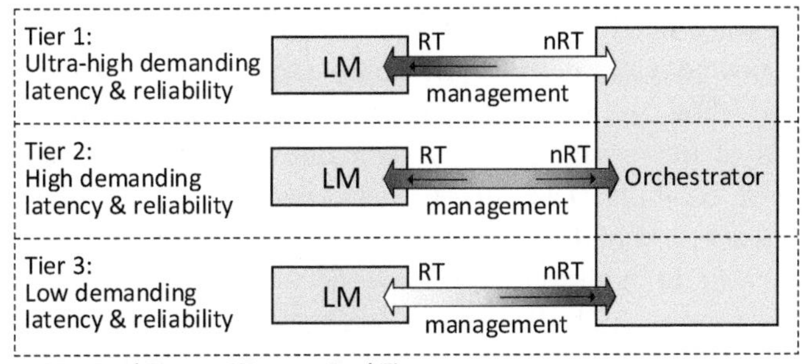

Figure 5.9 LM–Orchestrator interaction at different tiers of the management architecture.

underutilization of radio resources. The different requirements in terms of latency and reliability of the application supported by a cell also affect the exact locations where data should be stored and replicated. For example, in time-critical applications, the lower the data access latency bound is, the closer to the destination the data should be replicated.

The requirements of the nodes connected to a cell also influence the type of interactions between the LM of the cell and the Orchestrator. LMs of cells that support communication links with loose latency requirements can delegate some of their management functions to the Orchestrator. For these cells, a closer coordination between different cells could be achieved. Management decisions performed by LMs based on local information are preferred for applications with ultra-high demanding latency requirements (see Figure 5.9).

5.5.4 Architectural Enablers: Virtualization and Softwarization

Efficiency, agility, and speed are fundamental characteristics that future communication and networking architectures must accomplish to support the high diverging and stringent performance requirements of future communication systems (including but not limited to the industrial ones) [41]. In this context, the communication and data management architecture proposed within this chapter considers the use of RAN Slicing and Cloud RAN as enabling technologies to achieve the sought flexibility and efficiency.

5.5.4.1 RAN slicing

The proposed architecture considers the use of heterogeneous communication technologies. The assignment of communication technologies to industrial applications does not need to necessarily be a one-to-one matching. There is a clear trend nowadays in designing wireless technologies such that they can support more than one type of application even belonging to different "verticals", each of them with possibly radically different communication requirements. For example, LTE or 5G networks can be used to satisfy the ultra low-latency and high-reliability communications of a time-critical automation process. In addition, the same networks could also support applications that require high-throughput levels, such as virtual reality or 4K/8K ultra-high-definition video. This is typically achieved through network virtualization and slicing, to guarantee isolation of (virtual) resources and independence across verticals, or across applications in the same vertical.

In the proposed architecture, each cell can support several industrial applications with different communication requirements. The industrial applications supported by the same cell might require different management functions or techniques to satisfy their different requirements in terms of transmission rates, delay, or reliability. Moreover, it is important to ensure that the application-specific requirements are satisfied independently of the congestion and performance experienced by the other application supported by the same cell, i.e., performance isolation needs to be guaranteed between different applications. For example, the amount of traffic generated by a given application should not negatively influence the performance of the other application. In this context, we propose the use of RAN Slicing to solve the above-mentioned issues. RAN Slicing is based on SDN (Software-Defined Networking) and NFV (Network Function Virtualization) technologies, and it proposes to split the resources and management functions of an RAN into different slices to create multiple logical (virtual) networks on top of a common network [42]. Each of these slices, in this case, virtual RANs, must contain the required resources needed to meet the communication requirements of the application or service that such slice supports. As presented in [42], one of the main objectives of RAN Slicing is to assure isolation in terms of performance. In addition, isolation in terms of management must also be ensured, allowing the independent management of each slice as a separated network. As a result, RAN Slicing becomes a key technology to deploy a flexible communication and networking architecture capable of meeting the stringent and diverging communication requirements of industrial applications, and in particular, those of URLLC.

In the proposed architecture, each slice of a physical cell is referred to as virtual cell, as shown in Figure 5.10. Virtual cells resulting from the split of the same physical cell can be located at different levels of the

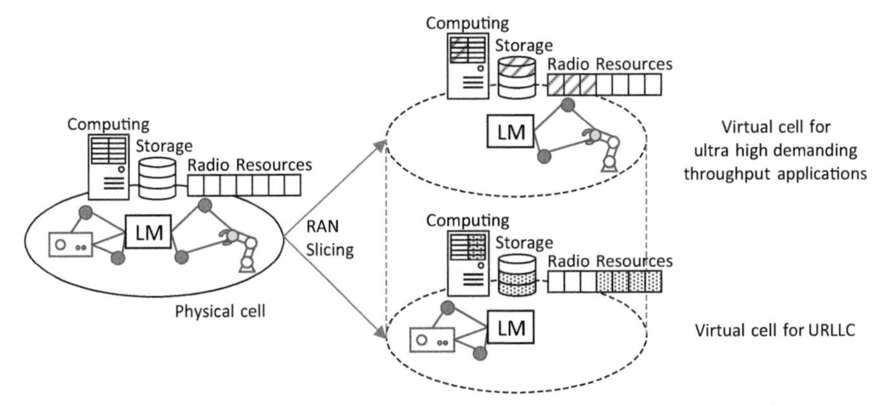

Figure 5.10 Virtual cells based on RAN Slicing.

multi-tier architecture depending on the communication requirements of the applications. Each virtual cell implements the appropriate functions based on the requirements of the application supported and must be assigned the RAN resources required to satisfy the requirements of the communication links it supports.

RAN resources (e.g., data storage, computing, radio resources, etc.) must be allocated to each virtual cell considering the operating conditions, such as the amount of traffic, the link quality, etc. The amount of RAN resources allocated to each virtual cell must be therefore dynamically adapted based on the operating conditions. Within the proposed reference architecture, the Orchestrator is the management entity in charge of creating and managing RAN slices or virtual cells. Thanks to the reports received from the LMs, the Orchestrator has a global view of the performance experienced at the different (virtual) cells. As a result, it is able to decide the amount of RAN resources that must be assigned to each virtual cell to guarantee the communication requirements of the applications.

With respect to data management functions, they will operate on top of the virtual networks generated by RAN Slicing. However, note that the requirements posed by data management will determine part of the network traffic patterns. Therefore, RAN Slicing defined by the Orchestrator might consider the traffic patterns resulting from data management operations, in order to optimize slicing itself.

5.5.4.2 Cloudification of the RAN

Cloud-based RAN (or simply Cloud RAN) is a novel paradigm for RAN architectures that applies NFV and cloud technologies for deploying

RAN functions [43]. Cloud RAN splits the base station into a radio unit, known as Radio Remote Head (RRH), and a signal-processing unit referred to as Base Band Unit (BBU) [44]. The key concept of Cloud RAN is that the signal processing units, i.e., the BBUs, can be moved to the cloud. Cloud RAN shifts from the traditional distributed architecture to a centralized one, where some or all of the base station processing and management functions are placed in a central virtualized BBU pool (a virtualized cluster which can consist of general purpose processors to perform baseband processing and that is shared by all cells) [43]. Virtual BBUs and RRHs are connected by a fronthaul network. Centralizing processing and management functions in the same location improves interworking and coordination among cells; virtual BBUs are located in the same place, and exchange of data among them can be carried out easier and with shorter delay.

We foresee Cloud RAN as the baseline technology for the proposed architecture, to implement hierarchical and multi-tier communication management. Cloud RAN will be a key technology to achieve a tight coordination between cells in the proposed architecture and to control inter-cell and inter-system interferences. As presented in [45] and [46], Cloud RAN can support different functional splits that are perfectly aligned with the foreseen needs of industrial applications; some processing functions can be executed remotely while functions with strong real-time requirements can remain at the cell site. In the proposed communication and data management architecture, the decision about how to perform this functional split must be made by the Orchestrator considering the particular communication requirements of the industrial applications supported by each cell (see Figure 5.11).

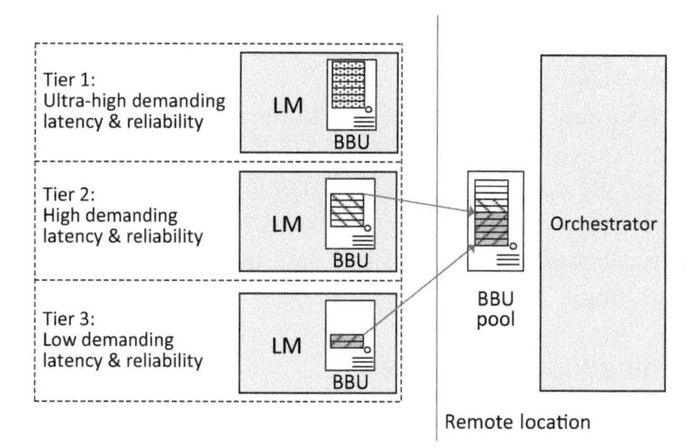

Figure 5.11 Cloudification of the RAN.

The Cloud RAN architectural paradigm allows for hardware resource pooling, which also reduces operational cost, by reducing power and energy consumption compared to traditional architectures [43], which results in an attractive incentive for industrial deployment. The cloudification of the RAN will also leverage RAN Slicing on a single network infrastructure and will increase flexibility for the construction of on-demand slices to support individual service types or application within a cell.

5.6 Hybrid Communication Management

Communication systems must be able to support the high dynamism of industrial environment, which will result from the coexistence of different industrial applications, different types of sensors, the mobility of nodes (robots, machinery, vehicles, and workers), and changes in the production demands. Industry 4.0 then demands flexible and dynamic communication networks able to adapt their configuration to changes in the environment to seamlessly ensure the communication requirements of industrial applications. To this end, communication management decisions must be based on current operating conditions and the continuous monitoring of experienced performance. The proposed hierarchical communication and data management architecture allows the implementation of hybrid communication management schemes that integrate local and decentralized management decisions while maintaining a close coordination through a central management entity (the Orchestrator in the reference AUTOWARE architecture) with global knowledge of the performance experienced in the whole industrial communication network. The hybrid communication management introduces flexibility in the management of wireless connections and increases the capability of the network to detect and react to local changes in the industrial environment while efficiently guaranteeing the communication requirements of industrial applications and services supported by the whole network.

In hybrid management schemes, management entities must interact to coordinate their decisions and ensure the correct operation of the whole network. Figure 5.12 represents the interactions between the management entities of the hierarchical architecture: the Orchestrator, LMs, and end-devices (as presented in Section 5.2, end-devices might also participate in the communication management). Boxes within each management entity represent different functions executed at each entity:

- Local measurements: This function measures physical parameters on the communication link, as for example, received signal level (received signal strength indication or RSSI), signal-to-noise ratio (SNR), etc. In addition, this function also measures and evaluates the performance experienced in the communication, as for example, throughput, delay, packet error ratio (PER), etc. This function is performed by each entity on its communication links.

- Performance gathering: This function collects information about the performance experienced at the different cells. This function is performed at the LMs, which collect performance information gathered by end-devices within its cell, and also at the Orchestrator, which receives performance information gathered by the LMs.
- Reasoning: The reasoning function processes the data obtained by the local measurements and the performance gathering functions to synthesize higher-level performance information. The reasoning performed at each entity will depend on the particular application supported (and the communication requirements of the application) and also on the particular management algorithm implemented. For example, if a cell supports time-critical control applications, the maximum value of latency experienced by the 99 percentile of packets transmitted might be of interest, while the average throughput achieved in the communication could be required to analyze the performance of a 3D visualization application.
- Reporting: This function sends periodic performance reports to the management entity in the higher hierarchical level. Particularly, end-devices send periodic reports to the LMs, which in turn report performance information to the Orchestrator.
- Global/local/communication management decision: This function executes the decision rule or decision policy. This function can be whatever of the communication management functions shown in Figure 5.8: for example, Admission Control or Inter-Cell Interference Coordination algorithms can be executed as the Global management decision function in the Orchestrator, Power Control or Radio Resource Allocation within a cell can be executed as the Local management decision function in the LMs, and Scheduling or Power Control can be executed as the Communication management decision function at the end-devices.

As shown in Figure 5.12, an end-device performs local measurements of the quality and performance experienced in its communication links. This local data (1) is processed by the reasoning function that provides high-level performance information (2a) that is reported to the LM in its cell (3). This high-level performance information can also be used by the end-device (2b) to get a management decision (4) and configure its communication

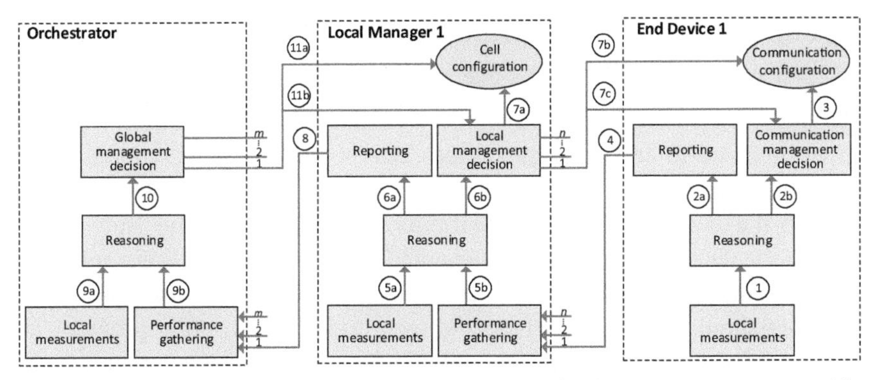

Figure 5.12 Hybrid communication management: interaction between management entities.

parameters in the case that the end-device has management capabilities. In this case, the management decisions taken by different end-devices in the same cell are coordinated by the LM in the cell, which can also configure some communication parameters of the end-devices (7b). Decisions taken by end-devices are constrained by the decisions taken by the LM (7c). If end-devices do not have management capabilities, the communication parameters for the end-devices are directly configured by the LM (8b). The Local management decisions taken by each LM are based on the performance information gathered by all end-devices in its cell (from 1 to *n* devices in the figure), and also on local measurements performed by the own LM. This data (5a and 5b) is processed by the reasoning function in the LM, and the resulting high-level performance information (6b) is used to take a local management decision and configure the communication parameters of the end-devices in its cell (7a, 7b, and 7c). Each LM also reports to the Orchestrator the processed information about the performance experienced in its cell (8). The Orchestrator receives performance information from all the LMs (from 1 to *m* LMs in the figure). The performance information gathered by the LMs (9b), together with local measurements performed by the Orchestrator in its communication links with the LMs (9a), is processed by the reasoning function in the Orchestrator. The high-level performance information (10) is used by the Orchestrator to achieve a global management decision and configure radio resources to use at each cell (11a). The global management decisions made by the Orchestrator constrain the local management decisions made by the LMs (11b) to guarantee the coordination among the different LMs in the network, and finally ensure the communication requirements of the industrial applications and services supported by the network.

5.7 Decentralized Data Distribution

The smart data management process provided by the architecture interacts with the underlying networking protocols. In order to provide both efficient data access and end-to-end delay guarantees, one of the technical components of the architecture is a dedicated decentralized data distribution. The main idea behind the decentralized data distribution is decoupling the Network plane from the Data plane. The data-enabled architecture functions selectively move data to different network areas and devise methods on how the data requests should be served, given a known underlying routing protocol. More specifically, the role of the decentralized data distribution component is three-fold:

1. It investigates where and when the data should be moved, and to which network areas.
2. It decides which network nodes can serve as special nodes and assume more responsibilities with respect to data management.
3. It indicates how the available data will be distributed and delivered to the individual network devices requesting it.

Note that the architecture enables the storing and replication of data between (i) (potentially mobile) nodes in the factory environment (e.g., the mobile nodes of the factory operators, nodes installed in work cells, nodes attached to mobile robots, etc.); (ii) edge nodes providing storage services for the specific (areas of the) factory; and (iii) remote cloud storage services. All the three layers can be used in a synergic way, based on the properties of the data and the requirements of the users requesting it. Depending on these properties, data processing may need highly variable computational resources. Advanced scheduling and resource management strategies lie at the core of the distributed infrastructure resources usage. However, such strategies must be tailored to the particular algorithm/data combination to be managed. Differently from the past, the scheduling process, instead of looking for smart ways to adapt the application to the execution environment, now aims at selecting and managing the computational resources available on the distributed infrastructure to fulfill some performance indicators.

The suggested architecture can be used in order to efficiently deploy the data management functions over typical industrial IoT networks. Initial results show that the decentralized data management scheme of the proposed architecture can indeed enhance various target metrics when applied to various industrial IoT networking settings. In the following subsections, we briefly review some recent examples, where the decentralized data distribution concepts resulted in an enhanced network performance.

5.7.1 Average Data Access Latency Guarantees

Assuming that applications in industrial IoT networks require that there is (i) a set of producers generating data (e.g., IoT sensors), (ii) a set of consumers requiring those data in order to implement the application logic (e.g., IoT actuators), and (iii) a maximum latency $L_{\max}$ that consumers can tolerate in receiving data after they have requested them; the decentralized data management module (DML) offers an efficient method for regulating the data distribution among producers and consumers. The DML selectively assigns a special role to some of the network nodes, that of the proxy. Each node that can become a proxy potentially serves as an intermediary between producers and consumers, even though the node might be neither a producer nor a consumer. If properly selected, proxy nodes can significantly reduce the average data access latency; however, when a node is selected as a proxy, it has to increase its storing, computational, and communication activities. Thus, the DML minimizes the number of proxies, to reduce as much as possible the overall system resource consumption. In [47], we have provided an extensive experimental evaluation, both in a testbed and through simulations, and we demonstrated that the proposed decentralized data management (i) guarantees that the access latency stays below the given threshold and (ii) significantly outperforms traditional centralized and even distributed approaches, in terms of average data access latency guarantees.

5.7.2 Maximum Data Access Latency Guarantees

Another representative example of decentralized data management is the exploitation of the presence of a limited set of pre-installed proxy nodes, which are more capable than resource-limited IoT devices in the resource-constrained network (e.g., fog nodes). Different to the previous example, here we focused on network lifetime and on maximum (instead of average) data access latencies. The problem we addressed in [48] is the maximization of the network lifetime, given the proxy locations in the network, the initial limited energy supplies of the nodes, the data request patterns (and their corresponding parameters), and the maximum latency that consumer nodes can tolerate since the time they request data. We proved that the problem is computationally hard and we designed an offline centralized heuristic algorithm for identifying which paths in the network the data should follow and on which proxies they should be cached, in order to meet the latency constraint and to efficiently prolong the network lifetime. We implemented the method and evaluated its performance in a testbed, composed of IEEE 802.15.4-enabled network nodes. We demonstrated that the proposed heuristic (i) guarantees data access latency below the given threshold and (ii) performs well in terms of network lifetime with respect to a theoretically optimal solution.

5.7.3 Dynamic Path Reconfigurations

As in the previous examples, we assume that applications require a certain upper bound on the end-to-end data delivery latency from proxies to consumers and that at some point in time, a central controller computes an optimal set of multi-hop paths from producers to proxies and from proxies to consumers, which guarantee a maximum delivery delay, while maximizing the energy lifetime of the network (i.e., the time until the first node in the network exhaust energy resources). In this example, we focus on maintaining the network configuration in such a way that application requirements are met after important network operational parameters change due to some unplanned events (e.g., heavy interference, excessive energy consumption), while guaranteeing an appropriate utilization of energy resources. In [49], we provided several efficient algorithmic functions that locally reconfigure the paths of the data distribution process, when a communication link or a network node fails. The functions regulate how the local path reconfiguration should be implemented and how a node can join a new path or modify an already existing path, ensuring that there will be no loops. The proposed method can be implemented on top of existing data forwarding schemes designed for industrial IoT networks. We demonstrated through simulations the performance gains of our method in terms of energy consumption and data delivery success rate.

5.8 Communications and Data Management within the AUTOWARE Framework

The reference communication and data management architecture of AUTOWARE supports the control plane of the communication network and the data management system. As shown in Figure 5.13, end (or field)-devices such as sensors, actuators, mobile robots, etc., are distributed throughout the factory plant participating in different industrial processes or tasks.

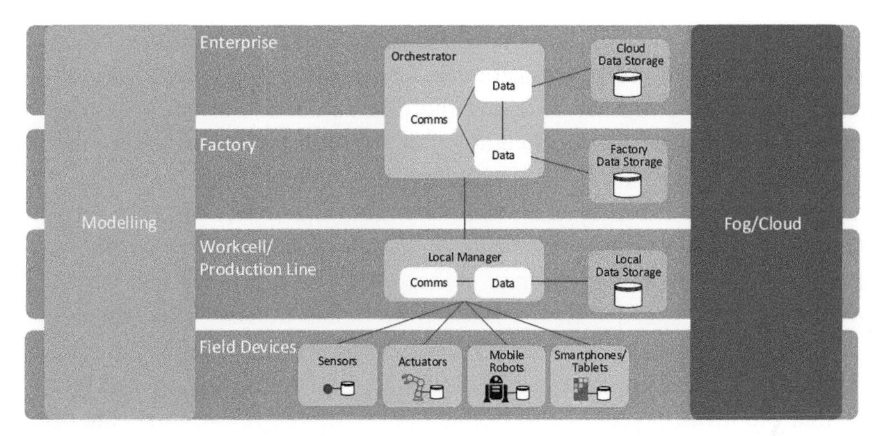

Figure 5.13 Integration of the hierarchical and multi-tier heterogeneous communication and data management architecture into the AUTOWARE Reference Architecture.

These field devices are then included within the Field Devices Layer of the AUTOWARE Reference Architecture defined in Chapter 10. Various LMs can be implemented at different workcells or production lines to locally manage the communication resources and data in the different communication cells deployed in the industrial plant. These management nodes are included in the Workcell/Production Line Layer, and they form a distributed management infrastructure that operates close to the field devices. As previously presented, both the Orchestrator and the LMs have communication and data management functionalities.

From the point of view of communications, the Orchestrator is in charge of the global management of the communication resources used by the different cells deployed within a factory plant. When there is only one industrial plant or when there are multiple but independent plants (from the communications perspective), the main communication functions of the Orchestrator are in the Factory Layer. However, if different industrial plants are deployed and they are close enough so that the operation of a cell implemented in a plant can affect the operation of a different cell in the other plant, then the Orchestrator should be able to manage the communication resources of the different plants. In this case, some of its communication functions should be part of the Enterprise Layer. Based on the previous reasoning, the Orchestrator and, in particular, the communication management function within the Orchestrator should be flexible and be able to be implemented in the Factory and the Enterprise Layer.

From the point of view of data storage, management, and distribution, the data can be circulated and processed at different levels of the architecture, depending on the targeted use case and the requirements that the industrial operator is imposing on the application. For example, if the requirements necessitate critical and short access latency applications (e.g., Table 5.5), such as condition monitoring, then imposing data transfers back and forth between the Field Layer, the Workcell/Production Line Layer, and the Factory Layer may lead to severe sub-optimal paths, which in turn negatively affect the overall network latency. At the same time, those transfer patterns will lead to poor network performance, as field devices often have to tolerate longer response times than necessary. In this case, the data can be stored and managed at the lower layers of the architecture, with the LMs in the role of the data coordinator. Another example is when the requirements necessitate the employment of computationally more sophisticated methods on larger volumes of data that can only be performed by stronger devices than those at the Field Layer, such as 3D object recognition or video tracking, which come with vast amounts of data. In this case, the data can be forwarded, stored, and processed in the higher levels of the architecture, the Factory Layer, or the Enterprise Layer, with the Orchestrator in the role of the data coordinator.

5.9 Conclusions

A software-defined heterogeneous, hierarchical, and multi-tier communication management architecture with edge-powered smart data distribution strategies has been presented in this chapter to support ubiquitous, flexible, and reliable connectivity and efficient data management in highly dynamic Industry 4.0 scenarios where multiple digital services and applications are bound to coexist. The proposed architecture exploits the different abilities of heterogeneous communication technologies to meet the broad range of communication requirements demanded by Industry 4.0 applications. Integration of the different technologies in an efficient and reliable network is achieved by means of a hybrid management strategy consisting of decentralized management decisions coordinated by a central orchestrator. Local management entities organized in different virtual tiers of the architecture can implement different management functions based on the requirements of the application they support. The hierarchical and multi-tier communication management architecture enables the implementation of cooperating, but distinct management functions to maximize flexibility and efficiency to meet the stringent and varying requirements of industrial applications. The proposed architecture considers the use of RAN Slicing and Cloud RAN as enabling technologies to meet reliably and effectively future Industry 4.0 autonomous assembly scenarios and modular plug & play manufacturing systems. The technological enablers of the communications and data management architecture were identified as part of the AUTOWARE framework, both in the user plane and in the control plane of the AUTOWARE reference architecture.

Acknowledgments

This work was funded by the European Commission through the FoF-RIA Project *AUTOWARE: Wireless Autonomous, Reliable and Resilient Production Operation Architecture for Cognitive Manufacturing* (No. 723909).

References

[1] V. K. L. Huang, Z. Pang, C. J. A. Chen and K. F. Tsang, "New Trends in the Practical Deployment of Industrial Wireless: From Noncritical to Critical Use Cases", in IEEE Industrial Electronics Magazine, vol. 12, no. 2, pp. 50–58, June 2018.

[2] T. Sauter, S. Soucek, W. Kastner and D. Dietrich, "The Evolution of Factory and Building Automation", in IEEE Industrial Electronics Magazine, vol. 5, no. 3, pp. 35–48, September 2011.

[3] *How Audi is changing the future of automotive manufacturing*, Feb. 2017. Available at https://www.drivingline.com/. Last access on 2017/12/01.

[4] C. H. Chen, M. Y. Lin and C. C. Liu, "Edge Computing Gateway of the Industrial Internet of Things Using Multiple Collaborative Microcontrollers", in IEEE Network, vol. 32, no. 1, pp. 24–32, January–February 2018.

[5] Plattform Industrie 4.0, "Network-based communication for Industrie 4.0", *Publications of Plattform Industrie 4.0*, April 2016. Available at http://www.plattform-i40.de. Last access on 2017/10/20.

[6] 5GPPP, *5G and the Factories of the Future*, October 2015.

[7] H2020 AUTOWARE project website: http://www.autoware-eu.org/.

[8] M. C. Lucas-Estañ, T. P. Raptis, M. Sepulcre, A. Passarella, C. Regueiro and O. Lazaro, "A software defined hierarchical communication and data management architecture for industry 4.0", *14th Annual Conference on Wireless On-demand Network Systems and Services* (WONS), Isola 2000, pp. 37–44, 2018.

[9] P. Zand, et al., "Wireless industrial monitoring and control networks: The journey so far and the road ahead", *Journal of Sensor and Actuator Networks*, vol. 1, no. 2, pp. 123–152, 2012.

[10] ETSI, "Technical Report; Electromagnetic compatibility and Radio spectrum Matters (ERM); System Reference Document; Short Range Devices (SRD); Part 2: Technical characteristics for SRD equipment for wireless industrial applications using technologies different from Ultra-Wide Band (UWB)", ETSI TR 102 889-2 V1.1.1, August 2011.

[11] E. Molina, et al., "The AUTOWARE Framework and Requirements for the Cognitive Digital Automation", in *Proc. of the 18th IFIP Working Conference on Virtual Enterprises (PRO-VE)*, Vicenza, Italy, September 2017.

[12] M.C. Lucas-Estañ, J.L. Maestre, B. Coll-Perales, J. Gozalvez, "An Experimental Evaluation of Redundancy in Industrial Wireless Communications", in *Proc. 2018 IEEE 23rd International Conference on Emerging Technologies and Factory Automation* (ETFA 2018), Torino, Italy, 4–7 September 2018.

[13] A. Varghese, D. Tandur, "Wireless requirements and challenges in Industry 4.0", in *Proc. 2014 International Conference on Contemporary Computing and Informatics (IC3I)*, pp. 634–638, 2014.

[14] A. Osseiran et al., "Scenarios for 5G mobile and wireless communications: the vision of the METIS project", *IEEE Communications Magazine*, vol. 52, no. 5, pp. 26–35, May 2014.

[15] Reference architecture model for the industrial data space, Fraunhofer, 2017. Available at https://www.fraunhofer.de. Last access on 2017/12/01.

[16] S. Montero, J. Gozalvez, M. Sepulcre, G. Prieto, "Impact of Mobility on the Management and Performance of WirelessHART Industrial Communications", in *Proc. 17th IEEE International Conference on Emerging Technologies and Factory Automation (ETFA)*, Krakw (Poland), pp. 17–21, September 2012.

[17] C. Lu et al., "Real-Time Wireless Sensor-Actuator Networks for Industrial Cyber-Physical Systems", *Proc. of the IEEE*, vol. 104, no. 5, pp. 1013–1024, May 2016.

[18] Xiaomin Li, Di Li, Jiafu Wan, Athanasios V. Vasilakos, Chin-Feng Lai, Shiyong Wang, "A review of industrial wireless networks in the context of Industry 4.0", *Wireless Networks*, vol. 23, pp. 23–41, 2017.

[19] J.R. Gisbert, et al., "Integrated system for control and monitoring industrial wireless networks for labor risk prevention", *Journal of Network*

and Computer Applications, vol. 39, pp. 233–252, ISSN 1084-8045, March 2014.

[20] R. Sámano-Robles, T. Nordström, S. Santonja, W. Rom and E. Tovar, "The DEWI high-level architecture: Wireless sensor networks in industrial applications", in *Proc. of the 2016 Eleventh International Conference on Digital Information Management (ICDIM)*, Porto, pp. 274–280, 2016.

[21] I. Aktas, J. Ansari, S. Auroux, D. Parruca, M. D. P. Guirao, and -B. Holfeld: "A Coordination Architecture for Wireless Industrial Automation", in *Proc. of the European Wireless Conference*, Dresden, Germany, May 2017.

[22] DEWI Project website. Available at http://www.dewiproject.eu/. Last access on 2017/05/26.

[23] KoI Project website. Available at http://koi-projekt.de/index.html. Last access on 2017/05/26.

[24] Mehmet Yavuz, *How will 5G transform Industrial IoT?*, Qualcomm Technologies, Inc., June 2018.

[25] Qualcomm, *Private LTE networks create new opportunities for industrial IoT*, Qualcomm Technologies, Inc., October 2017.

[26] P. Gaj, J. Jasperneite and M. Felser, "Computer Communication Within Industrial Distributed Environment-a Survey", in *IEEE Transactions on Industrial Informatics*, vol. 9, no. 1, pp. 182–189, February 2013.

[27] D. De Guglielmo, S. Brienza, G. Anastasi, "IEEE 802.15.4e: A survey", *Computer Communications*, vol. 88, pp. 1–24, 2016.

[28] T. Watteyne et al., "Industrial Wireless IP-Based Cyber –Physical Systems", *Proceedings of the IEEE*, vol. 104, no. 5, pp. 1025–1038, May 2016.

[29] J. Heo, J. Hong and Y. Cho, "EARQ: Energy Aware Routing for Real-Time and Reliable Communication in Wireless Industrial Sensor Networks", *IEEE Transactions on Industrial Informatics*, vol. 5, no. 1, pp. 3–11, February 2009.

[30] D. Kim et al., "Minimum Data-Latency-Bound k-Sink Placement Problem in Wireless Sensor Networks", *IEEE/ACM Transactions on Networking*, vol. 19, no. 5, pp. 1344–1353, October 2011.

[31] C. Antonopoulos, C. Panagiotou, G. Keramidas and S. Koubias, "Network driven cache behavior in wireless sensor networks", in *Proc. of the 2012 IEEE International Conference on Industrial Technology*, Athens, pp. 567–572, 2012.

[32] C. Panagiotou, C. Antonopoulos and S. Koubias, "Performance enhancement in WSN through data cache replacement policies", in *Proc. of 2012 IEEE 17th International Conference on Emerging Technologies & Factory Automation (ETFA 2012)*, Krakow, pp. 1–8, 2012.

[33] A. Saifullah, Y. Xu, C. Lu and Y. Chen, "End-to-End Communication Delay Analysis in Industrial Wireless Networks", *IEEE Transactions on Computers*, vol. 64, no. 5, pp. 1361–1374, 1 May 2015.

[34] J. Li et al., "A Novel Approximation for Multi-hop Connected Clustering Problem in Wireless Sensor Networks", in *Proc. of the 2015 IEEE 35th International Conference on Distributed Computing Systems*, Columbus, OH, pp. 696–705, 2015.

[35] 3GPP, Technical Specification Group Radio Access Network; Evolved Universal Terrestrial Radio Access (E-UTRA); Study on latency reduction techniques for LTE, 3GPP TS 36.881, version 14.0.0, 2016.

[36] ITU-R M.2083-0, IMT Vision – Framework and overall objectives of the future development of IMT for 2020 and beyond, September 2015.

[37] Qualcomm, *Private LTE networks create new opportunities for industrial IoT*, Qualcomm Technologies, Inc., October 2017.

[38] P. Gaj, A. Malinowski, T. Sauter, A. Valenzano, "Guest Editorial Distributed Data Processing in Industrial Applications", *IEEE Trans. Ind. Informatics*, vol. 11, no. 3, pp. 737–740, 2015.

[39] C. Wang, Z. Bi, L. Da Xu, "IoT and cloud computing in automation of assembly modeling systems", *IEEE Trans. Ind. Informatics*, vol. 10, no. 2, pp. 1426–1434, 2014.

[40] Z. Bi, L. Da Xu, C. Wang, "Internet of things for enterprise systems of modern manufacturing", *IEEE Trans. Ind. Informatics*, vol. 10, no. 2, pp. 1537–1546, 2014.

[41] Gary Maidment, "One slice at a time: SDN/NFV to 5G network slicing", *Communicate (Huawei Technologies)*, Issue 81, pp. 63–66, December 2016.

[42] J. Ordonez-Lucena, et al., "Network Slicing for 5G with SDN/NFV: Concepts, Architectures, and Challenges", *IEEE Communications Magazine*, vol. 55, Issue 5, pp. 80–87, May 2017.

[43] A. Checko et al., "Cloud RAN for Mobile Networks—A Technology Overview", *IEEE Communications Surveys & Tutorials*, vol. 17, no. 1, pp. 405–426, 2015.

[44] O. Chabbouh, S. B. Rejeb, Z. Choukair, N. Agoulmine, "A novel cloud RAN architecture for 5G HetNets and QoS evaluation", in *Proc. International Symposium on Networks, Computers and Communications (ISNCC) 2016*, pp. 1–6, Yasmine Hammamet, Tunissia, May 2016.

[45] *Cloud RAN*, Ericsson White Paper, September 2015.

[46] *5G Network Architecture. A High-Level Perspective*, Huawei White Paper, July 2016.

[47] T. P. Raptis and A. Passarella, "A distributed data management scheme for industrial IoT environments", in *Proc. of the IEEE 13th International Conference on Wireless and Mobile Computing, Networking and Communications* (WiMob), Rome, pp. 196–203, 2017.

[48] T. P. Raptis, A. Passarella and M. Conti, "Maximizing industrial IoT network lifetime under latency constraints through edge data distribution", in *Proc. of the IEEE Industrial Cyber-Physical Systems* (ICPS), Saint Petersburg, Russia, pp. 708–713, 2018.

[49] T. P. Raptis, A. Passarella and M. Conti, "Distributed Path Reconfiguration and Data Forwarding in Industrial IoT Networks", in *Proc. of the 16th IFIP International Conference on Wired/Wireless Internet Communications*, Boston, MA, (WWIC), 2018.

6

A Framework for Flexible and Programmable Data Analytics in Industrial Environments

Nikos Kefalakis[1], Aikaterini Roukounaki[1], John Soldatos[1] and Mauro Isaja[2]

[1]Kifisias 44 Ave., Marousi, GR15125, Greece
[2]Engineering Ingegneria Informatica SpA, Italy
E-mail: jsol@ait.gr; arou@ait.gr; nkef@ait.gr; mauro.isaja@eng.it

This chapter presents a dynamic and programmable distributed data analytics solution for industrial environments. The solution includes an edge analytics engine for analytics close to the field and in line with the edge computing paradigm. Each edge analytics engine instance is flexible and dynamically configurable based on an Analytics Manifest (AM). It is also based on distributed ledger technologies for configuring analytics tasks that span multiple edge nodes and instances of the edge analytics engine. In particular, it leverages ledger services for synchronizing and combining various AMs in factory wide analytics tasks. Based on these mechanisms, the presented distributed data analytics infrastructure is therefore flexible, configurable, dynamic and resilient. Moreover, it is open source and provides Open APIs (Application Programming Interfaces) that enable access to its functionalities. These features make it unique and valuable for vendors and integrators of industrial automation solutions.

6.1 Introduction

A large number of digital automation applications in modern shopfloors collect and process large amounts of digital data as a means of identifying the status of machines and devices (e.g., a machine's condition or failure mode) or the context of industrial processes (e.g., possible defects in an entire production process), including relevant events [1]. This context is accordingly used to support decision making, including decisions that drive automation and control operations on the shopfloor [2] such as the configuration of a production line or the operational mode of a machine. Therefore, data analytics operations are an integral element of most digital automation platforms [3], which is usually integrated within automation and simulation functionalities.

In this context, the automation platform that has been developed in the scope of the FAR-EDGE project includes also distributed data analytics functionalities. In particular, the FAR-EDGE platform offers functionalities in three distinct, yet complementary domains, namely Automation, Analytics and Simulation [4]. The Analytics domain provides the means for collecting, filtering and processing large volumes of data from the manufacturing shopfloor towards calculating indicators associated with manufacturing performance and automation. Analytics functions are offered by a Distributed Data Analytics (DDA) infrastructure, which enables the definition, configuration and execution of analytics functions at two different levels, namely:

- Local Level Analytics, i.e. at the edge of a FAR-EDGE deployment. These comprise typically analytics functions that are executed close to the field and have local/edge scope, e.g. they collect and process data streams from a part of a factory such as data streams associated with a station within the factory. Local Level Analytics in FAR-EDGE are configured and executed by means of an Edge Analytics Engine (EAE), which runs within an Edge Gateway (EG) and is a core part of the DDA.
- Global Level Analytics, i.e. concerning the factory as a whole and spanning instances of local level analytics. In FAR-EDGE, global level analytics combine information from multiple Edge Gateways (EGs) and instances of the Edge Analytics Engine. They can be configured and executed through an Open API. Global Level analytics are supported by the ledge and the cloud infrastructures of the FAR-EDGE platform.

The distinction between edge/local and global/cloud analytics is very common in the case of Big Data analytics systems (e.g. [5–7]). Moreover, there are different frameworks that can handle streaming analytics at the edge of the network, which is a foundation for edge analytics. The FAR-EDGE DDA infrastructure goes beyond the state of the art of these Big Data systems through employing novel techniques for the flexible configuration of edge analytics and the synchronization of multiple edge analytics deployments. In particular, the FAR-EDGE DDA includes an infrastructure for registering data sources from the plantfloor, as well as for dynamically discovering them.

Moreover, it includes a modular framework for the deployment of analytics functionalities based on a set of (reusable) processing libraries. The latter can be classified in three main types of data processing functions, which enable the pre-processing of data streams (i.e. pre-processing functions), their data analysis (i.e. analytics functions) and ultimately the storage of the analytics results (i.e. storage functions). In FAR-EDGE, edge analytics tasks are described as combinations of various instances of these three processing functions in various configurations, which are specified as part of relevant analytics workflows.

In this context, different edge analytics tasks can be described using well-defined configuration files (i.e. Analytics Manifests (AMs)), which reflect analytics workflows and are amenable by visual tools. This facilitates the specification and configuration of analytics tasks as part of the DDA. In particular, solution integrators and manufacturers can flexibly configure their analytics operations through defining proper AMs. Based on the use of proper visual tools, such definitions can be performed with almost zero programming, which is an obvious advantage of the FAR-EDGE DDA over conventional edge analytics frameworks. Furthermore, the DDA leverages several distributed ledger services for storing and configuring AMs across different edge nodes, which provides a novel, secure and resilient way for specifying and executing global analytics tasks.

This chapter is devoted to the presentation of the DDA infrastructure of the FAR-EDGE project, which has been briefly introduced in [4]. This chapter extends the work in [4] through providing more details on the design and implementation details of the DDA platform. Special emphasis is put in describing and highlighting the unique value propositions of the FAR-EDGE DDA in terms of configurability, programmability and resilience. The description includes dedicated parts for the Edge Analytics Engine (EAE) that enable edge scoped analytics and for the Ledger Services for data analytics configuration and synchronization that enable configurable global analytics. Note also that the DDA infrastructure complies with the overall FAR-EDGE reference architecture, which has been introduced in an earlier chapter, while leveraging digital models that are presented in a subsequent chapter. Hence, the present chapter does not detail the overall architecture of the FAR-EDGE platform and the digital models that are used as part of it, since they are both described in other parts of the book.

The structure of this chapter is as follows:

- Section 6.2 following the chapter's introduction presents the main drivers behind the development of a framework for DDA in industrial environments, through enhancing conventional and popular frameworks for Big Data analytics and streaming analytics.

- Section 6.3 presents the overall architecture of the DDA, including its main modules.
- Section 6.4 illustrates the edge analytics engine of the DDA, including the anatomy of the analytics workflows.
- Section 6.5 presents the ledger services that enable the synchronization of different manifests across edge nodes.
- Section 6.6 presents information about the open source implementation of the DDA, including information about the underlying technologies that have been (re)used.
- Section 6.7 is the final and concluding section of the chapter.

6.2 Requirements for Industrial-scale Data Analytics

As already outlined, most digital automation platforms need to process large volumes of data (including streaming data) as part of wider simulation, decision making and control tasks. Instead of implementing a data analytics function for every new use case, digital automation platforms can offer entire middleware frameworks that facilitate the distributed data analytics tasks (e.g., [8–10]). These frameworks offer facilities for dynamically discovering data sources and executing data processing algorithms over them. In principle, they are Big Data frameworks that should be able to handle large data volumes that features the 4Vs (volume, variety, velocity and veracity) of Big Data. Beyond these general and high-level requirements, the FAR-EDGE DDA infrastructure has been driven by the following principles:

- **High-Performance and Low-Latency:** The FAR-EDGE DDA enables the execution of data analytics logic with high performance, i.e. in a way that ensures low-overhead and low-latency processing of data streams. This is especially important towards handling high-velocity data streams i.e. data with very high ingestion rates such as data streams stemming from sensors attached to a machine.
- **Configurable:** The DDA is configurable in order to be flexibly adaptable to different business and factory automation requirements, such as the calculation of various KPIs (Key Performance Indicators) for production processes. Configurability should be reflected in the ability to dynamically select the data sources that should be used as part of a data analytics task.

- **Extensible:** The DDA provides extensibility in terms of the supported processing functions, i.e. to provide the ability to implement additional data processing schemes based on fair programming effort. In the case of FAR-EDGE, extensibility concerns the implementation of advanced processing capabilities in terms of pre-processing, analyzing and storing data streams.
- **Dynamic:** The DDA is able to dynamically update the results of the analytics functions, upon changes in its configuration. This is essential towards having a versatile analytics engine that can flexibly adapt to changing business requirements and production contexts in volatile industrial environments where data sources join or leave dynamically.
- **Ledger Integration:** One of the innovative characteristics of the DDA lies in the use of a distributed ledger infrastructure (i.e. blockchain-based services) [11] towards enabling analytics across multiple EGs, as well as towards facilitating the dynamic configuration of the data analytics rules that comprise these analytics tasks.
- **Stream Handling Capabilities:** The DDA can handle streaming data in addition to transactional static or semi-static data. This requirement has been considered in the design and the prototype implementation of the DDA infrastructure, which is based on middleware for handling data streams.

Table 6.1 associates these design principles with some concrete implementation examples and use cases.

Table 6.1 Requirements and design principles for the FAR-EDGE DDA

Design Principles and Goals	Examples and use Cases	DDA Implementation Guidelines
High performance and Lowlatency	Complex data analyses over real-time streams should be performed within timescales of a few seconds. As an example, consider the provision of quality control feedback about an automation process in a station, based on the processing of data from the station. The DDA support the collection and analysis of data streams within a few seconds.	Leverage high-performance data streaming technology as background for the EAE implementation (e.g. ECI's streaming technology)

(*Continued*)

Table 6.1 (Continued)

Design Principles and Goals	Examples and use Cases	DDA Implementation Guidelines
Configurable	A manufacturer needs to calculate multiple Key Performance Indicators (KPIs) such as indicators relating to quality control and performance of the automation processes. The DDA should flexibly support the on-line calculation of the different KPIs within the same instance of the EAE. To this end, the EAE should be easily configurable to support the calculation of all desired KPIs, ideally with minimal or even zero programming. Configurability can be gauged based on the time needed to set up and deploy a data analytics workflow comprising several processing functions. The use of EAE is destined to reduce this time, when compared to cases where data analytics are programmed from scratch (i.e. without support from the EAE middleware).	• Specify and implement DDA as a programmable & configurable engine, which executes analytics configurations specified in appropriate files ("manifests"). • Parse and execute the analytics rules of the configuration files, without a need for explicitly programming these rules
Extensible	The EAE should be extensible in terms of data processing, data mining and machine learning techniques. For example, in cases where deep learning needs to be employed (e.g., estimation of a failure mode in predictive maintenance), the EAE must support the execution of machine learning functions, including AI-based algorithms such as deep neural network. The latter can, for example, support the detection of complex patterns such as production quality degradation patterns.	• Provide a library of analytics functions/capabilities and integrate it within a directory. • Provide the means for discovering and using analytics functions from the library analytics configurations.
Dynamic	The EAE should be able to deploy on the fly (i.e. hot deploy) different data analysis instances. For example, when new KPIs should be calculated, calculation shall be done of the fly, without affecting the rest of deployed KPIs.	Leverage multi-threading and hot deployment capabilities of the selected implementation technologies.

Table 6.1 (Continued)

Design Principles and Goals	Examples and use Cases	DDA Implementation Guidelines
Ledger integration	The EAE must integrate functions from the Ledger Services in order to: (i) access configurations of analytics tasks through ledger smart contracts, such as a large scale distributed analytics tasks; (ii) collecting and analyzing data from multiple edge nodes/gateway through access to the publishing services. This can be, for example, the case there data analytics for calculating a product schedule must be computed, as this is likely to span multiple EGs.	• Represent analytics configurations as smart contracts. • Implement publishing services driven by the smart contracts and leveraging information from multiple edge nodes.
Stream handling capabilities	The EAE must be able to handle data-intensive data streams such as sensor data for predictive maintenance and data from other field devices for quality control in automation.	Leveraging streaming handling and management middleware of the ECI.

6.3 Distributed Data Analytics Architecture

A high-level overview of the DDA Infrastructure is provided in Figure 6.1. The DDA consists of wide range of components, which are described in the following subsections.

6.3.1 Data Routing and Preprocessing

The Data Routing and Pre-processing (DR&P) component is in charge of routing data from the data sources (i.e. notably industrial devices) to the Edge Analytics Engine (EA-Engine). The component includes a **Device Registry**, where the various device and data sources announce (i.e. "register") themselves, as well as the means to access their data (i.e. based on connectivity details such as protocol, IP address and port). The registry makes the system dynamic, as it ensures handling of all data sources that register with it. Moreover, the component provides pre-processing capabilities, which allow for transformations to data streams prior to their delivery to the EA-Engine. Note that the DR&P component is edge-scoped i.e. it is deployed at an Edge Gateway (EG). Likewise, the data sources that are registered and managed in the registry concern the devices that are attached to the specific edge gateway as well.

Along with the Device Registry, the DR&P provides a Data Bus, which is used to route streams from the various devices to appropriate consumers, i.e. processors of the EA-Engine. Moreover, the Data Bus is not restricted to routing data streams stemming directly from the industrial devices and other shopfloor data sources. Rather it can also support the routing of additional data streams and events that are produced by the EA-Engine.

6.3.2 Edge Analytics Engine

The EA-Engine is a runtime environment hosted in an EG, i.e. at the edge of an industrial FAR-EDGE deployment. It is the programmable and configurable environment that executes data analytics logic locally to meet stringent performance requirements, mainly in terms of latency. The EA-Engine is also configurable and comprises multiple analytics instances that correspond to multiple edge scoped analytics workflows.

As shown in Figure 6.1, the EA-Engine comprises several processors, which implement processing functions over the data streams of the Data Bus. As illustrated in a following paragraph, these processors are of three main

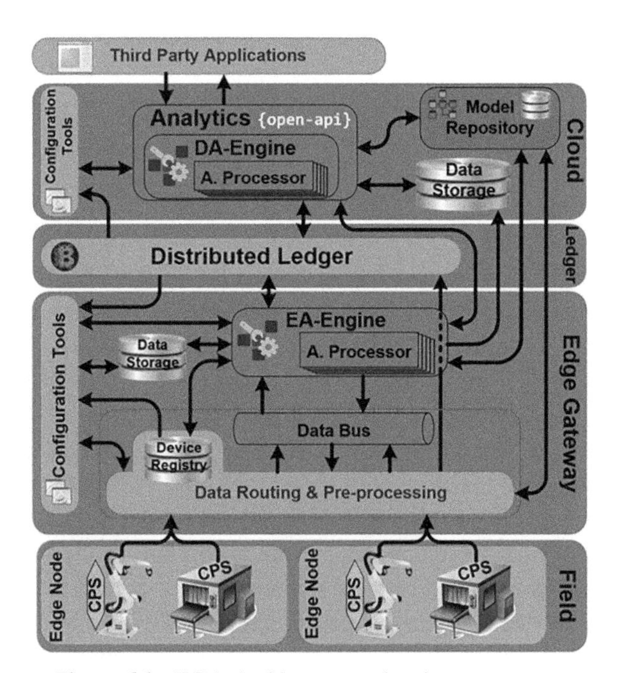

Figure 6.1 DDA Architecture and main components.

types, including processors that store/persist data streams, processors devoted to pre-processing functions, as well as processors in charge of data analytics. Furthermore, the outcomes of the EA-Engine can be written to the Data Bus in order to be consumed by other components and processing functions or even written at local/edge data storage.

6.3.3 Distributed Ledger

The Distributed Ledger is used to orchestrate analytics functionalities across multiple Edge Gateways. It is in charge of maintaining the configuration of different analytics tasks across multiple EGs, which at the same time keep track of their composition in factory-wide analytics tasks. Moreover, the distributed ledger is used to compute the outcomes of factory-wide analytics. Overall, the distributed ledger offers two kinds of services to the DDA, namely Data Publishing Services that synchronize the analytics computations and Configuration Services that synchronize the configuration of the analytics services.

6.3.4 Distributed Analytics Engine (DA-Engine)

While the EA-Engine is in charge of data analytics at edge scope, the DA-Engine is in charge of executing global analytics functions based on the analytics configurations that reside in the distributed ledger. The DA-Engine is configurable thanks to its interfacing with a set of data models that describe the configuration of the DDA infrastructure in terms of edge nodes, edge gateways, data sources and the processing functions that are applied over them as part of the DA-Engine. To this end, the DA-Engine interfaces to a models' repository, which comprises the digital representation of the devices, data sources and edge gateways that are part of the DDA. The Digital Models are kept up to date and synchronized with the status of the DDA's elements. As such, they are accessible from the DR&P and EA-Engine components, which make changes in the physical and logical configuration of the analytics tasks. Note also that the DA-Engine stores data within a cloud-based data storage repository, which is destined to persist and comprise the results of global analytics tasks.

6.3.5 Open API for Analytics

The Open API for Analytics enables external systems to take advantage of the DDA infrastructure functionalities, including both the configuration and execution of factory-wide analytics tasks, which span multiple edge gateways and take advantage of the relevant EA-Engine instances. Using the Open API any integrator of industrial solutions can specify and execute data processing functions over data streams stemming from the full range of devices that are registered in the device registries of the DR&P components of the DDA infrastructure. As illustrated in the figure, this gives rise to the use of the DDA infrastructure by third-party applications.

The following sections provide insights into the operation and novel features of the EA-Engine and the Distributed Ledger, which endows the DDA with modularity, extensibility and configurability.

6.4 Edge Analytics Engine

6.4.1 EA-Engine Processors and Programmability

One of the unique value propositions of the EA-Engine is that it is configurable and programmable. These properties stem from the fact that it is designed to handle analytics tasks that are expressed based on the combination of three types of processing functions, which are conveniently called "processors". The three types of processors are as follows:

- **Pre-processors**, which perform pre-processing (e.g. filtering) over data streams. In principle, pre-processors prepare data streams for analysis. A pre-processor interacts with a Data Bus in order to acquire streaming data from the field through the DR&P component. At the same time, it also produces and registers new streams in the same Data Bus, notably streams containing the results of the pre-processing.
- **Storage processors**, which store streams to some repository such as a data bus, a data store or a database.
- **Analytics processors**, which execute analytics processing functions over data streams ranging from simple statistical computations (e.g., calculation of an average or a standard deviation) to more complex machine learning tasks (e.g., execution of a classification function). Similar to pre-processors, analytics processors consume and produce data through interaction with the Data Bus.

Given these three types of "processors", analytics tasks are represented and described as combinations of multiple instances of such processing functions in the form of workflow or a pipeline. Such workflows are described through an Analytics Manifest (AM), which specifies a combination of the above processors. Hence, an AM follows a well-defined schema (as shown

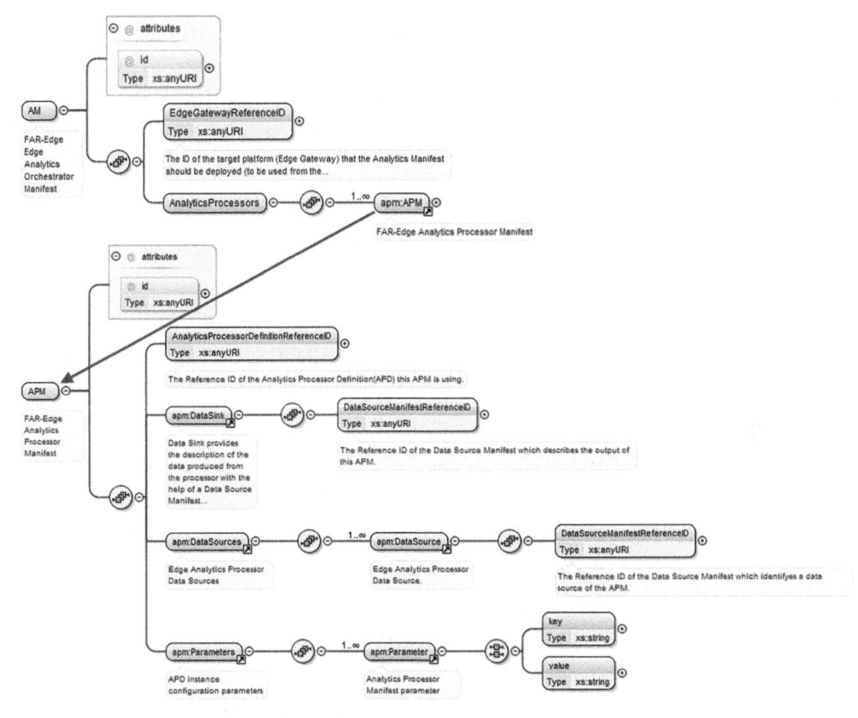

Figure 6.2 Representation of an Analytics Manifest in XML format (XML Schema).

in Figure 6.2), which specifies the processors that comprise the AM. In particular, an AM defines a set of analytics functionalities as a graph of processing functions that comprises the above three types of processors and which can be executed by the EA-Engine.

Note also that an AM instance is built based on the available devices, data sources, edge gateways and analytics processors, which are part of the data models of the DDA. The latter reflect the status of the factory in terms of available data sources and processing functions, which can be used to specify more sophisticated analytics workflows.

6.4.2 EA-Engine Operation

The EA-Engine provides the run-time environment that controls and executes edge analytics instances, which are specified in AMs. In particular, the EA-Engine is able to parse and execute analytics functions specified in an AM, based on the following processes:

- **Parsing:** The EA-Engine parses AMs and identifies the analytics pipeline that has to be executed.
- **Execution:** The EA-Engine executes (applied) the analytic functions that are identified following the parsing. Note that the EA-Engine is multi-threaded and enables the concurrent (parallel) execution of multiple analytics pipelines, which can correspond to different AMs.

Figure 6.3 illustrates an example topology and runtime operations for EA-Engine. In this example, two streams (CPS1 and CPS2) are pre-processed from Analytics Processor 1 (i.e. Pre-Processor) and Analytics Processor 2 (i.e. Pre-Processor) equivalently in order to enable the execution of an analytics algorithm that is in Analytics Processor 3, which is an Analytics Processor. Finally, the pipelines ends-up storing the result to a Data Store based on Analytics Processor 4, which is a Storage Processor. In this example, the EA-Engine is set up and runs based on the following steps:

- **Step 1 (Set-up):** Based on the description of the topology and required processors in the AM, the engine instantiates and configures the required Analytics Processors. Note that the AM is built based on real information about the factory, which is reflected in the digital models of the DDA infrastructure.
- **Step 2 (Runtime):** Analytics Processor 1 consumes and pre-processes streams coming from CPS1. Likewise, Analytics Processor 2 consumes

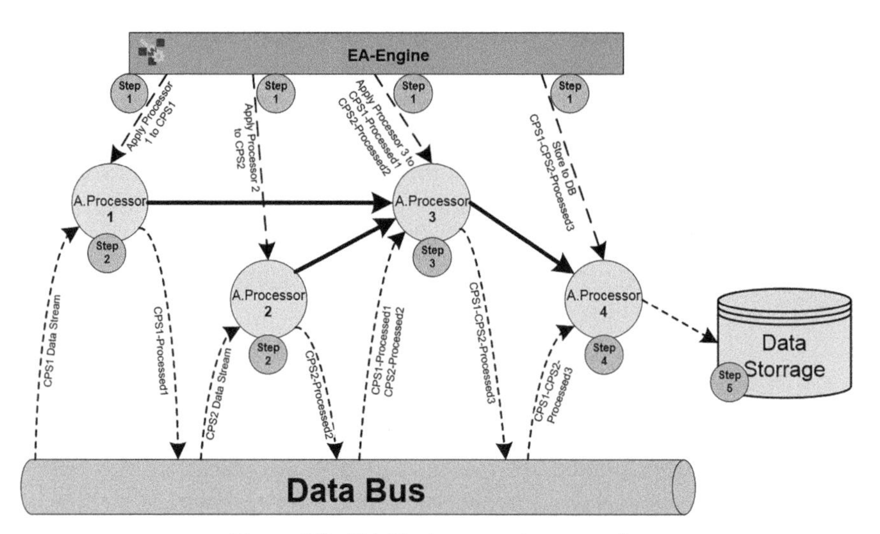

Figure 6.3 EA-Engine operation example.

and pre-processes streams coming from CPS2. In both cases, the streams are accessed through the Data Bus.

- **Step 3 (Runtime):** Analytics Processor 3 consumes the produced streams from Analytics Processor 1 and 2 towards applying the analytics algorithm. In this case, the analytics processor cannot execute without input for the earlier Analytics Processors.
- **Step 4 (Runtime):** Store Analytics Processor 4 consumes the data stream produced from Analytics Processor 3 and forwards it to the Data Store, which persists and data coming from Analytics Processor 4.

This is a simple example of the EA-Engine operation, which illustrates the use of all three types of processors in a single pipeline. However, much more complex analytics workflows and pipelines can be implemented based on the combination of the three different types of processors. The only limiting factor is the expressiveness of the AM, which requires that instances of the three processors are organized in a graph fashion, with one or more processors providing input to others.

Vendors and integrators of industrial automation solutions can take advantage of the versatility of the EA-Engine in two ways:

- First, they can leverage existing processors of the EA-Engine towards configuring and formulating analytics workflows in line with the needs of their application or solution.
- Second, they can extend the EA-Engine with additional processing capabilities, in the form of new reusable processors.

In practice, industrial automation solution integrators will use the EA-Engine in both the above ways, which are illustrated in the following paragraphs.

6.4.3 Configuring Analytics Workflows

Integrators can configure and execute edge-scoped analytics pipelines. The configuration of a new pipeline involves the following steps:

- **Discovery of Devices** and other data sources registered in the device registry. Analytics workflows can only take advantage of devices and data sources that are registered with the DR&P component.
- **Discovery of available processors**, a list of which is maintained in the EA-Engine. The rationale behind this discovery is to reuse existing processors instead of programming new ones. Nevertheless, in cases where the analytics workflow involves a processor that is not yet available,

this processor should be implemented from scratch. However, every new processor will become available for reuse in future analytics workflows.
- **Definition and creation of the Analytics Manifest**, based on the available (i.e. discovered) devices, data sources and processors. As already outlined, an AM comprises a graph of processors of the three specified types, defines the analytics results to be produced and specified where they are to be stored. The specification of the AM can take place based on the use of the Open API of the DDA. However, as part of our DDA development roadmap, we will also provide a visual tool for defining AMs, which facilitate zero-programming specification of the edge analytics tasks.
- **Runtime execution of the AM**, based on the invocation of appropriate functions of the EA-Engine's runtime. This step can be implemented based on the Open API of the DDA, yet it is also possible to execute it through a visual tool.

6.4.4 Extending the Processing Capabilities of the EA-Engine

Integrators can specify additional processing functions and make them available for use as part of the EA-Engine. The extension process involves the following steps:

- **Implementation of a Processor Interface:** In order to extend the EA-Engine with a new processor, an integrator has to provide an implementation of a specific interface i.e. the interface of the processor. In practice, each of the three processor types comes with its own interface.
- **Registration of the Processor to a Registry:** Once a new processor is implemented, it has to become registered to a registry. This will make it discoverable by solution developers and manufacturers that develop AMs for their needs, based on available devices and processors.
- **Using the processor**: Once a processor becomes available, it can be used for constructing AMs and executing analytics tasks that make use of the new processor.

6.4.5 EA-Engine Configuration and Runtime Example

In this section, we use the topology illustrated in Figure 6.3 above in order to provide a more detailed insight into the steps needed to configure the EA-Engine, but also in order to illustrate the interactions between the various components both at configuration time and at run time. As already outlined, the example involves two devices (CPS1, CPS2), which generate two data streams under a topic each one named after their ID. We therefore need to:

- Apply some pre-processing to each one of the two streams (by Processor 1 and Processor 2).
- Apply an Analytics algorithm (Processor 3) to the pre-processed streams.
- Persist the result to a Data storage (i.e. the Data Storage).

Figure 6.4 illustrates the steps required to register a new processor, build the Edge Analytics configuration (AM), register it to the EA-Engine and instantiate the appropriate Analytics Processors. In particular:

- The user of the EA-Engine (e.g. a solution integrator) registers new Processors required to the Model Repository. To this end, it can use an API or a visual tool.
- In order to set up an AM, all the available processors are discovered from the Model Repository and all the available Data Sources (DSMs) are discovered from the Distributed Ledger.
- The user has all the required information and with the help of the Configuration Dashboards can now set up a valid AM flow for the four Analytic Processors.
- The AM is set up based on a proper combination of devices data streams and processors. In this example, the AM includes the required configurations for Processor 1 (APM1), Processor 2 (APM2), Processor 3 (APM3) and Processor 4 (APM4).

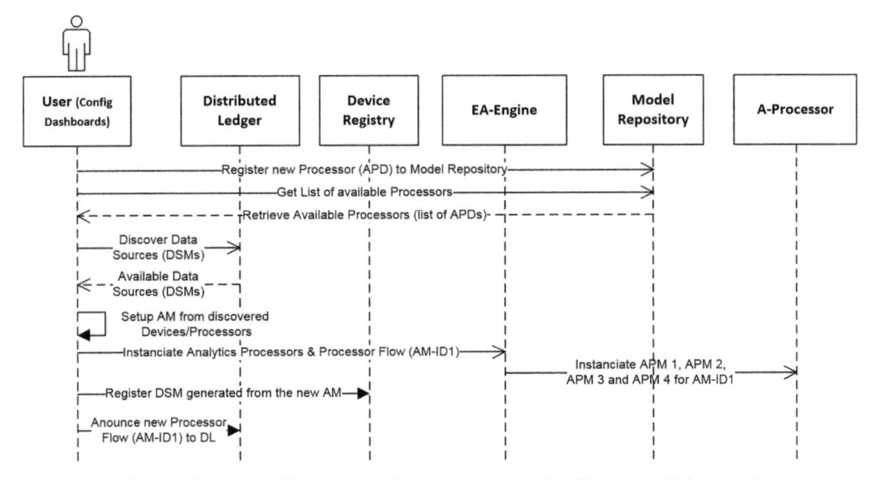

Figure 6.4 EA-Engine configuration example (Sequence Diagram).

- The AM is accordingly sent to the EA-Engine, which instantiates the four Analytic Processors.
- The output of the AM is automatically described in a new DSM, which is registered to the Device Registry as a new Data Source and synchronized with the Distributed Ledger through the Device Registry mechanisms.
- The capabilities of the new processor are also registered to the Distributed Ledger to enable the discoverability of the new processor for future use.

Figure 6.5 illustrates the interactions between the EA-Engine components, when the execution of the AM starts. These include:

- Instructing the EA-Engine to start the execution of the analytics task, as specified in the analytics manifest (AM1). To this end, the EA-Engine retrieves AM1 from the Distributed Ledger in order to instantiate the processors that AM1 comprises.
- The EA-Engine instantiates each of the four EA-Processors described in the AM1. Specifically:
    - As part of the instantiation of Processor 1 (pre-processor), its specification (APM1) contains the configurations of Processor 1,

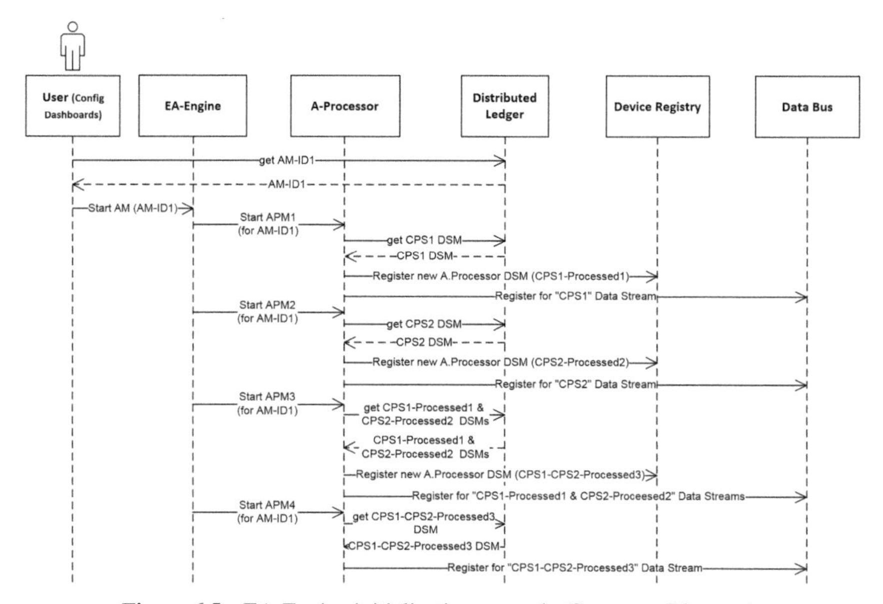

Figure 6.5 EA-Engine initialization example (Sequence Diagram).

which includes data inputs, data outputs and processor attributes required for the instantiation. The data type and data model of CPS1 are retrieved from the Ledger Service in order to apply the pre-processing properly. The processor data output description is provided within a new DSM that is registered to the Device Registry. Then, the EA-Processor (Processor 1) subscribes for the "CPS1" data stream of the Data Bus to apply the required pre-processing.

- As part of the instantiation of Processor 2 (pre-processor), its specification (i.e. APM2) contains the configurations of Processor 2, which includes data inputs, data outputs and processor attributes required for the instantiation. The data type and data model of CPS2 are retrieved from the Ledger Service. Also, the EA-Processor (Processor 2) subscribes for the "CPS2" data stream of the Data Bus in order to apply the required pre-processing.
- As part of the instantiation of Processor 3 (analytics processor), its specification (APM3) contains the configurations of Processor 3. Processor 3 subscribes to the topics named after the IDs of Processor 1 and Processor 2 ("CPS1-Processed 1" and "CPS2-Proceesed 2", respectively) in order to apply the required analytics.
- Finally, as part of the instantiation of Processor 4 (store processor), its specification (APM4) is retrieved from the EA-Storage. Processor 4 subscribes to the topics named after the ID of Processor 3 ("CPS1-CPS2-Processed 3") in order to store it to the data storage.

The runtime operation of the EA-Engine is further presented in Figure 6.6, which illustrates the sequence of runtime interactions of the components of the engine, following the conclusion of the above-listed configurations. At runtime, all the different processors run continuously in parallel until they are stopped from the end-user through a proper API command or based on the use of the visual tool. In particular:

- **Processor 1** gets notified every time new CPS1 data is published and collects it. It applies the required pre-processing and pushes the preprocessed data stream back to the data bus under the topic named after its own ID ("CPS1-Processed 1").
- **Processor 2** gets notified every time new CPS2 data is published and collects it. It applies the required pre-processing and pushes the preprocessed data stream back to the data bus under the topic named after its own ID ("CPS2-Processed 2").

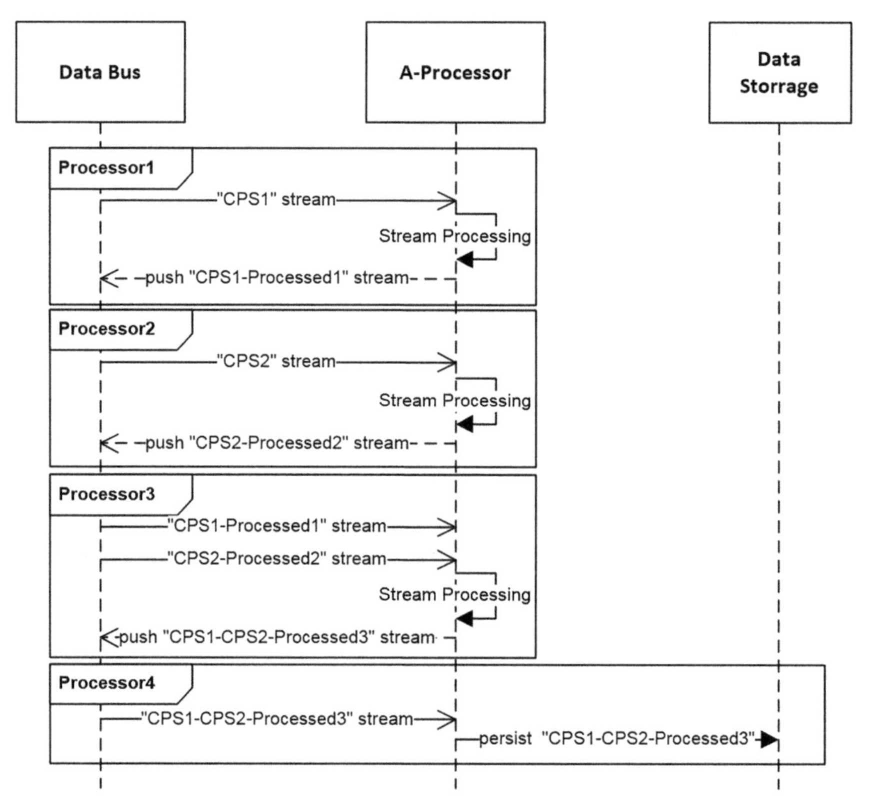

Figure 6.6 EA-Engine runtime operation example (Sequence Diagram).

- **Processor 3** gets notified every time new Processor 1 and Processor 2 data is published and collects it. It applies the required analytic and pushes the processed data stream back to the data bus under the topic named after its own ID ("CPS1-CPS2-Processed 3").
- **Processor 4** gets notified every time new Processor 3 data is published and collects it. It pushes the collected data to the EA-Storage to be persisted.

6.5 Distributed Ledger and Data Analytics Engine

6.5.1 Global Factory-wide Analytics and the DA-Engine

Given the presented functionalities of the EA-Engine, the DA-Engine enables the combination and synchronization of data from multiple edge analytics pipelines towards implementing factory-wide analytics. At a high level, the concept of global analytics workflows is similar to the one of edge analytics ones. In particular, an Analytics Manifest (AM) is used to express an analytics workflow based on the combination of analytics tasks that are configured and executed at edge gateways based on properly configured instances of the EA-Engine. To this end, a mechanism for constructing AMs that comprise global analytics tasks is provided through the Open API of the DDA. In particular, the Open API provides the means for creating, updating, deleting, managing and configuring global analytics tasks based on the combination and orchestration of edge analytics workflows.

At a lower level, the implementation of the AM configuration and execution mechanism is offered in two flavours:

- **A conventional edge computing implementation**, which is subject to conventional central control. It involves an analytics engine that combines edge analytics workflows to global ones for a central orchestration point. That is in line with the classical edge/cloud computing paradigm.
- **A novel distributed ledger implementation**, which is based on a disruptive cooperative approach without central control. This cooperative approach is based on the deployment and use of ledger services in each one of the edge nodes that participate in the DDA infrastructure. In particular, ledger services are deployed in each of the edge gateways in order to enable a consensus-based approach regarding the configuration of the global analytics task, as well as its execution based on publishing and combination of data from the edge gateways. Such a collaborative approach is fully decentralized and hence does not provide a single point of failure. Moreover, it can be generalized beyond edge gateways in order to enable data analytics workflows that comprise data from field objects (i.e. smart objects) and cloud nodes as well.

The next sub-section illustrates the scope and operation of these ledger services, which enable a novel and more interesting approach to supporting the functionalities of the DA-Engine.

6.5.2 Distributed Ledger Services in the FAR-EDGE Platform

For the implementation of the DA-Engine, we leverage the services of a permissioned blockchain, rather than of one of the popular public blockchains such as Bitcoin and Ethereum. The rationale behind this decision is that permissioned blockchains provide the means for controlling participation and authenticating participants to the blockchain network, while offering superior performance over public blockchains [12]. The latter performance is largely due to the fact that peer nodes (i.e. participants) in these blockchains need not employ complex Proof-of-Work (PoW) mechanisms. For these reasons, a permissioned blockchain is more appropriate for coordinating and synchronizing distributed processes in an industrial context.

In this context, a Ledger Service is a Chaincode program for IBM's Hyperledger Fabric, which uses some of the utility services that are provided by the FAR-EDGE platform. Chaincode is always designed to support a well-defined, application-specific process. Hence, the DDA implementation is not based on a generic Ledger Service implementation, but rather on application-specific Ledger Service. Nevertheless, four categories of abstract services are defined as part of the Ledger Tier of the FAR-EDGE Architecture, namely Orchestration, Configuration, Data Publishing and Synchronization. These categories are used to classify the application-specific implementations of Ledger Services rather than to denote some general-purpose framework services. In particular:

- **Orchestration Services** are related to edge automation workflows, aiming at synchronizing distributed edge automation tasks in factory-wide automation workflows.
- **Data Publishing Services** support edge analytics algorithms, through the combination of multiple edge analytics pipelines in factory-wide workflows.
- **Synchronization Services** enable the reconciliation of several independent views of the same dataset across the factory.
- **Configuration Services** support the decentralized system administration.

Overall, these four categories of Ledger Services cover all the mandatory platform-level functionality that is required for Edge Computing to deliver its promises in a manufacturing context. The Distributed Ledger of the FAR-EDGE platform can then be used to deploy any kind of custom Ledger Service that meets the secure state sharing and/or decentralized coordination requirements of user applications.

Any concrete Ledger Service implementation is responsible for three things:

- **Defining and managing a data model**. While the global state of the Ledger Service is automatically maintained in the background by the DL-Engine – which logs every state change in the Ledger that is replicated across all the peer nodes of the system – the data model of such state is shaped in code by the Ledger Service implementation itself. Practically speaking, the data store of a Ledger Service is initialized according to a specific data model by a special code section when the instance is first deployed. Once initialized, no structural changes in the data model occur.

- **Defining and executing business logic**. Application logic is coded in software and exposed on the network as a number of application-specific service endpoints, which can be called by clients. These service calls represent the API of the Ledger Service. Through them, callers can query and change the global state of the Ledger Service. The API can be invoked by any authorized client on the network following some well-documented calling conventions of the DL-Engine. Moreover, we have implemented an additional layer of software in order to simplify the development of client applications: each Ledger Service implemented in the project comes with its own client software library – called Ledger Client – which an application can embed and use as a local proxy of the actual Ledge Service API. The Ledger Client provides an in-process API, which has simple call semantics.
- **Enforcing (and possibly also defining) fine-grained access and/or usage policies**. This is optional one, as a basic level of access control is already provided by the DL-Engine, which requires all clients to have a strong digital identity and be approved by a central authority. When a more fine-grained control is required – e.g. an Access Control List (ACL) applied to individual service endpoints – the Ledger Service implementation is required to manage it as part of its code.

In the specific context of the FAR-EDGE Platform, peer nodes are usually – but not mandatorily – installed on Edge Gateway servers, together with Edge Tier components. This setup allows for DL clients that run on Edge Gateways, like the EA-Engine, to refer to a localhost address by default when resolving Ledger Service endpoints. However, this is not the only possible way to deploy the Ledger Tier in FAR-EDGE-enabled system: peer nodes can easily be deployed on the Cloud Tier to make them addressable from anywhere or even embedded in Smart Objects on the Field Tier to make them fully autonomous systems. In complex scenarios, peer nodes can actually be spread across all the three physical layers of the FAR-EDGE architecture (Field, Edge and Cloud), exploiting the flexibility of the DL enabler to its full extent.

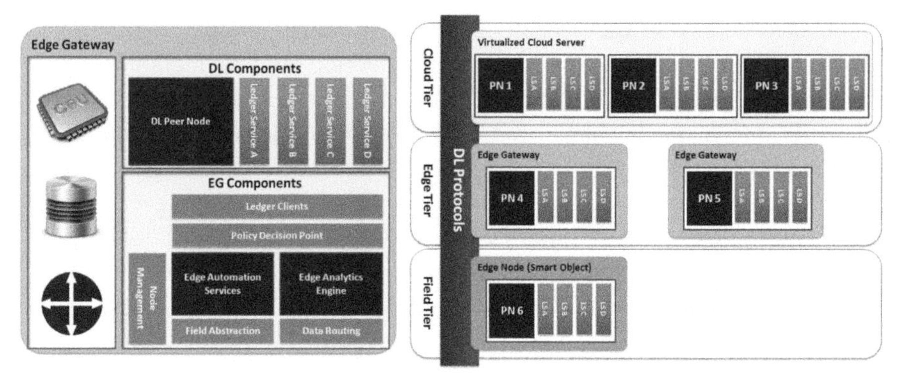

Figure 6.7 DL deployment choices (right) and EG deployment detail (left).

6.5.3 Distributed Ledger Services and DA-Engine

The DA-Engine takes advance of two of the above-listed types of Ledger Services, namely the Data Publishing and Configuration services. In particular, the DDA infrastructure implements Data Publishing and Configuration services at the Ledger Tier, in order to configure factory-wide AMs and to implement the respective analytics. In particular:

- **Configuration Services:** DDA configurations (i.e. AMs) are represented as smart contracts. Each smart contract is executed by the peers (notably edge gateway) that participate in the configuration and execution of the factory-wide AM. A set of Configuration services (Ledger Services) are used to ensure the configuration of the global analytics manifest based on consensus across the participating nodes. In this case, the distributed ledger is used as a distributed database that holds all the analytics configurations (in terms of manifests and their component). This allows the resilient configuration of global analytics without a need for centralized coordination and control from a single point of (potential) failure.
- **Publishing Services:** Publishing Services are implemented in order to compute factory-wide analytics tasks, based on data streams and analytics (i.e. processors) available across multiple instances of the EA-Engine, which are deployed in different Edge Gateways (EGs). The EGs act as peers in this case.

6.6 Practical Validation and Implementation

6.6.1 Open-source Implementation

The DA-Engine is implemented as open-source software/middleware, which is available at the FAR-EDGE github: https://github.com/far-edge/distributed-data-analytics. In the absence of general-purpose Ledger Services, the implementation includes the middleware for edge analytics framework of Section 6.3, as well as an Open API for creating Analytics Manifests for global, factory-wide analytics. Hence, a subset of the DDA architecture has been actually implemented, which is shown in Figure 6.8. As evident from the figure, the open-source implementation includes the EA-Engine and the DA-Engine, without however general-purpose ledger services, which is the reason why the Distributed Ledger database is not depicted in the figure. In a nutshell, the implementation includes and integrates the DR&P, the Data Bus, the Device Registry, the Data Storage (including both cloud and local data storage) and the Model Repository components.

The structure of the open-source codebase is as follows:

- **edge-analytics-engine**, which contains the source code of the EA-Engine component.

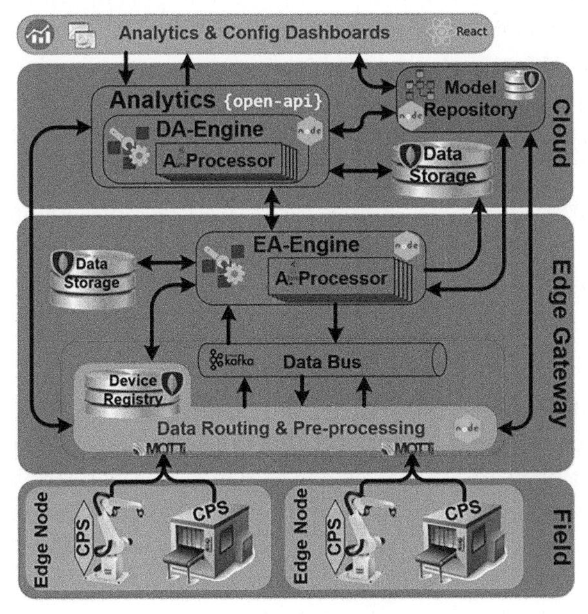

Figure 6.8 Elements of the open-source implementation of the DDA.

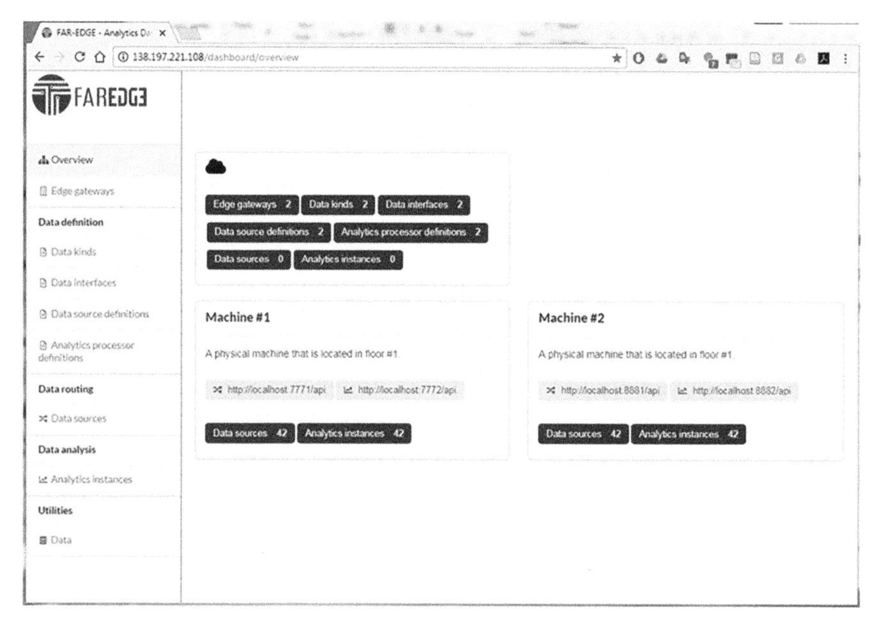

Figure 6.9 DDA Visualization and administration dashboard.

- **open-api-for-analytics**, which contains the component that implements and supports the Open API for Analytics.
- **mqtt-random-data-publisher**, which contains an application that simulates the functionality of DR&P component in order to facilitate the easier setup of simple demonstrators.

Furthermore, a set of administration dashboards that visualize the main entities of the DDA have been implemented. It allows the monitoring and the configuration of main entities like processors, data sources, devices and manifests (see Figure 6.9).

6.6.2 Practical Validation

6.6.2.1 Validation environment

The DDA Infrastructure has been also validated in a pilot plant and specifically in the pilot plant of SmartFactoryKL, which is a network with more than 45 industrial and research organizations that support and use an Industrie 4.0 testbed in Kaiserslautern, Germany. In particular, we set up a relatively simple analytics scenario over three Infrastructure Boxes (IB) of the pilot plant. Each Infrastructure Box (IB) provides energy sensors information through an MQTT interface (Broker), where Data are provided every 60 seconds. The available energy information provided includes data about the TotalRealPower, the TotalReactivePower, the TotalApparentPower, the Total-RealEnergy, the TotalReactiveEnergy and the TotalApparentEnergy that are consumed and used by the machine. The business rationale behind analyzing this data is to help the plant operator in finding anomalies during production. Indeed, with the power and energy values, it is possible to understand the machine behaviour as well as the "response time" of each business process. Moreover, the use of streaming processing and high-performance analytics enables the identification and understanding of abnormalities almost in real time.

The following components were deployed and used in the pilot plant:

- **The Data Routing and Pre-processing (DR&PP) Component** (including device registry service), which forwards data generated by Field sources.
- **The Edge Tier Data Storage**, which stores data stemming from the EA-Engine and provides a result storage repository.
- **The Model Repository**, which supports the sharing of common digital models, which are used from the various analytics components.
- **The EA-Engine**, which is the programmable and configurable environment that executes data analytics logic locally.
- **The Analytics Processor**, which implements the data processing functionalities for an edge analytics task.

The components are deployed in a Virtual Machine (VM) provided within the Smart Factory premises, which had access to data from the IB based on the MQTT protocol. The DDA has been tested and validated in two different scenarios, involving edge analytics and (global) distributed analytics. Various test cases have successfully run and analytics results have been correctly computed. The following subsections illustrate the setup of the EA-Engine and the DA-Engine in the scope of the two scenarios.

6.6.2.2 Edge analytics validation scenarios

For the Edge Analytics, we provide the hourly daily consumption from each Infrastructure Box for two parameters, namely TotalRealPower and TotalRealEnergy. The following steps have been followed for setting up and modelling the Edge Analytics scenario:

- **IB Modelling:** One Edge Gateway is built with each IB. The latter is modelled in line with the FAR-EDGE digital models for data analytics.

The respective data model is stored at the Data Model repository in the cloud.

- **IB Instantiation & Registration:** The specified Data models are used to generate the Data Source Manifest (DSM) and register it to each Edge Gateway.
- **Edge Analytics Modelling:**The required processor is modelled with the help of an Analytics Processor Definition (APD). In particular, the following processors are defined: (i) A processor for hourly average calculation from a single data stream and (ii) Processor for persisting results in a MongoDB. The above information is also stored at the Data Model repository in the cloud.
- **Edge Analytics Installation & Registration**: The specified Data models are used to generate the Analytics Processor Manifest (APM) for each required Processor, which is registered to the Edge Gateway. The following processors are set up: (i) A Processor for hourly average calculation from the TotalRealPower data stream; (ii) A Processor for hourly average calculation from the TotalRealEnergy data stream; (iii) A Processor for persisting results in the MongoDB of an EG in order to support edge analytics calculations; and (iv) A Processor for persisting results in a global (cloud) MongoDB in order to support (global) distributed analytics. Moreover, an AM is also created in order to combined values and data from the instantiated processors. The AM is registered and started through the API of the EG.

Following the setup and configuration of the system, runtime operations are supported, including the following information flows:

- IBs pushes the data to MQTT broker.
- The DR&P retrieves raw/text data from MQTT broker and pushes them to an Apache Kafka Data Bus.
- The data are retrieved and processed from the Analytics Engine.
- The data are finally stored to the local Data Storage repository.

6.6.2.3 (Global) distributed analytics validation scenarios

For the Distributed Analytics validation, we provide the hourly daily consumption from all IBs for the TotalRealPower and the TotalRealEnergy parameters. The following steps are also needed in addition to setting up the EA-Engine:

- **Distributed Analytics Modelling:** The required processors will be modelled with the help of an Analytics Processor Definition (APD) construct of the FAR-EDGE data models. The processors that are set up include: (i) A Processor for hourly average calculation for values from a MongoDB and (ii) A Processor for persisting results in a MongoDB. The above information is stored at the Data Model repository, which resides on the cloud.

- **Distributed Analytics Installation & Registration:** The specified data models are used to generate the Analytics Processor Manifest (APM) for each required Processor and are registered to the Cloud. The following processors are registered: (i) A Processor for hourly average calculation from the TotalRealPower parameters for all IBs based on information residing in the (global) MongoDB in the cloud; (ii) A Processor for hourly average calculation from TotalRealEnergy for all IBs based on information residing in the (global) MongoDB in the cloud; and (iii) A Processor for persisting results in the (global) MongoDB in the cloud. An Analytics Manifest (AM) will be generated for combining data from the instantiated Processors. The AM will be registered and started through the Open API of the DA-Engine.

6.7 Conclusions

Distributed data analytics is a key functionality for digital automation in industrial plants, given that several automation and simulation functions rely on the collection and analysis of large volumes of data (including streaming data) from the shopfloor. In this chapter, we have presented a framework for programmable, configurable, flexible and resilient distributed analytics. The framework takes advantage of state-of-the-art data streaming frameworks (such as Apache Kafka) in order to provide high-performance analytics. At the same time however, it augments these frameworks with the ability to dynamically register data sources in repository and accordingly to use registered data sources in order to compute analytics workflows. The latter are also configurable and composed of three types of data processing functions, including pre-processing, storage and analytics functions. The whole process is reflected and configured based on digital models that reflect the status of the factory in terms of data sources, devices, edge gateways and the analytics workflows that they instantiate and support.

The analytics framework operates at two levels: (i) An edge analytics level, where analytics close to the field are defined and performance and (ii) A global factory-wide level, where data from multiple edge analytics deployments can be combined in arbitrary workflows. We have also presented two approaches for configuring and executing global level analytics: One following the conventional edge/cloud computing paradigm and another that support decentralized analytics configurations and computations based on the use of distributed ledger technologies. The latter approach holds the promise to increase the resilience of analytics deployments, while eliminated single point of failure and is therefore one of our research directions.

One of the merits of our framework is that it is implemented as open-source software/middleware. Following its more extensive validation and the improvement of its robustness, this framework could be adopted by the Industry 4.0 community. It could be really useful for researchers and academics who experiment with distributed analytics and edge computing, as well as for solution providers who are seeking to extend open-source libraries as part of the development of their own solutions.

Acknowlegdements

This work was carried out in the scope of the FAR-EDGE project (H2020-703094). The authors acknowledge help and contributions from all partners of the project.

References

[1] H. Lasi, P. Fettke, H.-G. Kemper, T. Feld, M. Hoffmann, 'Industry 4.0', Business & Information Systems Engineering, vol. 6, no. 4, pp. 239, 2014.

[2] J. Soldatos (editor) 'Building Blocks for IoT Analytics', River Publishers Series in Signal, Image and Speech Processing, November 2016, ISBN: 9788793519039, doi: 10.13052/rp-9788793519046.

[3] J. Soldatos, S. Gusmeroli, P. Malo, G. Di Orio 'Internet of Things Applications in Future Manufacturing', In: Digitising the Industry Internet of Things Connecting the Physical, Digital and Virtual Worlds, Editors: Dr. Ovidiu Vermesan, Dr. Peter Friess. 2016. ISBN: 978-87-93379-81-7.

[4] M. Isaja, J. Soldatos, N. Kefalakis, V. Gezer 'Edge Computing and Blockchains for Flexible and Programmable Analytics in Industrial Automation', International Journal on Advances in Systems and Measurements, vol. 11 no. 3 and 4, December 2018 (to appear).

[5] T. Yu, X. Wang, A. Shami 'A Novel Fog Computing Enabled Temporal Data Reduction Scheme in IoT Systems', GLOBECOM 2017 - 2017 IEEE Global Communications Conference, pp. 1–5, 2017.

[6] S. Mahadev et al. 'Edge analytics in the internet of things', IEEE Pervasive Computing, vol. 14, no. 2, pp. 24–31, 2015.

[7] M. Yuan, K. Deng, J. Zeng, Y. Li, B. Ni, X. He, F. Wang, W. Dai, Q. Yang, "OceanST: A distributed analytic system for large-scale spatiotemporal mobile broadband data", PVLDB, vol. 7, no. 13, pp. 1561–1564, 2014.

[8] A. Jayaram 'An IIoT quality global enterprise inventory management model for automation and demand forecasting based on cloud', Computing Communication and Automation (ICCCA) 2017 International Conference on, pp. 1258–1263, 2017.

[9] J. Soldatos, N. Kefalakis, M. Serrano, M. Hauswirth, A. Zaslavsky, P. Jayaraman, and P. Dimitropoulos 'Practical IoT deployment on Smart Manufacturing and Smart Agriculture based on an Open Source Platform', in Internet of Things Success Stories, 2014.

[10] John Soldatos, Nikos Kefalakis et. al. 'OpenIoT: Open Source Internet-of-Things in the Cloud', OpenIoT@SoftCOM, 2014: 13–25, 2014.

[11] Z. Zheng, S. Xie, H. Dai, X. Chen, and H. Wan. 'An Overview of Blockchain Technology: Architecture, Consensus, and Future Trends', Proceedings of IEEE 6th International Congress on Big Data, 2017.

[12] Elli Androulaki et al. "Hyperledger Fabric: A Distributed Operating System for Permissioned Blockchains", Proceedings of the Thirteenth EuroSys Conference (EuroSys '18), Article No. 30, Porto, Portugal, April 23–26, 2018.

7

Model Predictive Control in Discrete Manufacturing Shopfloors

Alessandro Brusaferri[1], Giacomo Pallucca[1], Franco A. Cavadini[2], Giuseppe Montalbano[2] and Dario Piga[3]

[1]Consiglio Nazionale delle Ricerche (CNR),
Institute of Industrial Technologies and Automation (STIIMA),
Research Institute, Via Alfonso Corti 12, 20133 Milano, Italy
[2]Synesis, SCARL, Via Cavour 2, 22074 Lomazzo, Italy
[3]Scuola Universitaria Professionale della Svizzera Italiana (SUPSI),
Dalle Molle Institute for Artificial Intelligence (IDSIA),
Galleria 2, Via Cantonale 2C, CH-6928 Manno, Switzerland
E-mail: alessandro.brusaferri@itia.cnr.it; giacomo.pallucca@itia.cnr.it;
franco.cavadini@synesis-consortium.eu;
giuseppe.montalbano@synesis-consortium.eu; dario.piga@supsi.ch

This chapter describes the fundamental components of the Software Development Kit architecture developed in Daedalus and its integration in IEC-61499 paradigm, presenting the methodologies selected to face the issues related to the control of aggregated Cyber Physical System (CPS). The aim of the Software Development Kit is to help automation system engineers to synthesize Hybrid Model Predictive Control for aggregated CPS environment.

The guidelines of future development steps of the tool are described. The SDK is composed of three main parts: On-line System Identification (OIS), Online Control Modeller (OCM) and Online Control Solver (OCS). The first one is dedicated to automatically infer the system's model of aggregated CPS from input and output measurements. OIS absolves two functions: in a preliminary design phase, it is used in order to estimate a first model of the system; successively during execution, it works in real time for tuning the parameter of the system in relation to input and output measurements. The OCM is the main component of SDK and it contains direct interface to modify and customize the parameters of controller to be designed, like observer tuning, prediction horizon and so on. Moreover, the OCM is the synergic element that orchestrate the work flow of OCS, which performs the calculations during execution. The main computational aspects are related to the requirements of the solution of an optimization problem in the receding horizon fashion: in each step, an MIQP problem must be solved in the cycle time: an adequate solver is fundamental to realize Hybrid Model Predictive Control.

7.1 Introduction

Part of the Daedalus project is dedicated to the design and implementation of the Software Development Kit (SDK) that provides helpful tools to develop, implement and deploy advanced control system within a distributed IEC-61499-based control framework, dedicated to automation system engineers.

To such an aim, optimal orchestration of distributed IEC-61499 application is investigated and advanced control techniques as optimal control and model predictive control are considered.

The main features of aggregated Cyber Physical System (CPS) are evaluated to realize an advanced optimal control system: it exhibits, in particular, both continuous and discrete variables to represent the aggregated CPS. Straightforwardly, Hybrid system will be considered, and the various modelling techniques are investigated in Section 7.2.

Another important feature of optimal orchestration of aggregated CPS is the compliance with system constrains on both output variables, i.e. physical limits, and manipulated variables, e.g. actuators saturation and limits. The optimization of a measure of the performance of the system, i.e. the minimization of the cost function, is now a well-established approach in the academia and in certain industries like the chemical and aerospace industries, which have to be widespread in every industrial sector. Therefore, optimization-based control algorithms are investigated for the SDK. Among these, Model Predictive Control stands out as the most promising, considering that Receding Horizon approach offers a way to compensate for disturbances on the system and model mismatch.

Following the last decades of development of control theory, the most suitable solution for above requirements and objectives is Hybrid Model Predictive Control (Section 7.3.1). Indeed, this family of control method guarantees in an implicit manner the respect of constrains and manages multi-objectives control in an optimal way, thanks to Quadratic Programming solver (details will be reported in further sections).

The aim of this chapter is to introduce and carry out an in-depth analysis of the main components of the SDK of Daedalus. Figure 7.1 shows the idea of optimal hybrid orchestrator for aggregated CPS. It is divided into three main subcomponents: Online System Identification tool, Online Control Modeller and Online Control Solver, which are discussed in the following sections.

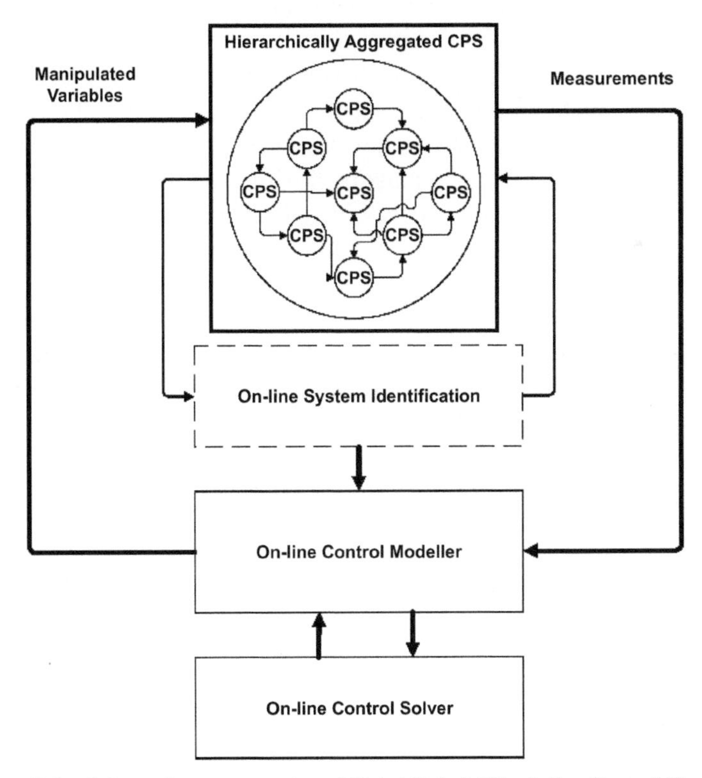

Figure 7.1 Schematic representation of Hybrid Model Predictive Control Toolbox.

7.1.1 Hybrid Model Predictive Control SDK

The proposed reference framework is composed of three main parts (shown in Figure 7.1). The first one is the On-line System Identification (OIS) tool, which is able to deduce the model of complex Multi-Input Multi-Output (MIMO) hybrid system. This data-driven tool uses input/output variables to extrapolate mathematical model of the system and it is based on iterative real-time procedure, and more details are reported in Section 7.4. The second block is the Online Control Modeller (OCM), where, given a model from the OIS, an optimal predictive controller able to orchestrate the aggregated Cyber-Physical Systems is synthesized. The OCM is developed based on latest paradigm of HMPC, explained in depth in Section 7.3. The last one is Online Control Solver (OCS) that is strictly related to OCM. This solver must be able to deal with Mixed-Integer Quadratic Problem (MIQP), to solve optimal predictive control problem for hierarchically aggregated CPS with quadratic function cost.

To such an aim, the proposed framework is developed to help control engineer to easily create an optimal controller for complex distributed CPS architecture. Each component will be developed with platform-independent software (see Section 7.1.2), which must be flexible and easy to use in order to create a standard procedure that deals with hybrid complex systems. Moreover, the resulting SDK will be integrated in a distributed IEC-61499-based control architecture (see Figure 7.2).

As analysed in Section 7.3.3, the computational aspect cannot be negligible; indeed, Mixed Integer Programming problem requires high computational power to be solved in runtime. This is more critical when complex systems require large controller bandwidth (Hz order): at 1 Hz, the OCS has to solve a Mixed Integer Problem in less than a second. An additional problem is the non-deterministic solving time of MIP. For the robustness of the modelled controller, it is important to evaluate in simulation the worst case of execution time and use a safety factor to evaluate a realistic and safety bandwidth of the controller. To face this problem, virtual commissioning is helpful: it is indeed possible to test control performance and its feasibility in a virtual environment and tune all control parameters.

7.1.2 Requirements

The investigation on orchestration of hierarchically aggregated CPS controller problems had led different needs. The basic development tools, to be compliant with IEC-61499 [1] and to have a platform-independent

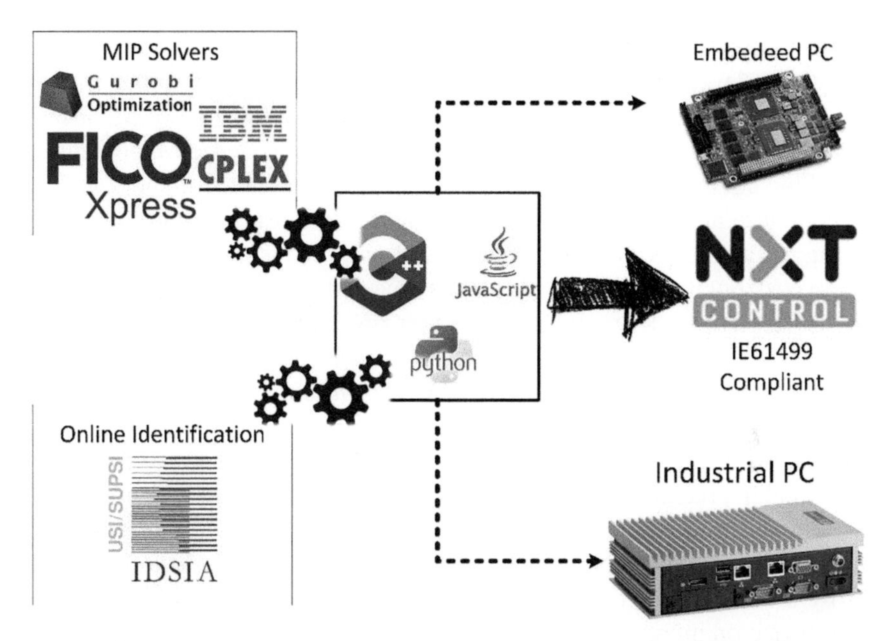

Figure 7.2 Conceptual map of used software. In the centre, there is object-oriented programming language that better supports an easy development and management between different application's needs.

toolbox, seem to be an object-oriented programming language used in cooperation with nxtControl. The nxtControl respects each paradigm of IEC-61499 and allows to build easily distributed control system using function blocks (for more details, see Section 7.5). The possible choice of object-oriented programming language allows to have a wide range of tool easily integrated in a single development environment. Object-oriented programming is easy to use for the purpose of this SDK, and this programming paradigm allows to develop effortlessly scalable and flexible software, independently from the application.

The investigated programming languages are Python, C++ and JavaScript. Even if the natural choice for a direct integration with nxtControl is C++, Python environment allows a better abstraction layer and enables easily the integration of a wide range of tools and libraries developed for optimization solver and control system. Moreover, nxtControl is able to compile Python with a wrapping toolkit, the computational time waste with the wrapper is negligible with respect to the computational time due to Quadratic Problem solver. This aspect conveys that choice of programming languages is not to be restricted to a specified one.

Another important benefit of possible Python's choice is availability of modelling and development environment of MIP solvers, both commercial and free-licence for it. Gurobi [2] and CPLEX [3] are the most powerful and optimized MIP commercial solvers [4], which have dedicated development and modelling environments for Python, also in C++. These environments are easy to configure and more important; they are easily integrable with hierarchically aggregated CPS controller. One limitation of industrial application is the license cost, but the difference of solving time and robustness respect freeware is not negligible. Regarding this, further investigation and benchmark will be done.

First release of the SDK will consider a centralized control scheme, where the on-line system identification tool returns the system's model. Straightforward Online control modeller builds up, based on identified model, a hybrid model predictive controller for the system with desired configuration. Finally, the proceeds controller sets up the online control solver and performs the desired performances respecting the tuning parameters chosen by the user, and moreover managing little modelling mismatching and disturbance on input and measurements.

Figure 7.2 shows the framework of the proposed toolbox. It is possible to see the different MIP solver and the Online Identification toolbox of the SDK; on the right, the different objective platforms where proposed Hybrid Model Predictive Controller will work are shown.

7.1.3 Hybrid System

The behaviour of physical phenomena can be represented by mathematical models. When these models exhibit continuous variable (like differential equation) and discrete/logical variables (like state machine), they are called Hybrid System Models. Every physical phenomenon can be described at different levels of detail; in applied science, it is possible to find various models of the same process, in relation of what the model had to describe. These models should not be too simple or too complicated. To formulate these models, we describe with sufficient level of details the behaviour of the physical phenomena efficiently by computational analysis point of view. In the following sections, the report analyzes the trade-off between simple and computational-light model with respect to more complex and computational-heavy model.

In the last three decades, several computer scientists and control theorists have explored models describing the interaction between continuous dynamics and logical components [5]. Such heterogeneous models

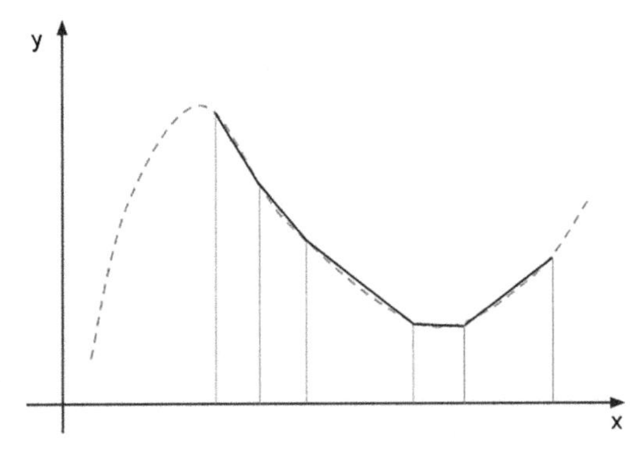

Figure 7.3 Subsequence approximation of a non-linear system.

are denoted as hybrid models; they switch among many operating modes described by differential equation, and mode transitions are triggered by events like states crossing pre-specified thresholds.

Another kind of system that is agreeably represented by hybrid model is non-linear system. Indeed, it is possible to represent non-linear system by a piece-wise linearized model, which consists in a sequence linearization of the system's model around consecutive operating points (see Figure 7.3). This kind of model representation is presented in Section 7.2.1, where its behaviour is also shown. Indeed, the relationship between every working mode is linear, whose slope changes in each region; this is called linearized model of non-linear system and can be represented like a Hybrid system that switches its operating mode.

7.1.4 Model Predictive Control

Model Predictive Control (MPC) arose in the late 1970s and has developed continuously since then. The term MPC does not correspond to specific control strategy, but fairly a wide range of control methods, which use mathematical model of the process to obtain control signal by minimizing an objective function.

Model Predictive Control is an advanced control technique that determinates the control action by solving on-line, at every sampling time k, an open-loop optimal control problem over a p-horizon (Equation (7.2)), based on the current state of the system at k-sample. The optimization generates an input sequence for the specified time horizon p. However, only the first calculated input is applied to the system (Figure 7.4).

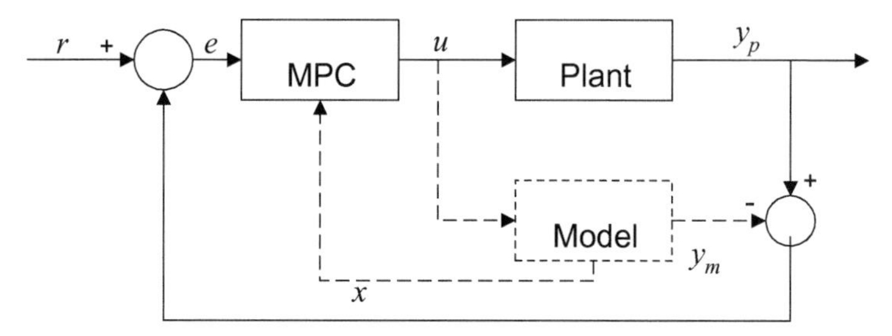

Figure 7.4 Model Predictive Control scheme.

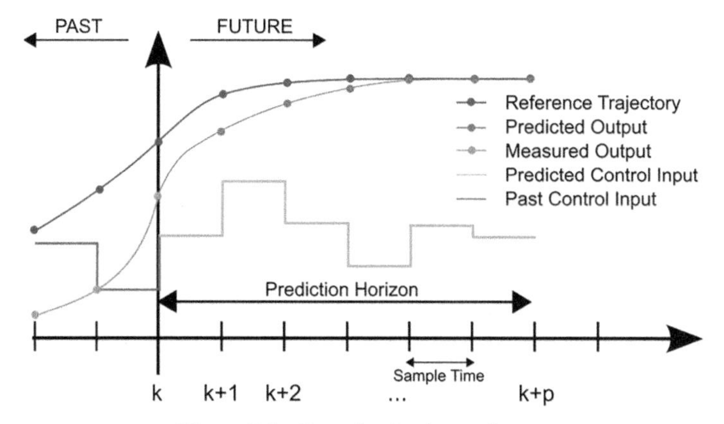

Figure 7.5 Receding horizon scheme.

The ideas at the basis of predictive control methods are:

- Explicit use of model to predict the process output evolution at future time instants (horizon).
- Calculation of control sequence minimizing an objective function.
- Receding strategy. As shown in Figure 7.5, at each sample time, the control computes the optimal sequence of control signal that minimizes the objective function along the horizon, but only the first control signal is applied to the system. This routine is called receding horizon strategy.

There are many successful applications of predictive control in use nowadays from process industry [6] to robots [7] through cement industry, chemical industry [8] or steam generation [9]. The good performance of these applications shows the capacity of the MPC to achieve highly durable and efficient control systems.

Moreover, MPC allows to adjust simultaneously all inputs to control all outputs, while accounting for all process interactions. As a result, MPC can take actions that improve plant performance that a more skilled and experienced operator can achieve.

Moreover, Model Predictive Control is able to consider limitations or constraints of the system, like saturation of actuators and/or physical constraints on output or state variables, directly in the problem formulation. This behaviour is a fundamental improvement that respects classical optimal control (like Linear Quadratic Regulator); in this way, the controller is able to calculate the optimal sequence of control actions that minimize a given cost function, respecting each specified constraint.

The most useful model formulation is the state-space form. This formulation is very helpful in both identification problem and optimal control problem. This modelling environment allows to easily relate inputs, outputs and states variable. In discrete time space for continuous variables, the formulation is (Equation (7.1)):

$$\begin{cases} x(k+1) = Ax(k) + Bu(k) \\ y(k) = Cx(k) + Du(k) \end{cases} \tag{7.1}$$

where $x(k)\ in\ \mathbb{R}^n$ is a vector of the state variables, $u(k)\ in\ \mathbb{R}^m$ are the input variables and $y(k)\ in\ \mathbb{R}^q$ are the output variables. The matrices $A, B, C\ and\ D$ have proper dimensions. In MPC framework, the control goals, such as the tracking of a reference or the satisfaction of constraints, are formulated as a numerical optimization problem. In most cases, this problem is represented as a Quadratic programming (QP) problem. For such an optimization problem, the cost function is the sum of individual terms that express various control requirements. The objective function is generally composed as follows (Equation (7.2)):

$$\begin{aligned} J \triangleq & \sum_{i=1}^{P} \|(y(k+i) - y_r)\|_{Q_y}^{N} + \sum_{i=1}^{P} \|(u(k+i) - u_r)\|_{Q_u}^{N} \\ & + \sum_{i=1}^{P} \|(\Delta u(k+i))\|_{Q_{\Delta u}}^{N} \end{aligned} \tag{7.2}$$

where $N = \{1, 2, \infty\}$ represents norm-type that defines the type of minimization problem. A linear problem is defined if $N = \{1, \infty\}$ and quadratic if $N = 2$. P is the prediction horizon that will be considered. $Q_{y,u,\Delta u}$ are positive defined matrices, also called weight matrices of different objectives

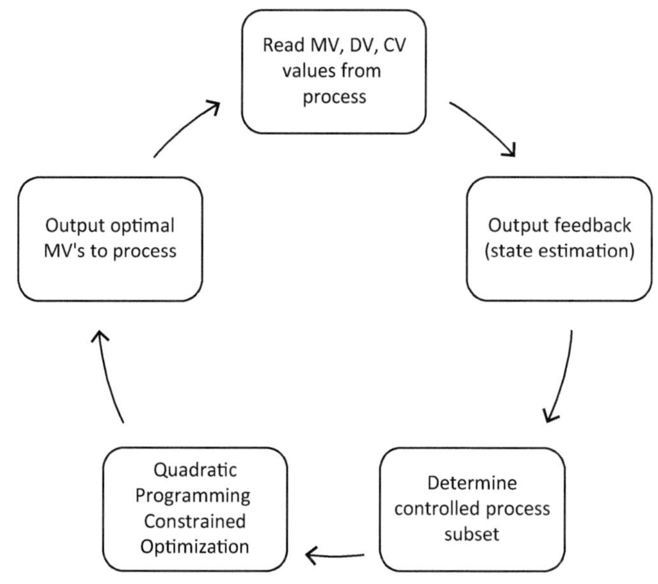

Figure 7.6 Flow of MPC calculation at each control execution.

of the controller: thanks to these parameters we can tune the controller. For example, if it is not important to control the first output y_1, it is possible to easily set $Q_{y_1} = 0$, and the same action will be applied for other weights.

Overall, the flow of computation for a typical MPC problem is represented in Figure 7.6.

7.2 Hybrid System Representation

During the last decades, Hybrid system arose naturally its interest in the scientific and research community. Many applications of hybrid system modelling in key areas were presented, such as automotive system [10] or power system [11].

A demonstration of considerable interest in hybrid system is the number of periodic conferences and entire session in major conferences completely devoted to them.

Moreover, this research field is relatively open to new advances. New approaches to mathematical representation of hybrid system have just appeared and a growing interest in applications is straightforward.

Hybrid systems are dynamic systems with both continuous states, discrete-states and event-variables. Consequently, a hybrid system provides a perfect structure to represent large plant of industrial process, which can be seen globally like an agglomeration of subsystems working in different modes, switching along the plant operation points. For example, the mathematical car's model with gear shift has different traction force curves related to selected gear [12]. To consider these different dynamics behaviour in a unique model, hybrid system modelling is mandatory. Moreover, hierarchical systems can be modelled as hybrid, in which lower components are described by continuous variables and higher-level blocks are governed by logic or decision modules.

Different kinds of models can be used to describe hybrid system. For control purpose, hybrid modelling techniques have to be descriptive enough to capture the behaviour of the interconnections between logic components (automata, switches, software code) and continuous dynamics (physical laws). Simultaneously, the model must to be simple enough to solve analysis and synthesis problems.

The state of the art of hybrid system modelling can be summarized in two main groups (Figure 7.7): the more used piecewise affine (PWA) system [13], mixed logical and dynamical (MLD) models [14] and hybrid automata (HA) [15]; and less used linear complementarity (LC), extended

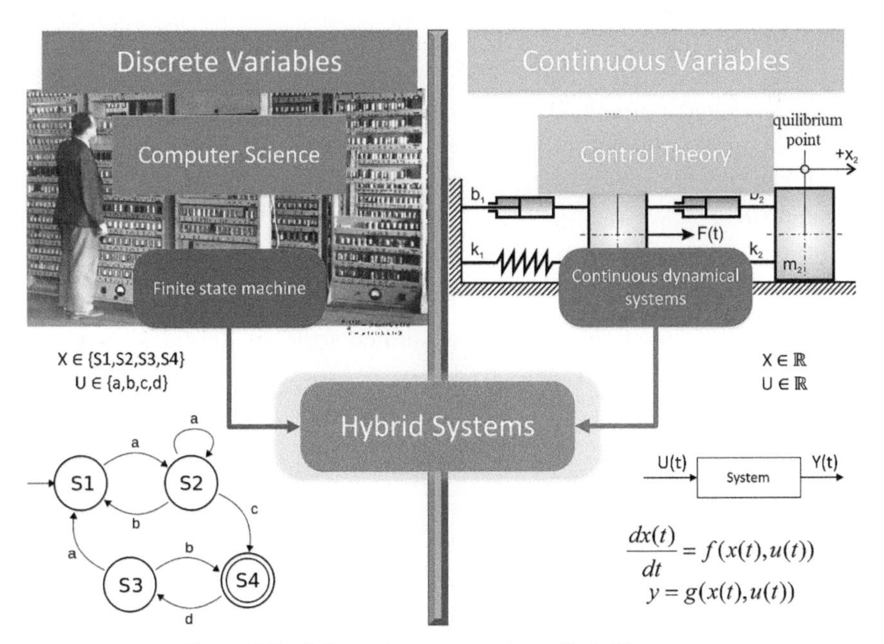

Figure 7.7 Schematic representation of hybrid system.

linear complementary (ELC) system and max-min-plus-scaling (MMPS) systems [16].

In detail, as proved in [16], all those modelling frameworks are equivalent and it is possible to describe the same system with models of each class. This characteristic is useful, for example, as each formulation offers some advantages in one particular situation: MLD framework is the best for the optimization of the system, while stability and robustness are more easily proved in a PWA formulation.

Hybrid system modelling allows to describe a variety of different kinds of systems, for example, it is possible to deal with complex system like switched dynamics system. Moreover, a hybrid model can describe the complete dynamics of the system and consider different aspects of the same system that works in different ways. For example, when a robot works in a cooperative environment, this type of modelling technique is able to consider each different dynamic, like free motion, contact with operator, different payloads applied at end-effector, etc.

Another kind of system that can be modelled as hybrid system is non-linear system. A common method to face non-linear system consists of piece-wise linearization around consecutive operating points. The output of this procedure is a PWA model (see Equation (7.3)).

The main advantage of using this kind of modelling system to synthesise a Model Predictive Control (MPC) is that the controller, when is calculating predicted outputs, is able to consider each different dynamics included in the model and optimize the control action in order to minimize the functional cost (i.e. minimize energy consumption, control action magnitude or tracking error).

7.2.1 Piece-Wise Affine (PWA) System

PWA systems representation is the most studied form of hybrid systems. A PWA system is defined as (Equation (7.3)):

$$\begin{cases} x(t+1) = A^i x(t) + B^i u(t) + f^i \\ y(t) = C^i x(t) + g^i \end{cases} \quad for\ [x(t), u(t)] \in \chi_i \tag{7.3}$$

where $x(t) \in \mathbb{R}^n$, $u(t) \in \mathbb{R}^m$ and $y(t) \in \mathbb{R}^r$ denote the state and the input and output vectors. $\{\chi_i\}_{i=1}^{s}$ is a convex polyhedral partition of the states and input space (i.e. see Figure 7.8). Each χ_i is given by a finite number of linear inequalities.

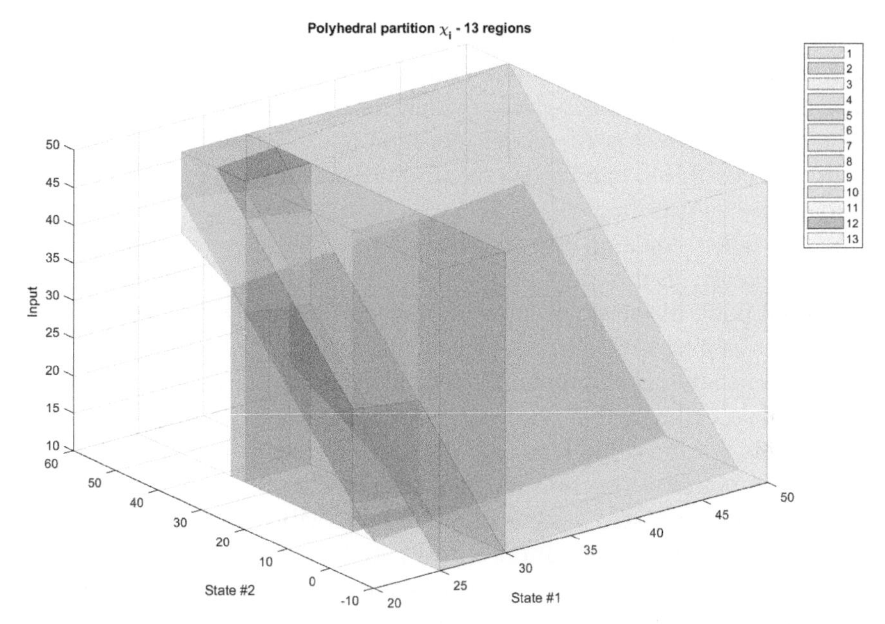

Figure 7.8 Polyhedral partition representation of a hybrid model. It is possible to see 13 partitions that divide the input state space into 13 pieces-wise sub-systems (using MatLab 2017b).

7.2.2 Mixed Logical Dynamical (MLD) System

In ref. [14], a new type of hybrid systems representation has been defined, in which logic, dynamics and constraints are integrated.

The MLD description is (Equation (7.4)):

$$\begin{cases} x(k+1) = Ax(k) + B_1 u(k) + B_2 \delta(k) + B_3 z(k) \\ y(k) = Cx(k) + D_1 u(k) + D_2 \delta(k) + D_3 z(k) \\ E_5 \geq E_1 x(k) + E_2 u(k) + E_3 \delta(k) + E_4 z(k) \end{cases} \tag{7.4}$$

where $x(k) = [x_r^T(k), x_b^T(k)]$ with $x_r(k) \in \mathbb{R}^{n_r}$ and $x_b(k) \in \{0,1\}^{n_b}$; $y(k) = [y_r^T(k), y_b^T(k)]$ with $y_r(k) \in \mathbb{R}^{m_r}$ and $y_b(k) \in \{0,1\}^{m_b}$; $u(k) = [u_r^T(k), u_b^T(k)]$ with $u_r(k) \in \mathbb{R}^{q_r}$ and $u_b(k) \in \{0,1\}^{q_b}$. $z(k) \in \mathbb{R}^{r_r}$ and $\delta(k) \in \{0,1\}^{r_b}$are auxiliary variables that are used to represent the switching between different operating modes.

The inequalities have to be interpreted component-wise, and they define the switching conditions of different operating modes. The construction of this inequality is based on tools able to convert logical facts involving continuous variables into linear inequalities (for more details, see [17]). This tool will be used to express relations describing the evolution of systems where physical laws, logic rules and operating constrains are interdependent.

Equation (7.4) commits linear discrete-time dynamics for the first two equations. It is possible to build up another formulation describing continuous time version by substituting $x(k+1)$ by $x(t)$ or a non-linear version by changing the linear equation and inequalities in (7.4) to more non-linear functions. However, in this way, the problem becomes hard tractable by a computational point of view, and more in general, the MLD representation allows to describe a wide range class of systems.

MLD models are successful thanks to good performance in computation aspect. The main claim of their introduction was the easy handling of non-trivial problems, for the formulation of Model Predictive Control for hybrid and non-linear system. This formulation performs well when it is used together with modern Mixed-Integer Programming (MIP) solver for synthesizing predictive controller for hybrid systems, as described in Section 7.4.1.

Note that the class of Mixed Logical Dynamical systems includes the following important system classes:

- Linear systems;
- Finite state machines;
- Automata;
- Constrained linear systems;
- Non-linear dynamic systems.

In fact, the next section introduces the equivalence between different hybrid system representations and it underlines the potential of MLD models (in Figure 7.9, it is possible to see the interconnection between MLD and other system representation models).

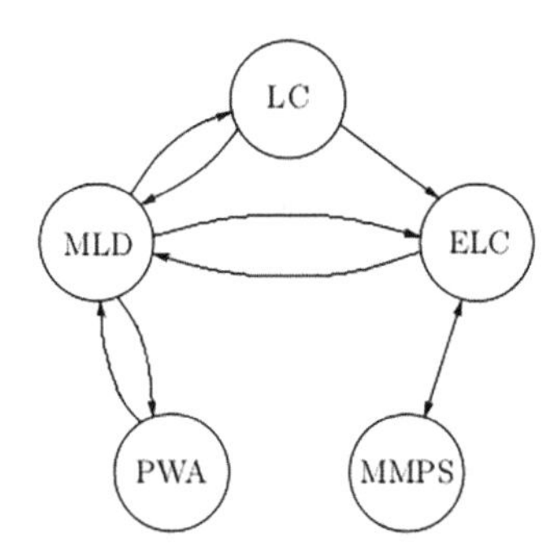

Figure 7.9 Graphic scheme of the links between the different classes of hybrid. The arrow from A to B classes shows that A is a subset of B.

7.2.3 Equivalence of Hybrid Dynamical Models

In ref. [16], there are different demonstrations of equivalence between each hybrid system model, summarized in Figure 7.9. For some transformations, additional conditions like boundedness of the state and input variables or well-posedness have to be made. Typically, the more frequent condition is that the polyhedral partition of input-state space must be univocally defined, i.e. with no overlapping between different χ_i. These requirements are fundamental in that case where, for example, in PWA or MLD, the modelling framework does not allow overlapping of sub-set of state-input space.

These equivalences are fundamental to demonstrate the properties of different hybrid models and commonly use stability analysis on a single representation, translating its effects on another modelling system.

7.3 Hybrid Model Predictive Control

Dealing with control of hybrid systems is an open field of research in both academia and industrial world. Model predictive control based its main advantage on the prediction of future outputs, which requires a model that considers the evolution of the system. In case of hybrid systems, discrete variables must be included. For this aim, the modelling frameworks described in Section 7.2 have to be considered.

7.3.1 State of the Art

Model predictive control was proposed for the first time in the late 1970s by Richalet et al. [9], who predicted future outputs in a heuristic manner. During that time, the application field of MPC was process industry, from chemical to oil and gas extraction through pharmaceutical industries.

Since then, model predictive control has been extended to a wide range of control problems. During the 1990s [18], the academics world was interested on stability analysis, because it is a very challenging problem not only for control engineers but also for mathematicians. Control engineers moved their focus to large systems, where both continuous and discrete variables describe the model of the system, therefore requiring a hybrid model predictive control solution [14]. HMPC consists in a repetitive solution of a Mixed-Integer Programming (MIP) problems, where variables could be both continuous and discrete. If the objective function is quadratic, these problems are classified as Mixed Integer Quadratic programming (MIQP) or Mixed Integer Linear programming (MILP), if a linear objective function is used.

MILP and MIQP problems are much more difficult to solve than a linear or quadratic programming problem (LP or QP), and some properties like convexity are lost (see ref. [19] for a more detailed description).

The computational load for solving an MIP problem is a key issue, as a brutal force approach consists of the evaluation of every possible combination. The optimal solution would be to solve every QP or LP related to all the feasible combinations of discrete decision variables. The solution is the minimum of all the computed solution of QP/LP problems. For example, if all the discrete decision variables are Boolean, then the number of possible LP/QP problems is 2^(n_b). Fortunately, there exists an entire research field on this topic and nowadays, there is a wide range of commercial solvers able to deal with MIP problem in a very fast way. These software are mainly based on branch and bound methods [20]; the most known and used are CPLEX (ILOG Inc. [3]), GLPK (Makhorin [21]) or GUROBI [2] for which APIs for many programming languages are available.

The application of the Model Predictive Control arose in the early 1990s. One of the first fundamental studies was made by Bemporad and Morari [14]: they proposed a rigorous approach to mathematical modelling of hybrid system where it is possible to obtain a compact representation of system called Mixed Logical Dynamical (MLD, see Section 7.3.2). Then, following the optimization step, it is possible to synthesize an optimal constrained receding horizon control. This methodology is helpful to optimize and orchestrate both large systems with mixed-variables and non-linear systems linearized around sequential operating points.

As in birth of MPC, the first implementation was in the field of refinery and chemical process. In these fields, Model Predictive Control was already a standard, and the possibility to build up a unique mathematical model that represents the whole system, like plant with all its components, and synthesize a unique controller able to find the optimal solution that respects every specified constrain was a revolution. In the next section, we deeply explore the issues and limits of Hybrid Model Predictive Control, which are roughly synthesizable in computational time and computational power. In that period, the solution of this problem was overcome by using off-line optimization, also called Explicit MPC. This control method is able to properly work only in a predetermined range of variable states: in fact, the on-line optimization was replaced by an off-line optimization, summarized in a lookup table. Using this methodology, the application of Hybrid MPC could be extended to mechanical and mechatronics system, where the cycle time can be very small. Some applications are summarized in refs. [10–12].

Indeed, in refinery and chemical process or more generally in process industry, the sampling time of the controller is in minutes-order. Since the solution of Mixed-Integer Programming problem is feasible, in these fields, it is used as industrial standard. However, in the last two decades, from ref. [14], the computational power of embedded micro-processor or Industrial PC has grown exponentially, as Moore's law said, and the commercial MIP solvers increase their "power" dramatically. These evolutions allow to rethink to Hybrid MPC with on-line optimization applied to fast system, with sampling time in the range of a few seconds. The aim of this study is to build a standard method to synthesize Model Predictive Control for hybrid system (aggregated CPS too) and have the opportunity to test a possible on-line execution of the controller, in order to understand the minimum sampling time of the controller. This possibility is a killer-feature in refinery and chemical process where Hybrid MPC already is in use, but there is not a powerful and standard tool able to help control's engineers to design HMPC for process industry. Otherwise, in the mechanical and mechatronics system control field, this tool can be revolutionary because it simplifies the design of the controller and standardizes it: in this way, the focus to realize a feasible controller is moved on MIP solving time. In addition, the designer can check in a meticulous, but fast, way the feasibility of the Hybrid Model Predictive Control and its performance.

7.3.2 Key Factors

In the last decades, since the introduction of MPC in control theory, a wide variety of application has been presented. All these applications are related to notable capabilities of fitting the control goals. Indeed, this methodology is able to realize very smooth and precise control. Moreover, MPC is capable of being tuned in a straightforward way in relation to desired performance of the system. As described in Section 7.2, a typical function cost contains different weights, which offer the possibility to tune the performance of the controller, easily to tune also for non-technical people. Moreover, the definition of constrains is direct in the optimization problem and it is simple to impose constraints on Manipulated Variables (MVs) and Output Variables (OVs), which means limits on actuator saturation, dynamical constrain on actuators and physic limits of the controlled system.

Summarizing the benefit of Model Predictive Control:

- Most widely used control algorithm in material and chemical processing industries [22];

- Increased consistency of discharge quality. Reduced off-specs products during grade changeover. Increased throughput. Minimizing the operating cost while meeting constrains (optimization, economic) [23];
- Superior for process with a large number of manipulated and controlled variables (multivariable, strong coupling) [24];
- Allows constraints to be imposed on both MVs and CVs. The ability to operate closer to constraints and over those (soft constraints);
- Allow time delays, inverse response, inherent non-linearities (difficult dynamics), changing control objectives and sensor failure (predictive);
- Optimal rejection to modelling error and disturbances;
- Multi-objectives control technique [25].

7.3.3 Key Issues

The basic issue of Hybrid MPC, and MPC in general, is related to the computational time needed to solve in real time the optimization problem. Indeed, when dealing with a large and fast system, the model of the system becomes really complex and the required closed loop time very precise and the online optimization is not achievable. In order to minimize the problem caused by large system, a pre-stored control allocation law can be used to avoid increased number of decision variables and increased solving time. This technique is known as Explicit Model Predictive Control [26], where the controller creates a look-up table during off-line simulation and uses it during the execution time. This method is able to avoid the main drawback of MPC removing the optimization procedure that is very time-consuming. This benefit enables the use of MPC, and mainly Hybrid MPC, inside application with very high sampling rates.

Another important issue is the difficulty to demonstrate the robustness of the control respect to the classical robust control technique like H_∞ [27]. A possible solution of this issue is to couple with the MPC controller an Online system identification tool, as it is shown in Errore. L'origine riferimento non è stata trovata., that is able to realize a more robust control. This is because the online system identification checks and tunes the system model recursively, compensating modellation errors.

7.4 Identification of Hybrid Systems

The design of a hybrid model predictive controller needs to describe the plant dynamics in terms of a hybrid linear model, which is used to simulate the plant behaviour within the prediction horizon. As known, there are basically two ways to construct a mathematical model of the plant:

- Analytic approach, where models are derived from first-principle physics laws (like Newton's laws, Kirchhoff's laws, balance equations). This approach requires an in-depth knowledge and physical insight into the plant, and in the case of complex plants, it may lead to non-linear mathematical models, which cannot be easily expressed, converted or approximated in terms of hybrid linear models;
- System identification approach, where models are derived and validated based on a set of data gathered from experiments. Unlike the analytic approach, the model constructed through system identification has a limited validity (e.g., it is valid only at certain operating conditions and for certain types of inputs) and it does not give physical insights into the system (i.e., the estimated model parameters may have no physical meaning). Nevertheless, system identification does not need, in principle, in-depth physical knowledge of the process, thus reducing the modelling efforts.

In this project, hybrid linear models of the process of interested will be derived via system identification, and physical insights into and knowledge of the plant will be used, if needed, to assist the whole identification phase, such as choosing the appropriate inputs to perform experiments, choosing the structure of the hybrid model (defined, for instance, in terms of number of discrete states and dynamical order of the linear subsystems), debugging the identification algorithms and assessing quality of the estimated model.

The following two classes of hybrid linear models will be considered, which mainly differ in the assumption behind the switches among the (linear/affine) time-invariant sub-models:

- Jump Affine (JA) models, where the discrete-state switches depend on an external signal, which does not necessarily depend on the value of the continuous state. The switches among the discrete states can be governed, for instance, by a Markov chain, and thus described in terms of state transition probabilities. Alternatively, in deterministic jump models, the mode switches are not described by a stochastic process, but they are triggered by or associated to determinist events (e.g. gear or speed selectors, evolutions dependent on if-then-else rules, on/off switches and valves). In this chapter, we will focus on the identification of deterministic jump models. Stochastic models might be considered at a later stage, only if necessary.

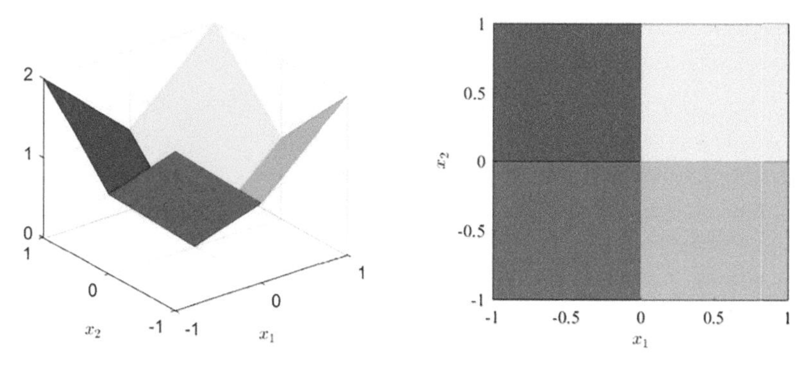

Figure 7.10 Example of a three-dimensional PWA function $y = f(x_1, x_2)$.

- Piece-Wise Affine (PWA) models, where the active dynamic affine sub-model at each time instant only depends on the value of the continuous state. More specifically, in PWA models, the (continuous) state space is partitioned into a finite number of polyhedral regions with non-overlapping interiors, and only one dynamical affine model is associated to each polyhedron. PWA models can be used to accurately describe dynamical systems that evolve according to different dynamics depending on the specific point in the state-input space (e.g. a bouncing ball or switching feedback control laws where the switches between the controllers depend on the state of the system). Furthermore, thanks to the universal approximation property of PWA maps, PWA models can be also used to approximate non-linear/non-smooth phenomena with an arbitrary degree of precision [28]. For the sake of visualization, an example of a three-dimensional PWA function, defined over four polyhedral regions of the state space, is plotted in Figure 7.10.

Note that Jump models and PWA models can be also combined to describe, for instance, finite state machines (with linear dynamics at each mode), where the mode transition depends on both an external event and the current value of the continuous state, input and output.

In the following, we formalize the hybrid system identification problem and discuss its main challenges. Finally, we provide an overview of the algorithm that will be used and implemented in the DAEDALUS platform, for the identification of both Jump Affine and PWA models.

7.4.1 Problem Setting

Let us consider a training dataset of input/output pairs $\mathcal{D} = \{u(t), y(t)\}_{t=1}^{N}$ (generated by the plant we would like to model), where t denotes the time index, $u(t) \in Rnu$ and $y(t) \in Rny$ are the input and output of the system at time t, respectively, and N is the length of the training set. Our goal is to estimate, from the considered training set D, a hybrid linear dynamical model approximating the input/output relation of the system and described in the input/output Auto-Regressive with Exogenous input (ARX) form (Equation (7.5)):

$$\hat{y}(t)= \Theta_{s(t)}x(t) \tag{7.5}$$

where $\hat{y}(t) \in Rny$ is the output of the estimated model, $s(t) \in \{1,\ldots,\overline{s}\}$ is the active mode at time t (i.e. the value of the discrete state at time t) and $x(t) \in X \subset Rnx$ is the regressor vector containing past values of the input and of the output (Equation (7.6)), i.e.

$$x(t) = [1\ y(t-1)' \ldots y(t-n_a)'u(t)'u(t-1)'u(t-n_b)']' \tag{7.6}$$

for some fixed values of n_a and n_b, and $\Theta_s \in Rny, nx$ (with $s = 1,\ldots,\overline{s}$) is the parameter matrix describing the linear sub-model associated to the discrete state s.

The identification of a hybrid linear dynamical model (Equation (7.5)) thus requires: (i) choosing the number $\overline{s}$ of modes (i.e. size of the discrete state); (ii) computing the parameter matrices Θ_s (with $s = 1,\ldots,\overline{s}$) characterizing the affine sub-models; (iii) finding the hidden sequence of discrete states $\{s(t)_{t=1}^{N}\}$ and (iv) in the case of PWA model identification, finding the polyhedral partition of the regressor space X where the affine sub-models are defined.

When choosing the dimension $\overline{s}$ of the discrete state, one must take into account the trade-off between data fitting and model complexity. For small values of $\overline{s}$, the hybrid model cannot accurately capture the non-linear and time-varying dynamics of the system. On the other hand, increasing the number of modes also increases the degrees of freedom in the description of model, which may cause overfitting and poor generalization to unseen data (i.e., the final estimate is sensitive to the noise corrupting the observations), besides increasing the complexity of the estimation procedure and of the resulting model. In the identification algorithms, which will be developed during the project, we will assume that $\overline{s}$ is fixed by the user. The value of $\overline{s}$ (as well as the values of the parameters na and nb defining the dynamical order of the affine sub-models) will be chosen through cross-validation, with a possible upper-bound dictated by the maximum tolerable complexity of the estimated model or by some physical insight into the system.

Fitting-Error Minimization

The hybrid linear model structure in Equation (7.5) suggests to formulate the identification of the hybrid models as the following fitting-error minimization problem

$$\min_{\substack{\{\Theta_s\}_{s=1}^{\overline{s}} \\ \{s(t)\}_{t=1}^{N}}} \frac{1}{N} \sum_{t=1}^{N} \|y(t) - \Theta_{s(t)} x(t)\|_2^2 \tag{7.7}$$

which aims at minimizing, over the parameter matrices Θ_s (with $s = 1, \ldots, \overline{s}$) and the discrete state sequence $\{s(t)\}_{t=1}^{N}$, the power of the error between the measured output $y(t)$ and the model output $\hat{y}(t) = \Theta_{s(t)} x(t)$.

In the cases where the discrete state sequence $\{s(t)\}_{t=1}^{N}$ is exactly known (e.g. when $s(t)$ is associated to the gear number in a car or to an external switching signal controlled by the user, or, for PWA models, the partition of the regressor space X is fixed a priori), the fitting-error minimization problem (7.7) becomes a simple linear regression problem, and the parameter matrices Θ_s (with $s = 1, \ldots, \overline{s}$) defining the affine sub-models can be easily estimated through standard least squares, i.e.

$$\hat{\Theta}_s = arg \min_{\Theta_s} \frac{1}{N} \sum_{t=1}^{N} \mathbb{I}\{s = s(t)\} \|y(t) - \Theta_{s(t)} x(t)\|_2^2 \tag{7.8}$$

with $I\{s{=}s(t)\}$ denoting the indicator function, i.e.

$$I\{s{=}s(t)\} = \begin{cases} 1 & if \; s{=}s(t) \\ 0 & otherwise \end{cases} \tag{7.9}$$

Namely, in computing an estimate of Θ_s through Equation (7.8), only the regressor/output pairs $(x(t), y(t))$ such that $s{=}s(t)$ are considered.

In the more general case, where the discrete state sequence $\{s(t)\}_{t=1}^{N}$ is not available, the identification of hybrid models becomes NP hard (strictly speaking, Equation (7.8) is a mixed-integer quadratic programming problem, which might be computationally intractable, except for small-scale problems). Furthermore, besides reconstructing the discrete1 state sequence $\{s(t)\}_{t=1}^{N}$ and estimating the parameter matrices Θ_s (with $s = 1, \ldots, \overline{s}$), the identification of PWA models also requires to compute a polyhedral partition of the regressor space X.

7.4.2 State-of-the-Art Analysis

Several heuristics have been proposed in the literature to overcome the challenges encountered in hybrid system identification (see [29, 30] for an exhaustive overview of algorithms for identification of Jump Affine and PWA models). Among the proposed algorithms, we have analyzed:

- the bounded-error approach [31], which addresses the identification of Jump Affine models under the assumption that the noise corrupting the output observations $y(t)$ is norm-bounded (with known bound). The goal is to estimate the set of all model parameters Θ_s, which are compatible with the a-priori assumptions on the noise bound, the chosen model structure and the observations. A polynomial optimization problem is formulated, whose solution is approximated through convex-relation techniques based on the theory of moments [32]. This approach turns out to be very sensitive to outliers (i.e. noise outside the supposed bounds) and conservative if a large bound on the noise is assumed. Furthermore, it suffers from high computational complexity because of the high computational burden of the employed theory-of-moment-based relaxation;
- the sparse optimization-based approaches [33] and [34], which address the segmentation of linear models by formulating an optimization problem penalizing the fitting error and the number of switches among the affine sub-models. Therefore, these methods are suited only for Jump Affine systems with infrequent switches;
- the mixed-integer quadratic programming approach [35], which addresses the identification of PWA systems using hinging-hyperplane ARX models and piecewise affine Wiener models. A mixed-integer quadratic programming problem is formulated (similar, but not exactly equal to (6.3)) and solved through brunch-and-bound. Unfortunately, the number of integer variables increases with the number of training samples, limiting the applicability of the method to small-/medium-scale problems;
- the two-stage clustering based approach [36], which can be used for both Jump Affine and PWA model identification. At the first stage, the regressor observations are clustered by assigning each data-point to a sub-model through a k-means-like algorithm, and the affine sub-model parameters Θ_s are estimated at the same time. In the case of PWA identification, a second stage is performed to compute a partition of the

regressor space X. Although ref. [36] is able to handle large training sets, poor results might be obtained when the affine local sub-models are over-parameterized (i.e. large values of the parameters na and nb in the definition of the regressor (6.2) are used), since the distances in the regressor space (namely, the only criterion used for clustering) turns out to be corrupted by redundant, thus irrelevant, information;

- the recursive two-stage clustering-based approach [37], which is based on the same two-stage clustering philosophy of [36], is suited for both Jump Affine and PWA model identification. The proposed approach consists of two stages: (S1) simultaneous clustering of the regressor vector and estimation of the model parameters Θ_s ($s = 1, \ldots, \overline{s}$). This step is performed recursively by processing the training regressor/output pairs sequentially; (S2) computation of a polyhedral partition of the regressor space through efficient multi-class linear separation methods. This step is performed either in a batch way (i.e. offline) or recursively (i.e. online). Note that stage S2 is required only for PWA system identification. Because of its computational efficiency and the possibility to be used both for batch and recursive identification, we have decided to use and implement this algorithm in the DAEDALUS project. Further details on this algorithm are discussed below.

7.4.3 Recursive Two-Stage Clustering Approach

The main ideas behind the recursive two-stage clustering approach proposed in ref. [37] are presented in this section. As mentioned in the previous paragraph, the hybrid system identification problem is tackled in two stages: S1 (iterative clustering and parameter estimation) and S2 (polyhedral partition of the regressor space, necessary only for PWA model estimate).

Stage S1 is carried out as described in Algorithm 1, where clusters and sub-model parameters are updated iteratively, making the algorithm suitable for online applications, when data are acquired in real time.

Algorithm 1 Recursive clustering and parameter estimation

Input: Observations $\{x(t), y(t)\}_{t=1}^{N}$, desired number $\overline{s}$ of affine submodels, initial condition for model parameter matrices $\Theta_1, \ldots, \Theta_{\overline{s}}$.

---

1. **let** $\mathcal{C}_s \leftarrow \emptyset$, $s = 1, \ldots, \overline{s}$;

2. **for** $t = 1, \ldots, N$ **do**
   2.1. **let** $e_s(t) \leftarrow y(t) - \Theta_s x(t)$,
   2.2. **let** $s(t) \leftarrow \arg\min_{s=1,\ldots,\overline{s}} \|e_s(t)\|_2^2$;
   2.3. **let** $\mathcal{C}_{s(t)} \leftarrow \mathcal{C}_{s(t)} \cup x(t)$;
   2.4. **update** $\Theta_{s(t)}$ using recursive least-squares;
3. **end for**;
4. **end.**

Output: Estimated matrices $\Theta_1, \ldots, \Theta_{\overline{s}}$, clusters $\mathcal{C}_1, \ldots, \mathcal{C}_{\overline{s}}$, sequence of active modes $\{s(t)\}_{t=1}^{N}$.

The main idea of Algorithm 1 is to compute, at each time instant t, the fitting error $es(t) = y(t) - \Theta_s x(t) (s \in \{1, \ldots, \overline{s}\})$ achieved by all the $\overline{s}$ local affine sub-models, and select the local model that "best fits" the current output observation $y(t)$ (Steps 2.1 and 2.2). The regressor $x(t)$ is then assigned to the cluster $C_s(t)$ (Step 2.3) and the parameter matrix $\Theta_s(t)$ associated to the selected submodel is updated using recursive least squares (Step 2.4).

Due to the greedy nature of Algorithm 1, the estimates of the model parameters Θ_s and the clusters Cs are influenced by the initial choice of the parameters Θ_s. A possible initialization for the parameter matrices is to take $\Theta_1, \ldots, \Theta_{\overline{s}}$ all equal to the best linear model, i.e.

$$\Theta_s = \arg\min_{\Theta} \frac{1}{N} \sum_{t=1}^{N} \|y(t) - \Theta x(t)\|_2^2, \quad s = 1, \ldots, \overline{s}.$$

Moreover, the estimation quality can be improved by reiterating Algorithm 1 multiple times, using its output as an initial condition for the following iteration. This can be performed only if the algorithm is executed in a batch mode (offline). Alternatively, a subset of data can be processed in a batch mode to find proper initial conditions. Then, Algorithm 1 is executed in real time to iteratively process data streaming.

7.4.4 Computation of the State Partition

If a PWA identification problem is addressed, besides estimating the model parameters $\{\Theta_s\}_{s=1}^{\overline{s}}$ and the sequence of active modes $\{s(t)\}_{t=1}^{N}$, also a polyhedral partition of the regressor space X should be found. More specifically, let X_s (with $s = 1, \ldots, \overline{s}$) be a collection of polyhedra which form a complete polyhedral partition[1] of the regressor space X. Each polyhedron Xs is defined as:

$$X_s = \{x \in Rnx : Hsx \leq B_s\}, \tag{7.10}$$

for some matrix Hs and vector Bs of proper dimensions. The goal is thus to estimate Hs and Bs (with $s = 1, \ldots, \overline{s}$) defining the polyhedron Xs, where the s-th local affine submodel is active. Two approaches can be followed:

- according to the idea discussed in [11], the Voronoi diagram generated by the clusters' centroids can be used as a polyhedral partition of the regressor space X. Specifically, let cs be the centroid of cluster Cs. Then, the polyhedron Xs associated to cluster Cs (Equation (7.11)) is the set of all the values of the continuous state x such that cs is the closest centroid to x among all the other centroids c_j (with $j \neq s$), i.e.,

$$\mathcal{X}_s = \{x \in \mathbb{R}^{n_x} : \|x - c_s\|_2 \leq \|x - c_j\|_2, \quad j = 1, \ldots, \overline{s}, j \neq s\}, \tag{7.11}$$

Through simple algebraic manipulations, Xs can be expressed in a form like Equation (7.10), i.e.

$$\mathcal{X}_s = \{x \in \mathbb{R}^{n_x} : -2(c_s' - c_j')x \leq c_j' c_j - c_s' c_s, \quad j = 1, \ldots, \overline{s}, j \neq s\}. \tag{7.12}$$

Note that the definition of the polyhedron Xs (Equation (7.12)) only depends on the clusters' centroids, which can be easily updated recursively once the cluster Cs is updated (Step (2.3) of Algorithm 1). This makes the use of the Voronoi diagram particularly suited for real-time applications, where data are processed iteratively. However, a limitation of the Voronoi diagram is that it does not take into account how much the points are spread around the clusters' centres, making the state-space partition less flexible than general linear separation maps. In order to overcome this limitation, the approach described below can be followed.

- separate the clusters $\{C_s\}_{i=1}^{\overline{s}}$ provided by Algorithm 1 via linear multi-category discrimination (see, e.g. [37–39]). In the following, we briefly describe the algorithm used in [37], which is suited for both offline and online computations of the state partition.
  The linear multi-category discrimination problem is tackled by searching for a convex piecewise affine separator function φ: $Rnx \rightarrow$

---

[1] A collection $\{\mathcal{X}_s\}_{s=1}^{\overline{s}}$ is a complete partition of the regressor domain $\mathcal{X}$ if $\bigcup_{s=1}^{\overline{s}} \mathcal{X}_s = \mathcal{X}$ and $\mathcal{X}_s^\circ \cap \mathcal{X}_j^\circ = \emptyset$, $\forall s \neq j$, with $\mathcal{X}_s^\circ$ denoting the interior of Xs.

R discriminating between the clusters $C_1, \ldots, C_{\overline{s}}$. The separator φ (Equation (7.13)) is defined as the maximum of $\overline{s}$ affine functions $\{\phi_i(x)\}_{i=1}^{\overline{s}}$, i.e.

$$\phi(x) = \max_{s=1,\ldots,\overline{s}} \phi_s(x) \tag{7.13}$$

with $\varphi_{s(x)}$ described as (Equation (7.14))

$$\varphi_s(x) = x'\omega^s \tag{7.14}$$

where $\omega s \in Rnx$ $(s = 1, \ldots, \overline{s})$ are the parameters to be computed.

For $s = 1, \ldots, \overline{s}$, let Ms be an $ms \times nx$ dimensional matrix (with ms denoting the cardinality of cluster Cs) obtained by stacking the regressors $x(t)'$ belonging to Cs in its rows. If the clusters $\{C_s\}_{s=1}^{\overline{s}}$ are linearly separable, the piecewise-affine separator φ satisfies the conditions:

$$Msss \geq Ms\omega j + 1ms, \quad s, j = 1, \ldots, \overline{s}, s \neq j \tag{7.15}$$

where $1ms$ is an ms-dimensional vector of ones.

The piecewise-affine separator φ thus satisfies the conditions (Equation (7.16)):

$$\begin{cases} \varphi(x) = x'\omega^s & \forall x \in C_s,\ s = 1, \ldots, \overline{s} \\ \varphi(x) \geq x'\omega^j + 1 & \forall x \in C_s,\ s \neq j \end{cases} \tag{7.16}$$

From (7.16), the polyhedra $\{\mathcal{X}_s\}_{s=1}^{\overline{s}}$ are defined as

$$\mathcal{X}_s = \{x \in \mathbb{R}^{n_x} : (\omega^s - \omega^j)'x \leq -1, \quad j = 1, \ldots, \overline{s}, j \neq s\}.$$

The condition (7.15) thus suggests computing the parameters $\{\omega^s\}_{s=1}^{\overline{s}}$ by minimizing the convex cost

$$\min_{\omega^1,\ldots,\omega^{\overline{s}}} \sum_{s=1}^{\overline{s}} \sum_{\substack{j=1 \\ j \neq s}}^{\overline{s}} \frac{1}{m_s} \|([M_s - \mathbf{1}_{m_s}](\omega^j - \omega^s) + \mathbf{1}_{m_s})_+\|_2^2, \tag{7.17}$$

with $(\cdot)+$ defined as $f+ = \max\{0, f\}$. Problem (7.17) minimizes the averaged squared 2-norm of the violation of the inequalities in Equation (7.15). The solution of the convex problem (7.17) can be then computed numerically in two ways: (i) offline through a Regularized Piecewise-Smooth Newton method or (ii) online through a Stochastic Gradient Descent method, as explained in [10].

7.5 Integration of Additional Functionalities to the IEC 61499 Platform

The DAEDALUS automation platform is built on top of the IEC-61499 standard and makes it the main core technology to enable the implementation of industrial grade applications in distributed control scenarios. The function block (FB) is one of the base elements of this standard. Function blocks are a concept to define solid, reusable software components in industrial automation systems. They allow the encapsulation of algorithms in an easy, understandable, even for newcomer, and usable form. Each function block has defined inputs, which are read and processed from an internal algorithm. The result will be outputted at defined outputs. Whole applications can be created out of various function blocks by connecting their inputs and outputs. Concretely, each function block consists of a head, a body, input/output events and input/output data.

The IEC 61499 standard defines various kinds of function blocks:

- Basic Function Blocks. Basic function blocks are used to implement basic functionalities of applications. Basic function blocks include internal variables, one or more algorithms and an “Execution Control Chart”, to define the processing of the algorithms;
- Service Function Blocks. Service function blocks represent the interfaces to the hardware;
- Composite Function Blocks. Several basic, service or composite function blocks as well can be grouped to form a composite function block. The composite FB presents itself as a closed function block with a clearly defined interface.

7.5.1 A Brief Introduction to the Basic Function Block

Basic function blocks are the atomic units of execution in IEC 61499. A basic FB consists of two parts, i.e. a function block interface and an execution control chart (ECC) that operates over a set of events and variables. The execution of a basic FB entails accepting inputs from its interface, processing the inputs using the ECC and emitting outputs.

A basic FB is encapsulated by a function block interface, which exposes the respective inputs and outputs using ports. These input and output ports may be classified as either event or data ports.

Figure 7.11 shows the interface of the function block that implements a valve control logic. This interface exposes input events (INIT,

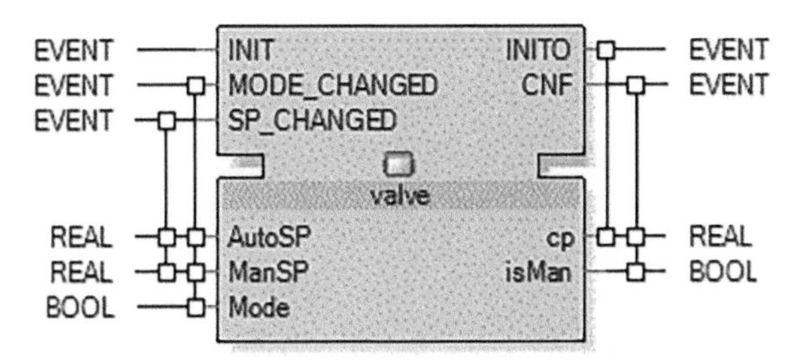

Figure 7.11 Valve: an example of basic function block.

MODE_CHANGED, SP_CHANGED), output events (INITO, CNF), as well as input variables (AutoSP, ManSP, mode) and output variables (cp, isMan).

Event ports are specialized to accept or emit events, which are pure signals that represent status only, i.e. they are either absent or present. On the other hand, data ports can accept or emit valued signals that consist of a typed value, such as integer, string or Boolean. Variable ports of a special type Any can accept data from a range of typed values. In addition, a concept of multiplicity is also applicable to data ports, which allows accepting or emitting arrays of values. A data port can be associated with one or more event ports.

As shown in Figure 7.11, for example, Mode is associated with MODE_CHANGED.

However, this association can only be defined for ports of the matching flow direction, e.g. input data ports can only be associated with input event ports. This event–data association regulates the data flow in and out of a basic FB, i.e. new values are loaded or emitted from the data ports on the interface when an associated event is present.

The behaviour of a basic FB is expressed as a Moore-type state machine, known as an ECC. An ECC reacts to input events and performs actions to generate the appropriate outputs.

Figure 7.12 shows the ECC of the valve basic function block, which consists of four states: START, INIT, exec_SPChange and exec_ModeChange.

States in ECCs have provision to execute algorithms and emit output events upon ingress, which are represented as ordered elements in their respective action sets.

As an example, in Figure 7.12, the algorithm exec_SPChange is executed (represented as a gray label), and the CNF event is emitted upon entering the exec_SPChange state (represented as a blue oval).

The execution of an ECC starts from its initial state (START in Figure 7.12) and progresses by taking transitions, which are guarded by an

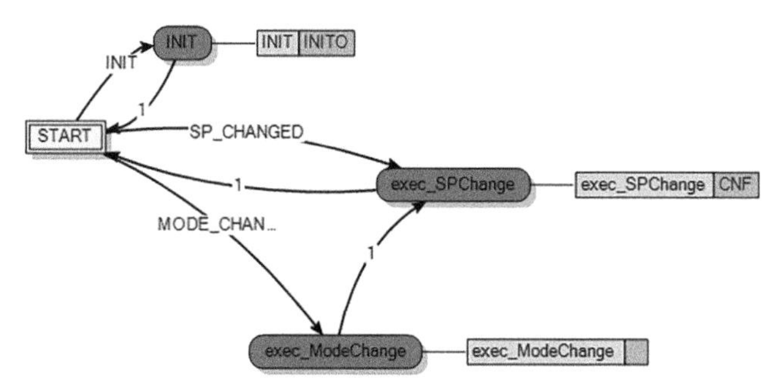

Figure 7.12 Example of execution control chart (ECC).

```
1   ALGORITHM exec_SPChange IN ST:
2   (* Add your comment (as per IEC 61131-3) here
3
4   *)
5   IF isMan THEN
6    cp := ManSP;
7   ELSE
8    cp := AutoSP;
9   END_IF;
10  END_ALGORITHM
```

Figure 7.13 exec_SPChange algorithm from the valve basic FB.

input event and an optional Boolean expression over input and/or internal variables. Upon evaluation, a transition is considered to be enabled if the respective guard condition evaluates to true. The ECC will then transition to the next state by taking the enabled egress transition from the source state to the corresponding target state.

An algorithm is a finite set of ordered statements that operate over the ECC variables. Typically, an algorithm consists of loops, branching and update statements, which are used to consume inputs and generate outputs. The IEC 61499 standard allows algorithms to be specified in a variety of implementation-dependent languages. As an example, the implementation from nxtControl allows the development of custom algorithms in Structured Text (ST).

The exec_SPChange algorithm from the valve basic FB is presented in Figure 7.13 that uses the ST language as defined in IEC-61131-3. Here, the IF–THEN–ELSE construct is used to update the output value of cp based on the value of the input isMan.

7.5.2 A Brief Introduction to the Composite Function Block

Composite function blocks facilitate the representation of structural hierarchy. Composite FBs are similar to basic FBs in the sense that they too are encapsulated by function block interfaces. However, unlike a basic FB, the behaviour of a composite FB is implemented by a network of function blocks.

Basic and composite function blocks characterize different types of specifications, which are referred to as function block types (FBTypes). A function block network may consist of instances of various FBTypes, where any given FBType may be instantiated multiple times. This concept is very similar to the object-oriented programming paradigm, which contains classes (analogous to FBTypes) and their instances, namely objects (analogous to FB instances). These FB instances connect and communicate with each other using wire connections, and with external signals via the encapsulating function block interface. This facilitates the structural hierarchy, i.e. a given function block network may contain instances of other composite FBs that encapsulate sub-FBNs.

Figure 7.14 shows a function block network with three function block instances that communicate with each other using wire connections, e.g. a Real output value SetPoint of the AutoCommand instance can be read as AutoSP by the valve instance.

Furthermore, some signals directly flow from the interface of the top-level composite FB into the encapsulated function block network, e.g. the event MODE_UPDATED is read from an external source and made available to the MODE_CHANGED input event of both the AutoCommand and valve instances. However, only compatible signals flow in this manner, meaning that an input event on a composite FB interface can only flow into an input event of nested FB interfaces. Similarly, data flow in this manner must also conform to data-type compatibility, e.g. a Boolean input on the composite FB interface cannot flow into a string type input of the nested FB interface. One exception to this rule is the Any type, which, as the name suggests, can accept any data type. This mode of signal flow is thus directly responsible for effecting the interface definition of a composite FB, i.e. if a nested FB needs an input from an external source, there must be an input defined on the composite FB interface, which flows into the said nested FB. This encapsulation of nested FBs from external sources simplifies the reuse of FBTypes.

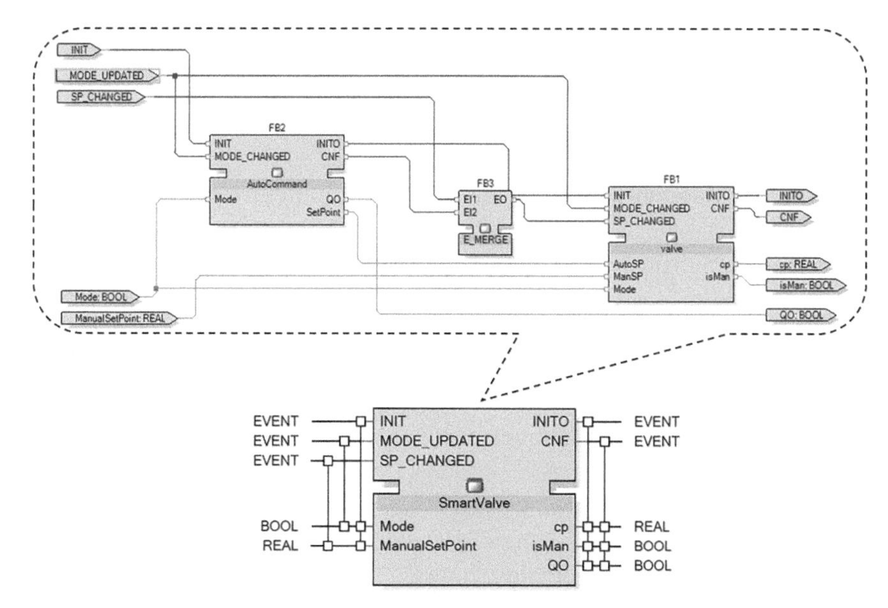

Figure 7.14 A composite function block with an encapsulated function block network.

7.5.3 A Brief Introduction to the Service Interface Function Block

Service interface function blocks (SIFB) can be considered as device drivers that connect the external environment with function block applications. These blocks are used to provide services to a function block application, such as the mapping of I/O pin interactions to event and data ports and the sending of data over a network.

There are two categories of SIFBs described in the standard, namely communication function blocks and management function blocks. While composite FBs capture centralized entities, resources are reminiscent of tasks and devices represent PLCs. Hence, both resources and devices need specific entities that facilitate either task-level (inter-resource) or distributed (inter-device) communication.

Communication function blocks are SIFBs providing interfaces that enable communication between IEC 61499 resources. Within the context of IEC 61499, a resource is a functional unit contained in a device that has independent control of its operations, so it may be created, configured, parameterized, started up, deleted, etc., without affecting other resources. The goal of a resource is to accept data and/or events from one or more interfaces, elaborate them and return data and/or events to some interfaces.

For the sake of completeness, it is worth mentioning that an IEC 61499 device contains one or more interfaces and those interfaces can be of two different types: communication and process. While communication interfaces provide a mapping between resources and the information exchanged via a communication network, a process interface provides a mapping between the physical process (e.g. analog measurements, discrete I/O, etc.) and the resources. Different types of communication function blocks may be used to describe a variety of communication channels and protocols.

On the other hand, management function blocks are SIFBs that are used to coordinate and manage application-level functionalities by providing services, such as starting, stopping, creating and deleting function block instances or declarations. They are somewhat analogous to a task manager in a traditional operating system. Unlike basic FBs, where the behaviour is specified using an ECC, SIFBs are specified using time-sequence diagrams.

7.5.4 The Generic DLL Function Block of nxtControl

The IEC 61499 software tool engineered by nxtControl provides a mechanism to integrate custom code in an IEC 61499 application. The mechanism is called Generic DLL function block and enables the exploitation of custom IEC 61499 function blocks interfaced by means of an abstract interface layer.

It provides the possibility to implement basic or service IEC 61499 function blocks in a custom programming language that are compiled in a dynamical loadable library (DLL) and then loaded and bound to the IEC 61499 runtime at the execution phase.

The Generic DLL function block mechanism builds on top of two components:

- a DLL that exposes a C interface where a predefined number of functions and data structures (embedded in a prototype which follows a well-defined template) implement the custom functionalities to be integrated in the distributed control application;
- a graphical representation of the custom function block, whose FBType is FB_DLL, and which is used in the nxtControl's engineering software environment to instantiate as many FBs as needed.

Such a mechanism enables the development of customized FB, providing:

- a representation of the IEC 61499 simple data types (as well as one-dimensional arrays of them) and plain C types;
- an input/output interface for passing these data between the IEC 61499 runtime software and the DLL implementation;

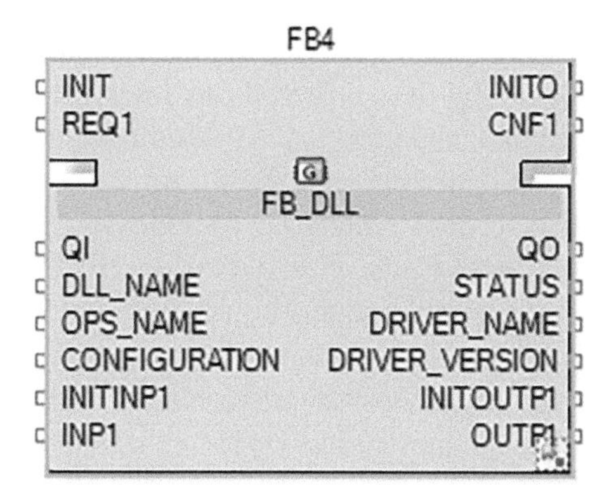

Figure 7.15 Example of FB_DLL function block.

- an interface for a custom function block where one initialization event and an arbitrary number of input events can be fed;
- the possibility to generate output events asynchronously;
- an interface to register and unregister a function block with the custom DLL;
- a way to query the provided data interface, so it is possible to implement consistency checks or to implement operations on different data types by one implementation;
- the possibility to implement several function blocks through a single DLL.

More than one instance of the Generic DLL function block (FB_DLL, Figure 7.15) can be instantiated in an IEC 61499 application, and the parameters provided as input to those FBs are exploited to select the appropriate DLL. All the FB_DLL instances are characterized by an INIT input event that is used to load the DLL: in particular, when the INIT event of any FB_DLL is received for the first time, the associated DLL is loaded and the IEC 61499 runtime registers the function block with that DLL. Furthermore, if the constructor is implemented in the custom code, then it is run afterward.

To leverage this flexible customization mechanism for implementing distributed automation applications, the custom code has to expose a data structure whose specification is detailed in the nxtControl's documentation material. That interfacing structure defines different elements that characterize the generic DLL function block, like:

- the number of input and output events;

- the number of data values that are associated to the input and output events;
- the data type associated to data values.

In addition to the description of the input/output events and data, the custom code used in a generic DLL function block has to define a precise set of functions that the IEC 61499 runtime uses to interact with the DLL when the distributed control application needs to execute the custom code. The most relevant of such functions are those used to register/unregister an FB_DLL with the appropriate DLL, the one used to execute the code associated to a specific input event, as well as the one dedicated to signal the triggering of an output event. In addition to those, there is also a function dedicated to the log information that can be used by the code in the DLL to report diagnosis information to the IEC 61499 runtime.

7.5.5 Exploiting the FB_DLL Function Block as Interfacing Mechanism between IEC 61499 and External Custom Code

Leveraging the generic DLL function block it is possible to extend functionalities available in the nxtControl automation platform with additional features that can be integrated in a seamless manner into an IEC 61499 control application.

That possibility opens the opportunity to integrate in an engineering software tool, designed to develop IEC 61499 applications, features that are not strictly related to the standard itself but that are interesting for implementing advanced distributed control applications. Actually, this can be leveraged to integrate the advanced functionalities that characterize a CPS that conforms to the DAEDALUS' vision, as for example, the integration of the "simulation dimension" and advanced MPC algorithms.

The possibility to extend the type of elaborations that can be performed within a function block in a distributed control application based on IEC 61499 enables the possibility to introduce new functionalities. Furthermore, it enables to test new features while respecting the normative rules and constraints of the standard and, as a consequence, allows to keep a high level of portability of the solution developed by means of this mechanism.

Since the DLL code is developed and compiled outside the classic development toolchain that is normally used for a plain IEC 61499 application (i.e. leveraging the development environment from nxtControl), the DLL has to be compiled by means of appropriate software tools to address the specific platform where the DLL will run. This means that an appropriate software toolchain is needed to generate a binary code that can run on the controller platform selected.

The main constraints that characterize this approach are:

- All the algorithms that define the behaviour of the FB_DLL have to be compiled as a dynamic loadable library (DLL) with a binary format compatible with the architecture of the controller, where the DLL will have to be installed;
- The DLL has to expose a C interface corresponding to the template imposed by the generic DLL function block mechanism;
- In the case where the FB_DLL is conceived to provide an output event to confirm the completion of the elaboration performed by the FB before a new input event can be processed by the FB, the elaboration performed by the DLL has not to take too much time before generating the output event. Otherwise that elaboration can affect negatively the controller's real-time performance;
- When the elaboration to be performed takes many computational resources and a lot of time to generate a result from the elaboration, another approach should be used: for example, the approach to run elaborations in parallel and generate output events asynchronously is a valid alternative;
- One of the aspects that needs to be considered at design is that a DLL can be shared by all the FB_DLL instances that make use of that library. This means, as a consequence, that the current number of function blocks registered with a DLL have to be managed appropriately, in order to keep track of the code portions that need to be executed for each FB_DLL instance.

The compact approach

The first approach enabled by the use of the generic DLL function block consists in exploiting the mechanism to implement a basic function block fully customized, where the constraint of using an execution control chart (ECC) is no more effective. In this case, the developer can freely design the finite state machine for government of the function block's logic states by exploitation of any preferred development tool (Figure 7.16).

By means of this approach, the logic algorithms that need to be executed when the associated input events are received by the FB_DLL instance can be designed and implemented following a customized approach that satisfies the developer's preferences and needs. At the same time, this mechanism enables

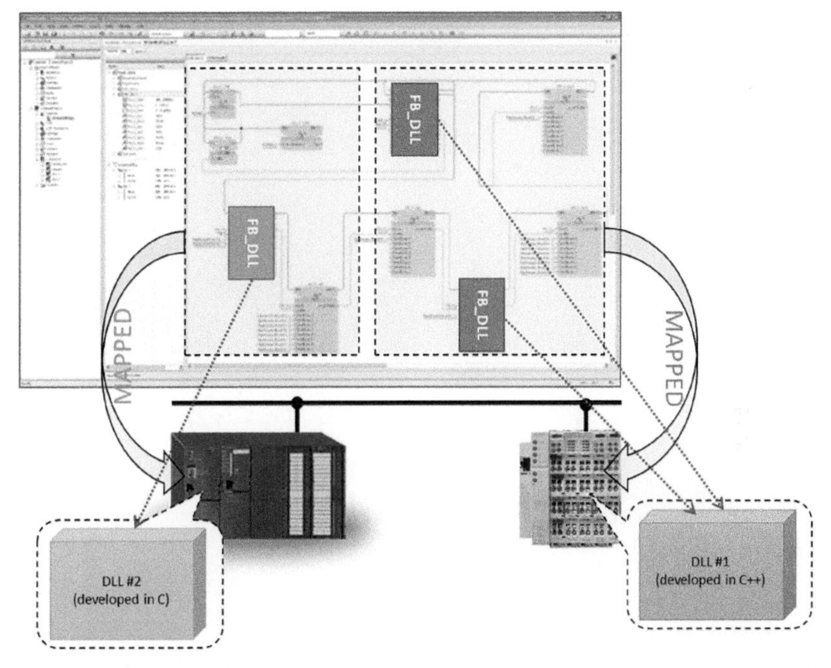

Figure 7.16 Illustration of the compact approach based on exploitation of generic DLL FBs.

to leverage other programming languages to implement the algorithms of the basic function block, in addition to the structured text (ST) language currently supported by the nxtControl software development tool.

The extended approach

A generalization of the previous approach consists in leveraging one or more additional DLLs when implementing the code associated to the FB_DLL instance. This basically means that the dynamic loadable library associated to the generic DLL function block is linked, in turn, to one or more other DLLs (Figure 7.17).

In such a case, it is possible that the exploitation of third parties' libraries implements customized function blocks usable in an IEC 61499 distributed control application. In this way, it is possible to develop custom service interface function blocks, making use of operating system function calls to access low-level hardware features or input/output data via interfacing devices.

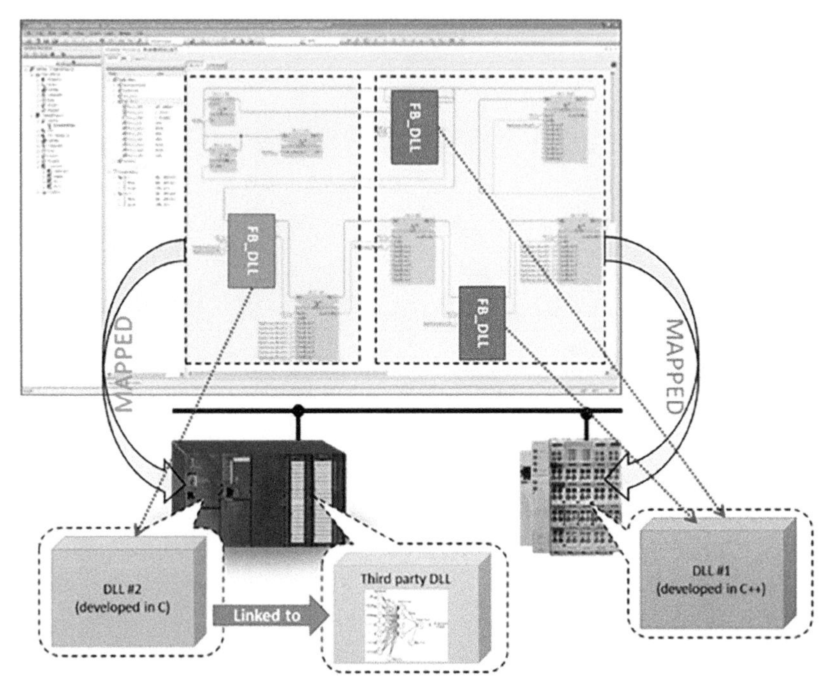

Figure 7.17 Illustration of the extended approach based on exploitation of generic DLL FBs.

In order to make this approach applicable, all the DLLs that are going to be exploited within the code of a general DLL function block have to be compiled for the specific architecture of the controller that will run that code.

That constraint can be limiting in certain scenarios, where the DLLs referenced by the custom code are not available for the platform selected and therefore it makes the use of those libraries impossible in such a scenario. On the other hand, that limitation has not to be ascribed to the generic DLL function block mechanism but to the lack of a compatible version of third party's libraries.

All the considerations that have been done for the basic approach of exploiting the FB_DLL are valid also for this extended case.

The distributed approach

The most general and flexible exploitation approach of the generic DLL function block mechanism consists not only in leveraging the FB_DLL FBs to integrate custom made and/or third-party software algorithms, but also in expanding the distributed computational network with additional

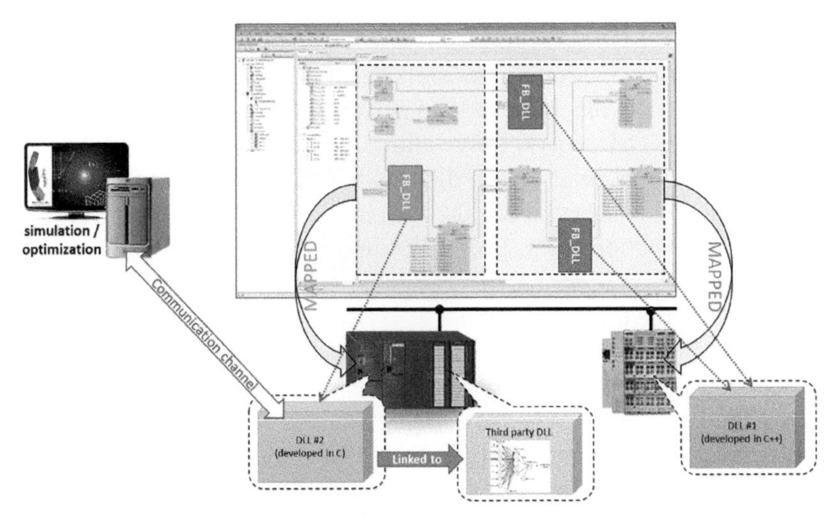

Figure 7.18 Illustration of the distributed approach based on exploitation of generic DLL FBs.

elaboration devices via interfacing mechanisms that can co-exist in parallel to the IEC-61499 communication interface (Figure 7.18).

This means that in addition to custom and advance algorithms embedded in DLLs that run locally in the controller where the FB_DLL instance is mapped, we can leverage the computational resources of other devices, in which specific data processing is allocated.

In such a scenario, the dynamic loadable library associated to an FB_DLL instance is used to open appropriate communication channels toward other computational nodes of the network where the data elaboration is actually performed. The FB_DLL has to leverage the asynchronous generation of events and appropriate mechanism to accept new requests in order to manage appropriately the elaboration and communication time without affecting negatively on the responsiveness of the IEC 61499 controller where the FB-DLL instance is running.

7.6 Conclusions

A deeply review of state of the art regarding solutions for controlling aggregated CPS has been carried out: the focus has been pointed on Model Predictive Control, especially on Hybrid Model Predictive Control. The analysis delves into Hybrid System representation and modelling, showing different techniques, mainly PWA and MLD. The advantages of PWA representation is related to the presence of numerous tools developed in the control system and identification fields, which are able to perform stability proof of system, convergence analysis. Moreover, PWA allows to build up an easier-to-use interface for SDK Interface in future development step. On the other hand, the MLD representation allows a deeply computational cost reduction for the solver as shown in Section 7.2.2. Both PWA and MLD are used in the SDK of Daedalus and they will work synergistically to improve the performance and the usability of the toolbox (SDK).

A review of the literature on data-driven modelling of hybrid systems has been carried out, with emphasis on PieceWise Affine (PWA) models and Jump Affine models, where the switches among the discrete state are triggered by deterministic events (e.g. if–then–else rules). These two models will be combined in the future stages of the project to arrive at Jump Piece-Wise Affine (JPWA) models, where the PieceWise Affine part will be used to describe the non-linear dynamics of the continuous (physical) states of the CPS, while the Jump part will be used to describe the time-evolution of the discrete (logical) states.

As a next step, a user-friendly software toolbox for identification of hybrid systems will be developed and the software functions will be integrated in the Daedalus' platform. This toolbox for on-line identification will contain the algorithm in ref. [37]. If necessary, improvements and/or extensions of this identification algorithm will be proposed and implemented in the toolbox. Benchmark examples available in the literature and case studies proposed by the project's partners will be used to test the implemented identification algorithms.

The IEC-61499 standard defines a technology for the implementation of distributed control applications applicable on several industrial scenarios. Many are the key aspects that make such a technology a valid solution for the development of the new generation of industrial control systems, leveraging networks of interacting CPSs.

The modularity that characterizes the control software design approach, which builds on the concept of function block, and the event-based execution paradigm are, just as an example, two of the core architectural aspects of the IEC-61499 standard that provide an effective development tool for complex control applications.

Advanced control software can be implemented exploiting the hierarchical development approach based on nesting of different types of function blocks. Custom algorithms can be implemented both through the composition of function blocks and by the development of Basic Function Blocks, leveraging the programming languages supported by the selected software development toolkit.

Acknowledgements

This work was achieved within the EU-H2020 project DAEDALUS, which has received funding from the European Union's Horizon 2020 research and innovation programme, under grant agreement No. 723248.

References

[1] V. Vyatkin, "The IEC 61499 standard and its semantics", IEEE Industrial Electronics Magazine, vol. 3, 2009.
[2] G. Optimization et al., "Gurobi optimizer reference manual", URL: http://www.gurobi.com, vol. 2, pp. 1–3, 2012.
[3] I. L. O. G. CPLEX, Reference Manual, 2004, 2011.
[4] B. a. T. M. Meindl, "Analysis of commercial and free and open source solvers for linear optimization problems", Forschungsbericht CS-2012-1, 2012.
[5] J. Lunze, F. Lamnabhi-Lagarrigue, Handbook of hybrid systems control: theory, tools, applications, Cambridge University Press, 2009.
[6] S. A. Nirmala, B. V. Abirami, and D. Manamalli, "Design of model predictive controller for a four-tank process using linear state space model and performance study for reference tracking under disturbances", in Process Automation, Control and Computing (PACC), 2011 International Conference on, 2011.
[7] J. G. Ortega, E. F. Camacho, "Mobile robot navigation in a partially structured static environment, using neural predictive control", Control Engineering Practice, vol. 4, pp. 1669–1679, 1996.
[8] M. Kvasnica, M. Herceg, L. irka, M. Fikar, "Model predictive control of a CSTR: A hybrid modeling approach", Chemical papers, vol. 64, pp. 301–309, 2010.
[9] J. Richalet, A. Rault, J. L. Testud, J. Papon, "Model predictive heuristic control: Applications to industrial processes", Automatica, vol. 14, pp. 413–428, 1978.

[10] F. Borrelli, A. Bemporad, M. Fodor, D. Hrovat, "An MPC/hybrid system approach to traction control", IEEE Transactions on Control Systems Technology, vol. 14, pp. 541–552, 2006.

[11] G. Ferrari-Trecate, E. Gallestey, P. Letizia, M. Spedicato, M. Morari, M. Antoine, "Modeling and control of co-generation power plants: a hybrid system approach", IEEE Transactions on Control Systems Technology, vol. 12, pp. 694–705, 2004.

[12] D. Corona, B. De Schutter, "Adaptive cruise control for a SMART car: A comparison benchmark for MPC-PWA control methods", IEEE Transactions on Control Systems Technology, vol. 16, pp. 365–372, 2008.

[13] E. Sontag, "Nonlinear regulation: The piecewise linear approach", IEEE Transactions on automatic control, vol. 26, pp. 346–358, 1981.

[14] A. Bemporad, M. Morari, "Control of systems integrating logic, dynamics, and constraints", Automatica, vol. 35, pp. 407–427, 1999.

[15] R. Alur, C. Courcoubetis, T. A. Henzinger, P.-H. Ho, "Hybrid automata: An algorithmic approach to the specification and verification of hybrid systems", in Hybrid systems, Springer, pp. 209–229, 1993.

[16] W. P. M. H. Heemels, B. De Schutter, A. Bemporad, "Equivalence of hybrid dynamical models", Automatica, vol. 37, pp. 1085–1091, 2001.

[17] H. P. Williams, Model building in mathematical programming, John Wiley & Sons, 2013.

[18] J. H. Lee, "Model predictive control: Review of the three decades of development", International Journal of Control, Automation and Systems, vol. 9, no. 3, pp. 415–424, 2011.

[19] C. A. Floudas, Nonlinear and mixed-integer optimization: fundamentals and applications, Oxford University Press, 1995.

[20] R. Fletcher, S. Leyffer, "Numerical experience with lower bounds for MIQP branch-and-bound", SIAM Journal on Optimization, vol. 8, pp. 604–616, 1998.

[21] A. Makhorin, "GLPK", GNU Linear Programming Kit, 2004.

[22] E. F. a. C. B. Camacho, Model predictive control in the process industry, Springer Science & Business Media, 2012.

[23] E. B. E. Y. a. I. E. G. Perea-Lopez, "A model predictive control strategy for supply chain optimization", Computers & Chemical Engineering 27.8 (2003), vol. 27, no. 8–9, pp. 1201–1218, 2003.

[24] J. e. a. Łirokɩ, "Experimental analysis of model predictive control for an energy efficient building heating system", Applied energy, vol. 89, no. 9, pp. 3079–3087, 2011.

[25] S. e. a. Li, "Model predictive multi-objective vehicular adaptive cruise control", IEEE Transactions on Control Systems Technology, vol. 19, no. 3, pp. 556–566, 2011.
[26] A. A. a. A. Bemporad, "A survey on explicit model predictive control". Nonlinear model predictive control, in Non Linear Model Predictive Control, Springer Berlin Heidelberg, pp. 345–369, 2011.
[27] S. Y. Xu, T. W. Chen et al., "Robust H-infinity control for uncertain stochastic systems with state delay", IEEE Transactions on Automatic Control, vol. 47, pp. 2089–2094, 2002.
[28] L. Breiman, "Hinging hyperplanes for regression, classification, and function approximation", IEEE Transactions on Information Theory, vol. 39, pp. 999–1013, 1993.
[29] S. Paoletti, A. L. Juloski, G. Ferrari-Trecate, R. Vidal, "Identification of hybrid systems a tutorial", European journal of control, vol. 13, pp. 242–260, 2007.
[30] A. Garulli, S. Paoletti, A. Vicino, "A survey on switched and piecewise affine system identification", in 16th IFAC Symposium on System Identification, Brussels, 2012.
[31] N. Ozay, C. Lagoa, M. Sznaier, "Set membership identification of switched linear systems with known number of subsystems", Automatica, vol. 51, pp. 180–191, 2015.
[32] J. B. Lasserre, "Global optimization with polynomials and the problem of moments", SIAM Journal on Optimization, vol. 11, pp. 796–817, 2001.
[33] H. Ohlsson, L. Ljung, S. Boyd, "Segmentation of ARX-models using sum-of-norms regularization", Automatica, vol. 46, pp. 1107–1111, 2010.
[34] D. Piga, R. Tth, "An SDP approach for 0-minimization: Application to ARX model segmentation", Automatica, vol. 49, pp. 3646–3653, 2013.
[35] J. Roll, A. Bemporad, L. Ljung, "Identification of piecewise affine systems via mixed-integer programming", Automatica, vol. 40, pp. 37–50, 2004.
[36] G. Ferrari-Trecate, M. Muselli, D. Liberati, M. Morari, "A clustering technique for the identification of piecewise affine systems", Automatica, vol. 39, pp. 205–217, 2003.
[37] V. Breschi, D. Piga, A. Bemporad, "Piecewise Affine Regression via Recursive Multiple Least Squares and Multicategory Discrimination", Automatica, vol. 73, pp. 155–162, 2016.

[38] K. P. Bennett, O. L. Mangasarian, "Multicategory Discrimination via Linear Programming", Optimization Methods and Software, vol. 3, pp. 27–39, 1994.

[39] Y. J. Lee, O. L. Mangasarian, "SSVM: A Smooth Support Vector Machine for Classification", Computational Optimization and Applications, vol. 20, pp. 5–22, 2001.

8

Modular Human–Robot Applications in the Digital Shopfloor Based on IEC-61499

Franco A. Cavadini[1] and Paolo Pedrazzoli[2]

[1]Synesis, SCARL, Via Cavour 2, 22074 Lomazzo, Italy
[2]Scuola Universitaria Professionale della Svizzera Italiana (SUPSI), The Institute of Systems and Technologies for Sustainable Production (ISTEPS), Galleria 2, Via Cantonale 2C, CH-6928 Manno, Switzerland
E-mail: franco.cavadini@synesis-consortium.eu; paolo.pedrazzoli@supsi.ch

This chapter presents the results of the conception effort done under Daedalus to transfer the technological results of IEC-61499 into the industrial domain of Human–Robot collaboration, with the aim of deploying the concept of mutualism in next-generation continuously adaptive Human–Machine interactions, where operators and robots mutually complement their physical, intellectual and sensorial capacities to achieve optimized quality of the working environment, while increasing manufacturing performance and flexibility. The architecture proposed envisions a future scenario where Human–Machine distributed automation is orchestrated through the IEC-61499 formalism, to empower worker-centred cooperation and to capitalize on both worker's and robot's strengths to improve synergistically their integrated effort.

8.1 Introduction

Personnel costs in Europe are higher compared to other industrial regions; hence, EU industry today competes in the global market by offering high added-value products. This is possible thanks to the extreme qualification level and know-how of its 17 million shopfloor workers.

To keep this true, European manufacturing industry needs to adopt a new production paradigm, focusing on processes where robots collaborate with humans with mutual benefits in terms of skill growth and support. The problem is that current automation approaches in Europe have disregarded the importance of added value of workers, enhancing de-skilling of European workforce and labour shedding.

Future European value-adding manufacturing industries will have to rely more and more on the virtuous combination of machines and operators [1], to increase the standard of quality of their shopfloors while remaining competitive with low-wage countries. To exploit new synergies between operators and machines, future manufacturing processes will have to exhibit a dynamically reconfigurable overall behaviour, through continuous physical interactions and bidirectional exchange of information between them.

A possible solution towards a comprehensive management of this highly integrated collaboration, by pushing the boundaries of the topic of human–robot Mutualism, is to apply the IEC-61499 standard to orchestrate their joint behaviour.

In fact, in collaborative tasks, the overall dynamics of the interactions between a robot and a human is currently an emergent property that implicitly arises from their individual behaviours. On the contrary, the aim should be of making these implicit properties explicit, by representing, standardizing and orchestrating the overall dynamics of human–robot interaction. Achieving human–robot mutualism through such orchestration will dramatically improve the transparency and acceptance of robots for users, as well as substantially increase the ergonomy and efficiency of collaborative tasks.

The main bottleneck in orchestration for mutualism is that current scheduling and planning algorithms, powerful as they may be, are limited by the expressiveness and fidelity of the representations they operate upon. Therefore, the need of the market (partially tackled in Daedalus) is to develop and standardize such representations for open manufacturing processes and ensure their compatibility with existing norms, with the IEC-61499 standard for industrial distributed automation being the cornerstone. This can be achieved only by working on three high-level objectives:

- Standardizing and homogenizing the way intelligent agents, both human and robotics, are represented and orchestrated, from design to runtime stage, in a team with multiple dynamically varying objectives;

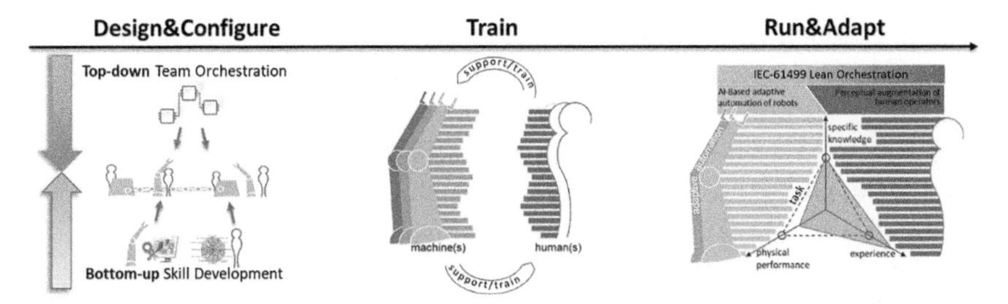

Figure 8.1 Life-cycle stages to achieve human–robot symbiosis from design to runtime through dedicated training.

- Engineering a new generation of mechatronic intelligent devices for human–robot collaboration, to facilitate and augment bidirectional interactions with operators while exhibiting inherently safe behaviours;
- A widespread application of AI-based techniques, from semantic planning for orchestration and task planning to deep learning for human intent recognition. Applying AI at different functional levels and life-cycle stages is essential to go beyond current rigidity of robotics systems and, thus, to increase by-design compatibility with human operators.

What European manufacturing processes need is a more effective and extensive symbiosis between humans and robots in the work environment, achieved by proposing new technological solutions but also reshaping the way those processes (and the systems executing them) are conceived, designed, run and reconfigured (Figure 8.1).

8.2 Human and Robots in Manufacturing: Shifting the Paradigm from Co-Existence to Mutualism

European manufacturing industry, with more than 2M enterprises that employ 30M people, must face two main challenges: (i) reduction of product life cycles, with a corresponding reduction in the amortization time for the investments; (ii) increase of customization from the market, that requires flexibility [2] and adaptability for manufacturing smaller batch sizes with constant product changes [3]. EU industry therefore competes in the global market by offering high-quality products, thanks to the high qualification level and know-how of its 17 million shopfloor workers. Sustaining it requires to maintain as much time as possible these high skilled workers, postponing their retirement age and ensuring a smooth transition by incorporating younger workers. However, the negative impression of manufacturing due to its negative impact on health [4] and low attractiveness (monotonous and boring work), combined with the aging of European population, will provoke a lack of qualified shop floor workers in 2030 [5].

The problem is that current automation approaches in Europe have disregarded the importance of added value of workers, enhancing de-skilling process in European workforce and labour shedding [6]. However, Japan, one of the key manufacturers of robotics, has shown that other manufacturing models where robots and technologies support and improve worker, instead of substituting them, are possible [7].

As explored in the literature and in previously approved EC-funded projects (e.g. MAN-MADE [8]), the workplaces of the future are expected to be worker-centric (as opposed to task-centric), with an increased role of workers in pursuing production performances and personal well-being. In this new paradigm, it is the task that suits skills, experience, capacities and needs of the worker, here turned from a passive constraint to a variable opportunity. Previous research and pilot implementations have in fact demonstrated that collaborative human–robot workspaces, knowledge networks and augmented reality support [9] can improve productivity and workers' well-being, reducing 50% the time of first time assembly [10] or bringing near novice workers to experienced ones [11].

The paradigm of human–machine symbiosis suggested by Tzafestas [12] and revised (with a focus on assembly systems) in Ferreira et al. [13] appoints advanced human–machine symbiotic workplaces as the foundation for human-centric factories of the future.

Ultimately, such a positive and human-centric vision still lacks in actual instantiations. The reasons are several:

- A new conceptual framework, intended to deploy symbiotic human–robot ecosystems, needs to be created for the manufacturing environment, effectively describing workers, intelligent machine and, especially, their interactions and collaborative tasks;
- Advanced algorithms and tools based on artificial intelligence are missing to "augment" the mutual perception and understanding of the behaviour of robots and operators, for an effective synergistic approach in executing joint tasks;

- Trust-creation environments, where human–machine team can orchestrate their respective actions in a controlled situation and where machines are aware of not only the human worker's physical but also mental state. In fact, human–machine interaction is usually conceived in one way, with one of the elements providing and the other receiving support;
- Worker-centric human–machine symbiotic interaction in real industrial environments is currently limited by available legislation (e.g. EU Machinery Directive 2006/42/EC, ISO 10218 standard) and insufficient empirical data, so that only sequential and parallel cooperation as forms of co-work are possible;
- Configuration of human–machine interaction is currently performed at its setting-up and, then, seldom updated. But human operators (and, just partially, machines) modify their behaviour and mood daily: a manufacturing symbiosis needs to be constantly adapted and tuned, considering actual symbionts' behaviours;
- A unified definition of what is "good" for the worker still lacks: a new approach combining subjective and objective measures is needed for this evaluation.

8.3 The "Mutualism Framework" Based on IEC-61499

Advanced human–machine interaction is the most promising approach to enable worker-centric manufacturing in the factory of the future. Some authors [13, 15] have recently suggested the implementation of a "symbiotic system" paradigm, where human and robotic operators cooperate for an effective accomplishment of manufacturing tasks.

What has been conceived aims at embracing this approach and proposes a more complete and concrete interpretation based on the biological concept of Mutualism, a peculiar relationship between two organisms of different species where each individual benefit from the activity of the other. Mutualism is a specific instance of symbiosis, establishing a win–win interaction.

This paradigm is adopted as basis for an innovative methodological framework that supports the effective integration and implementation of collaborative robotics technology over the life cycle of the plant, from conceptual stage to runtime and re-configuration. The objective is to sustain a deeper and more extensive collaboration between humans and robots, as intelligent agents orchestrated to achieve dynamically varying objectives, while mutually compensating for their limits through their respective strengths.

Members of this orchestrated manufacturing team are called symbionts and they are either humans or robots/intelligent devices. Regardless of their nature, symbionts are all able to provide and receive support (i.e. giving or receiving a quantifiable benefit thanks to their interaction).

Symbionts are intended to operate in real manufacturing environments, and the effectiveness of their Mutualism is assessed continuously (from design to runtime) considering a holistic worker-centric perspective of "well-being" and psychological safety. Clearly, Symbionts are "living" entities (both humans and robots) in that they adjust their behaviour on a task-wise basis and modify their performances according to changing exogenous elements.

These qualitative characteristics result in the following concepts composing the so-called "Mutualism Framework".

8.3.1 "Orchestrated Lean Automation": Merging IEC-61499 with the Toyota Philosophy

When dealing with classical industrial automation use-cases, a core concept is the real-time orchestration of automation tasks – provided by various subsystems, machines or robotic manipulators – to guarantee the exact execution of a well-specified behaviour, usually within the constraints of a time cycle. On the other hand, complex production processes, especially where the intervention of highly qualified human operators is essential, have been optimized throughout the last decades with the lean manufacturing approach, mostly under the Toyota philosophy.

For a more holistic integration of humans and robots, it has been introduced the concept of hybrid orchestration (at runtime stage) of IEC-61499 automation tasks executed by both types of Symbionts but framed in a lean methodological and engineering context (in the design stage). The synthesis of these two elements (Figure 8.2) requires adaptation and evolution of both, where orchestration will have to consider the inevitable variability (and, partly, unpredictability) of human tasks (thanks to the support of artificial intelligence), while the lean concepts will have to be extended considering the very peculiar capacities of intelligent mechatronic systems.

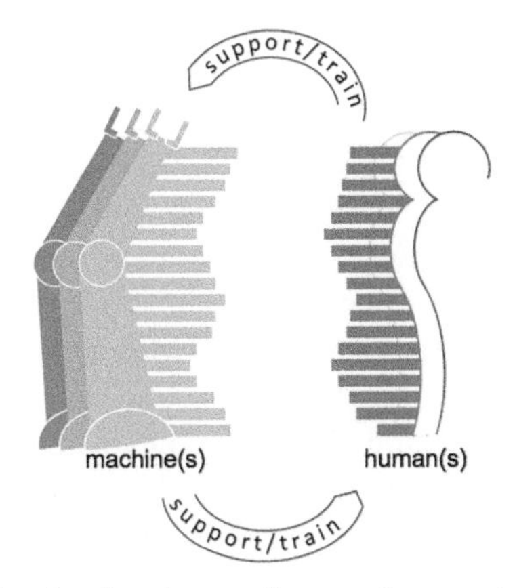

Figure 8.2 Bidirectional exchange of support between humans and robots.

8.3.2 A Hybrid Team of Symbionts for Bidirectional Mutualistic Compensation

Latest studies [13, 15] addressing human–machine interaction propose to establish symbiotic environments for optimal collaboration between human and machine. Nevertheless, proposed symbioses are always unidirectional, namely pursuing the benefit for just the human component. This confines the machine to a purely servant role: they actually co-exist with humans in the same working environment (under more flexible safety constraints, thanks to the advancement of collaborative robotics), but just giving them the required support in very specific and limited situations.

The Mutualism Framework aims at overstepping this concept by proposing that both robots and human operators are treated as intelligent agents (Symbionts), each with its own special traits, exchanging bidirectionally physical support, information and even knowledge of the process. In this new vision, also robots (machines) and, more in general, intelligent devices can be the addressees of support and training actions, with human operators transmitting knowledge and experience to them.

Making the symbiosis mutualistic provides two innovative ways of exploiting the role of machines:

i. human–robot automation tasks can be designed exploiting real collaboration, thanks to the continuous exchange of orchestration signals between symbionts;
ii. intelligent mechatronic systems are transformed into active repositories of manufacturing knowledge.

8.3.3 Three-Dimensional Characterization of Symbionts' Capabilities

Worker characterization is traditionally performed considering just a subset of all the possible describing dimensions: vital statistics, ergonomics and anthropometry; functional capacities; knowledge and experience. In fact, these are rarely included at the same time in the creation of a dynamic worker profile where all these elements holistically concur to a unique profiling strategy.

On the other hand, the ongoing transition towards the concept of Cyber-Physical Systems (CPS) for mechatronics is still incomplete, meaning that we have not yet reached a level of maturity where CPS are treated as intelligent agents that may evolve in time (i.e. machine learning) and, as such, require an equally dynamic characterization strategy.

The Mutualism Framework aims at providing a comprehensive assessment and characterization approach of Symbionts (both humans and robots) operating in its shop floors, to find valuable win–win combinations for effective execution of collaborative tasks.

Under a common overarching interpretation methodology, all the considered and monitored characteristics will be used to define, on a task-wise basis, a three-dimensional picture representing a generic symbiont profile (Figure 8.3) in terms of:

- Experience, which indicates the symbiont level of practice in executing that tasks. A proper taxonomy is thus required to quantify and qualify the level of experience also considering the practice gained using edutainment tools and virtual/augmented reality environments;

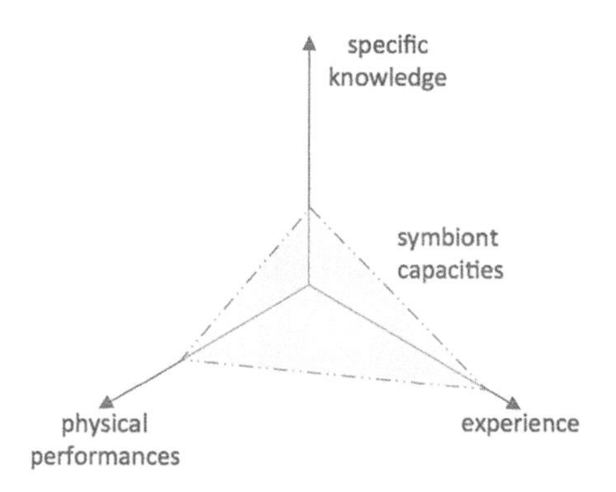

Figure 8.3 Three dimensions of characterization of Symbionts.

- Specific knowledge, related to the level of knowledge to correctly understand and handle a specific subject. It refers to a theoretical or practical understanding of the topic which can be either implicit or explicit;
- Physical performances, to include the quantification of the physical characteristics (such as movement capacities, strength in different positions and operations, sensorial capabilities, etc.) that are needed to execute that task.

8.3.4 Machine Learning Applied to Guarantee Dynamic Adherence of Models to Reality

Kruger et al. [14], in a review study on human–machine cooperation types in manufacturing systems, state that cooperation is an important aspect for flexibility, adaptability and reusability. As also stated by Ferreira et al. [15], manufacturing systems have been pressed in recent years to provide highly adaptable and quickly deployable solutions to deal with unpredictable changes following market trends. But what about the adaptability (short term) and evolution (mid/long term) of the capacities of both human operators and robots? In fact, workers' performances may importantly vary even on a day-by-day basis, while learning-augmented mechatronic systems are conceived to improve over their life cycle.

The Mutualism Framework puts the dynamicity of a Symbiont's characterization as one of its cornerstones, proposing to integrate this sort of "live portrait" into a so-called Virtual Avatar, that is, a digital representation of the Symbiont that implements the characterization data model and is fed of real-time information coming from the shopfloor, where humans and robots operate and interact. Avatars will not be simply passive containers of information concerning symbionts; they will also be able to represent that part of their dynamics (= behaviours) which is needed to design and then run an adequate orchestration towards the common manufacturing goals

8.4 Technological Approach to the Implementation of Mutualism

To transform the aforementioned concepts into a valid implementation of the Mutualism Framework, several technological contributions are needed, integrated in a coherent functional architecture, as represented in Figure 8.4.

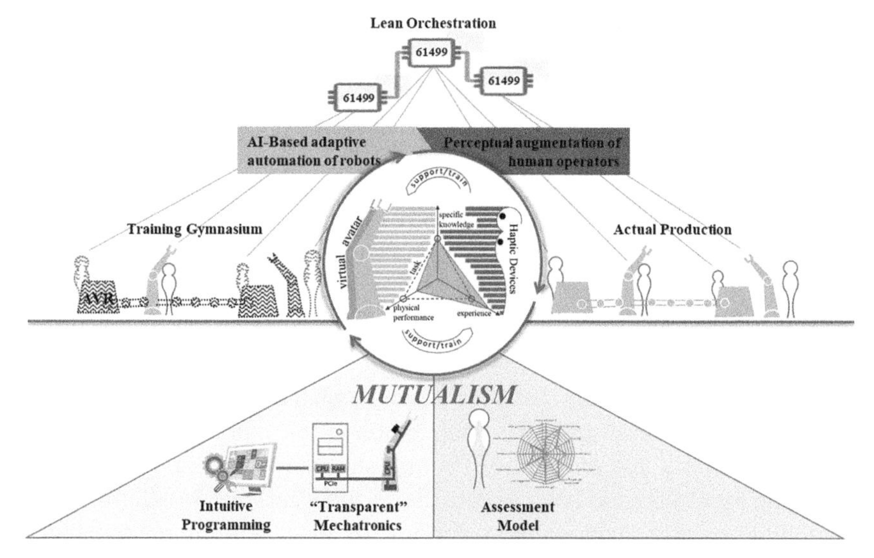

Figure 8.4 Qualitative representation of the technological key enabling concepts of the Mutualism Framework.

8.4.1 "Mutualism Framework" to Sustain Implementation of Symbionts-Enhanced Manufacturing Processes

Mutualistic symbiosis constitutes the fundamental brick to boost worker centrality in real production environments, where human–robot collaboration is put as a cornerstone. For this to become a new design and runtime paradigm, a "Mutualism Framework" must be established, that is, a sound methodological characterization of what the Symbionts are and of how their mutualistic interactions are modelled and exploited.

The Mutualism Framework (MF) explores four major aspects of the above-mentioned functional architecture:

- A dedicated semantic data model to describe all the elements of the Mutualism concept;
- An assessment model to capture the dynamics of their characteristics (slow) and behaviours (fast);
- The functional mapping, on a per-task basis, between the dynamic representation of Symbionts and their 3-dimensional (experience, specific knowledge and physical performances) profile;
- An overarching set of indicators to evaluate the multi-dimensional performance of Symbionts.

Clearly, for what concerns the intelligent mechatronic systems (= robotic symbionts), the assessment model is directly linked to their functional and non-functional specifications. On the other hand, characterization of human symbionts must consider in a holistic approach all its major "dimensions":

i. Vital statistics, ergonomics and anthropometry, filling the existing gaps of the state of the art, especially for what concerns the variability of these aspects due to the physical and non-physical exposure at the collaborative workplace;
ii. Functional capacities, to consider the relation between workers' abilities and the potentially assigned tasks, and to sustain the corresponding AI learning of robots;
iii. Knowledge and expertise, to effectively and systematically enable knowledge sharing and transfer among workers but also between humans and robots.

The second key focus of the Mutualism Framework is about quantification of performance of collaborating human and robots (measured and/or foreseen), on a per-task basis, but under different perspectives. In fact, the 3D profiling previously introduced provides a simplified but effective way of evaluating how much different Symbionts are "fitting" to execute a specific activity, independently from how such level of adequacy is achieved (i.e. differently between operators and machines); this enables the adaptable orchestration of collaboration based on IEC-61499.

Contemporary, performance of the mutualism must be assessed, to define and then evaluate the achievement of specific process objectives. The MF puts worker's safety (Physical AND Psychological) and well-being as cornerstone of its set of KPIs, without forgetting about manufacturing sustainability (economic, environmental and social) and reconfigurability, and compliance with regulations in force.

8.4.2 IEC-61499 Engineering Tool-Chain for the Design and Deployment of Real-Time Orchestrated Symbionts

During the ongoing transition towards Industry 4.0, the concept of highly distributed intelligent mechatronic devices (also called CPS) is emerging, whose joint behaviour satisfies the production objectives, while guaranteeing much higher flexibility and reconfigurability. This convergence of the industrial world towards an agent-based paradigm for industrial automation asks for adequate new methodologies and technologies for the so-called (real-time) Orchestration of these distributed systems and, thanks to the support of

Daedalus, the open and interoperable IEC-61499 standard is taking the lead in solving this need.

With the Mutualism approach, it is recognized that a collaborating team of human and robotic symbionts is, in fact, an extension of the above-mentioned concept of distributed intelligence, to encompass the hybrid nature of shopfloors where operators and machines work shoulder-to-shoulder. This means that the concept of Mutualism must be developed towards its technical dimension of (soft) real-time orchestration of Symbionts, designed, deployed and then executed at runtime thanks to the usage of an IEC-61499-based engineering tool-chain.

The design stage of a Mutualistic manufacturing process will consider both the conceptual definition of the specifications of the process itself, and the engineering of the automation logics (through the IEC-61499 formalism and programming language) that will control orchestration of the distributed intelligence of Symbionts.

Conception of the Mutualism is where the principles of Lean Manufacturing are applied towards a new production model that considers the opportunities of human–robot collaboration and therefore exploits them. It originates from the key principles of the "Toyota Way" (especially the kaizen) to implement Mutualism keeping the Human operator at the centre of the process.

Leveraging on the state of the art of R&D on "Lean Automation", it is possible to focus mostly on the implementation of those design-support tools that can simplify the definition of requirements for Mutualistic tasks, help assembling the most appropriate team of Symbionts for those tasks and support the generation of specifications for the corresponding IEC-61499 orchestration.

For what concerns the engineering of orchestration logics, the usage of the IEC-61499 IDE and runtime developed in Daedalus allows to: (i) Guarantee ease of interfacing and functional wrapping of lower-level automation architecture of specific robotic symbionts; (ii) Use 61499 formalism to consider the 3D performance of Symbionts and (iii) Integrate with an adequate perceptual learning platform.

8.4.3 AI-Based Semantic Planning and Scheduling of Orchestrated Symbionts' Tasks

Complementary to the IEC61499-based design stage of Mutualism is the development of the Mutualism Execution Platform (MEP), that is, a set of

IEC-61499-Compliant runtime modules to adapt continuously the orchestration of human and robotics symbionts. In fact, the management of the runtime stage of the orchestration of distributed symbionts is where the flexibility and adaptability enabled by the mutualistic symbiosis of humans and robots is achieved effectively.

The collaboration between operators and machines designed within the IEC-61499 IDE is then operatively achieved by the MEP, deployed within the shopfloor and connected in real time to both machines and workers through the IEC-61499 runtime framework. In fact, it is responsible for defining the most suitable optimization pattern(s) to be executed over a specific time horizon and adapting coherently with the variability of production and workers' needs.

In practice, the MEP can exploit availability of ready-to-use AI-based techniques to tackle the operational challenges of robust planning and scheduling (over a distributed agents' functional architecture) the team of Symbionts needed to reach specific production objectives, considering that:

- First, P/S human activities are complex by itself, since people have different skills, attitudes and preferences; moreover, working shifts and rosters are subject to strict union and legislative regulations that cannot be violated, but force severe restrictions on working plan feasibility.
- Second, although machines do not have "personal" preferences and they are not subject to "union" regulations, they must undergo precise and rigorous maintenance plans, which must ensure that the operating conditions of each machine meet very high operating standards and security rules.
- Third, humans and machines are called to cooperate in a highly dynamic and uncertain environment where the tasks to be executed constantly evolve and their distribution (over different symbionts) and scheduling must be very robust.

The second key aspects of dealing with the runtime management of Mutualism is to provide intuitive programming tools, which enable advanced users and shopfloor workers to program novel robot skills that are verifiably compliant with IEC-61499. This requires the analysis of robotic skills with linear time temporal logic model checking. These programming tools can be based, for instance, on RAFCON [16], a visual programming tool for robotics skills that enables logging with semantic labels, so that the context in which data was logged is known. Having contextual knowledge greatly facilitates the data-driven analysis of skills [17], and the application of data mining techniques.

8.4.4 Modular Platform for Perceptual Learning and Augmentation of Human Symbionts

Intelligent mechatronic systems exchange (potentially strict real time) I/O signals to sustain high-performance interactions. Humans do not; in fact, despite having an incredibly sophisticated and flexible set of sensing and actuating apparatus, their capacity of receiving and transmitting complex and structured information is very limited. This (very simply introduced) specific issue is also one of the major reasons for which, until now, human–robot collaboration has been severely hindered: orchestrating physical tasks accomplished by distributed agents requires a bi-directional information flow continuously exchanged WHILE those tasks are being executed.

This aspect of the Mutualism is tackled from a systemic point of view, aware of the fact that what is needed today is not so much a new, very specific smart sensing or interfacing solution (those are already developed and released continuously at market level) as rather a HW/SW abstraction layer to:

i. Simplify the aggregation and elaboration of several and heterogeneous signals coming from and going into the shopfloor (through multiple devices) and
ii. Provide already integrated interpretation and machine learning functionalities, then available to the orchestration layer for increased flexibility and adaptability.

Following the natural distinction induced by the bi-directionality of signals to be exchanged between human and robotic Symbionts, it is correspondingly possible to identify two major contributions: holistic monitoring and learning of Virtual Avatars, and adaptive cognitive interfacing.

The first component is a hierarchical functional architecture composed of three layers: A modular monitoring system composed of distributed smart sensors; a data-interpretation layer and a machine learning middleware to provide high-level tasks. To assure an adaptive manufacturing environment, data gathering for worker profile adaptation will be continuous, performed both during everyday manufacturing operations and within properly designed training sessions.

Complementary to this is the possibility of sending feedback to an operator, directly from a robot, to enable the coordination of their respective activities during the execution of joint tasks (which may involve also several symbionts). We call this an adaptive cognitive interfacing, achieved through the augmentation of the capabilities of the worker through dedicated smart devices. Augmented/Mixed/Virtual Reality will be the most important approach to tackle this challenge, knowing that the area of human–robot collaboration is still developing and novel interfaces supporting in an effective way the collaboration need to be designed, implemented and evaluated.

8.4.5 Training Gymnasium for Progressive Adaptation and Performance Improvement of Symbionts' Mutualistic Behaviours

All biological symbioses go through a preliminary training phase, where symbionts know each other and measure up. This step is fundamental also for Mutualism. In fact, improper human–robot interaction may cause counter-effects, such as misuse of machine and/or safety issues.

Because the trust of human to robots will directly affect the degree of autonomy of the industrial robot, which is related to the efficiency of manufacturing processes, trust is a critical element in HRI when a human worker observes a discrepancy between his/her performance and what he/she expects from the robot partner, his/her trust to the robot decreases accordingly. When the robot performance matches human expectation, the human's trust to robot increases.

The solution to this is the so-called Training Gymnasium, where:

- A task recording & displaying facility will enable the recording and retrieval of working parameters, machines movements, machine and worker roles, etc. According to worker literacy rate, the training facility may record or retrieve the above-mentioned parameters;
- Virtual reality environments and devices will be implemented thanks to the augmentation platform. These solutions are especially intended for machines still under design to assure rich workers' experience and value-adding data gathering since the design of new workplaces (with a closed loop with the other phases of the life cycle of the manufacturing process);
- Augmented reality will be used to guide the less skilled workers in interacting with robotics Symbionts and other physical machines.

8.5 The Potential to Improve Productivity and the Impact this Could Have on European Manufacturing

According to Holdren [18], manufacturing has a larger multiplier effect than any other major economic activity: every euro spent in manufacturing drives an additional €1.35. EC data show that in 2012 the manufacturing sector in the EU employed 30 million persons directly and provided twice as many jobs indirectly, manufactured goods amount to more than 80% of total EU exports and manufacturing accounted for 80% of private Research & Development expenditure. This notwithstanding, for diverse and widely discussed reasons, EU manufacturing has slightly declined in the last few years.

European Commission made huge investments in manufacturing topics to reverse this trend, and Advanced Manufacturing has been identified as the major driving force for improving competitiveness of the European Industry, namely through [19] ICT-enabled intelligent manufacturing, more sustainable technologies and processes, and high-performance production. To guarantee flexibility and even total flexibility [20], the European industry needs to adopt new paradigms of production models, focusing on human–machine collaboration, more than what is currently done in fully automatized islands, where humans are out of the decision loop [21], to take benefit of workers abilities, fostering human skills and human motivation [22].

It is therefore necessary to move to a new concept of human-centred automation [23], and it is under this conceptual framework that the Mutualism Framework seeks for an effective symbiosis between human and robots, to overcome the challenges that the European manufacturing industry must face in the years to come:

- To promote value-adding non-repetitive non-alienating jobs in the manufacturing industry;
- Claiming for R&D investments in a wide plethora of knowledge fields;
- Making economically sustainable high-quality production processes that targets overall sustainability;
- Supporting lifelong learning in the shopfloor exploiting AVR;
- Promoting re-shoring of sustainable businesses.

This is clearly aligned with the Europe 2020 policy framework targets for Smart, Sustainable and Inclusive Growth, towards its five major objectives of Employment (75% of the 20–64 year olds to be employed), R&D (3% of the EU's GDP to be invested in R&D), climate change and energy sustainability (greenhouse gas emissions 20% lower than that in 1990; 20% of energy from renewables; 20% increase in energy efficiency), education (reducing the rates of early school leaving below 10%; at least 40% of 30–34 year olds completing third-level education) and fighting poverty and social exclusion (at least 20 million fewer people in or at risk of poverty and social exclusion).

As for different application domains, manufacturing is facing a bend where traditional production processes and work methods must evolve to be more flexible and adaptive to a quick changing context. Many manufacturing companies experience unpredictable and dynamic production environment due to increased customization in their production, low product life cycles and increased competition from low labour countries. To remain competitive in such globalized market, they must adapt their production systems accordingly and create flexible automatic solutions.

Adaptability in manufacturing can be defined as the ability of the production system to alter itself efficiently to changed production requirements. According to Järvenpää [24], manufacturing system adaptability can be achieved either statically (i.e.: while the system is not operating) or dynamically (while the system is running) and working on: (a) physical adaptation (of layout, machines and machine elements); (b) logical adaptation (re-routing, re-planning, re-scheduling and re-programming) and (c) parametric adaptation (changing machine settings).

To fulfil these requirements, reconfigurable manufacturing systems have been proposed as a set of possible solutions, "designed at the outset for rapid change in structure, as well as in hardware and software components, to quickly adjust production capacity and functionality within a part family in response to sudden changes in market or regulatory requirements" [25]. In these systems, human operators remain an invaluable resource, by being superior to robots at rapidly interpreting unplanned tasks and situations and handling flexibility and complexity.

Six core properties of reconfigurable manufacturing systems impact the overall time and cost of reconfiguration and for each of them the Mutualistic Framework based on IEC-61499 is capable of providing a concrete technological and methodological answer:

- **Scalability** (design for capacity changes): The dynamic creation of Mutualistic teams, thanks to the IEC-61499 orchestration layer, allows to manage rapid changes of production capacity by simply adding or removing new symbionts from the shopfloor event at runtime stage. Two major technological innovations guarantee this level of reconfigurability: (i) The IEC-61499 platform support to plug & produce, integrating functionalities of auto-discovery and auto-configuration of symbionts; (ii) The dynamic planning and scheduling of mutualistic tasks, looking for the best coupling of characteristics and skills of Symbionts, which guarantees to exploit the introduction (or removal) of one member of the team in the most appropriate way.

- **Convertibility** (design for functionality changes): the Mutualism Framework brings ease of programmability directly at the hands of human operators, thanks to its dedicated intuitive interfaces. To achieve a greater flexibility along with a more efficient production, traditional industrial robots are substituted by flexible and autonomous robotic systems with an intuitive on-the-fly programming, enabling an ideal batch size of 1. This translates into a faster and less costly management of the re-conversion of functionalities of robotic symbionts, even over the very short term. This aspect is further enhanced by machine learning capabilities of robots, which enables a direct adaptation (without programming actions) of the behaviour with respect to minor changes of requested functionalities.
- **Diagnosability** (design for easy diagnostics): An effect of the human–robot symbiosis is that the monitoring and inspection capabilities of robots can be directly used by human operators to augment their perception of the working environment and of the production process. Thanks to an appropriate AVR layer, it is possible to provide to human Symbionts contextual and dynamic information, leading to much better error diagnosis.
- **Customization** (flexibility limited to part family): This is one of the drivers most impacted by a more extensive usage of HRC, since it is where manufacturing activities are currently done mostly by human operators. Through its IEC-61499-based approach to orchestration, the framework guarantees a new degree of collaboration in executing joint tasks, especially those which require variations within the same part family. The flexible implementation of the automation logics for these mutualistic tasks considers, from design stage, how they will be executed by hybrid human–robot teams, where single behaviours will be dynamically adapted. This means that operators will be able to exploit their natural cognitive flexibility to modify the activity with respect to the single part, while robots will consequently adapt to this human-induced changed.
- **Modularity** (modular components): the whole concept of Symbionts is targeted towards an extreme modularization of functional units but guaranteeing their ease of interchange through an elevated degree of decentralized intelligence. While this is natural for human operators, the innovation of the Mutualistic Framework is to bring the same approach to robots and to then bring physical and functional modularity into a dynamic integration with the unique IEC-61499 orchestration layer.

This means proposing a systematic way to design its orchestration in a robustly reconfigurable way.

- **Integrability** (interfaces for rapid integration): this key driver of reconfigurability is where the interoperable and open nature of IEC-61499 maximizes its impact. The approach of proposing this integration and orchestration layer is a major enabler to implement real plug & produce functionalities for the intelligent symbionts of Mutualism. The 61499 platform acts both as abstraction layer with respect to the lower levels of automation and, at runtime stage, as manager of the dynamic integrability of new Symbionts into a specific mutualistic task.

8.6 Conclusions

During the coming decades, the whole European manufacturing sector will have to face important social challenges. In fact, while shop floor operators are usually considered a “special” community, with their job being considered one of the most disadvantaged in terms of workplace healthiness and safety [26], Europe’s issue of an ageing population will lead inevitable workers to postpone their retirement age. With this prospect, without a concrete solution, the European industry is condemned to lose qualified workers who are needed for manufacturing high-quality products, while national assistance frameworks of EU-27 will have to assist retired workers who need to be kept active to balance the new demographic distribution of population.

Through the Mutualism Framework based on the IEC-61499 platform of Daedalus, we answer back to the popular belief that automation wipes out many jobs and that is currently under the attack of many Industry 4.0 opposers. Indeed, recently, several experts have repeatedly proven this thesis as groundless, demonstrating the mutually virtuous coexistence of humans and machines interacting in industrial environments [27]. Even in advanced automated scenarios, where machine learning can support adaptation to variable and unpredictable situations, interaction with humans is still essential in the process of reacting to contextual information (thus, machines need workers). Contemporarily, automation encompasses not only repetitive tasks, but also sophisticated and high-performance functionalities that human’s senses and capacities are not keen to; moreover, machines could compensate human knowledge gaps, thus extending the opportunity to have actual support for junior workers (thus, workers can benefit from machines).

The deployment of these technologies may improve the mental and physical strain on human operators, reducing the number of injuries related to manufacturing work. In the medium term, this will also improve the perception of shopfloor workers about how their job negatively influences their health status (currently 40% [28]). This will be mirrored within the society, improving the general perception that population has about shopfloors and increasing social acceptance of this profession.

At the same time, the opportunity is to bring new skills to the role of the shopfloor worker, increasing the reputation of operators at social level. The new task typology that operators will have to perform will create jobs opportunities at shop floor level for more qualified profiles like technicians, increasing appealing for younger workers. This is in line with current FOF Roadmap 2020 to achieve sustainable and social acceptance of this sector and to strengthen the global position of the EU manufacturing Industry.

Finally, the implementation of Mutualism distributes dynamically tasks between operators and robots and coordinates their collaborative execution according to their strengths and weaknesses. This approach offers new job opportunities to people with disabilities, as the automation can overcome functional limitations, facilitating inclusion of this community.

Acknowledgements

The work hereby described was achieved within the EU-H2020 project DAEDALUS, which was funded by the European Union's Horizon 2020 research and innovation programme, under grant agreement No. 723248.

References

[1] Brown, A. S. "Worker-friendly robots", Mechanical Engineering 136.9, pp. 10–11, September 2014.

[2] Spring et al. Product customisation and manufacturing strategy. Int. J. of Operations & Production Management, 20(4), pp. 441–467, 2000.

[3] EFRA. Factories of the Future. Multi-annual roadmap for the contractual PPP under Horizon 2020, 2013.

[4] European Working Conditions Surveys, http://www.eurofound.europa.eu/surveys

[5] EUROSTATS: ec.europa.eu/eurostat/statistics-explained/index.php/Population_structure_and_ageing

[6] Buffington, J. The Future of Manufacturing: An End to Mass Production. In Frictionless Markets (pp. 49–65). Springer International Publishing, 2016.
[7] Sawaragi et al. Human-Robot collaboration: Technical issues from a viewpoint of human-centred automation. ISARC, pp. 388–393, 2006.
[8] MANufacturing through ergonoMic and safe Anthropocentric aDaptive workplacEs for context aware factories in EUROPE, FP7-2013-NMP-ICT-FOF(RTD), prj. ref.: 609073
[9] Krüger et al. Cooperation of Human and Machines in Assembly Lines. Annals of the CIRP, 58/2, pp. 628–646, 2009.
[10] Duan et al. Application of the Assembly Skill Transfer System in an Actual Cellular Manufacturing System. Autom. Sce & Eng., 9(1), pp. 31–41, 2012.
[11] Ong et al. Augmented reality applications in manufacturing: a survey, Int. J. of Production Research, 46 (10), pp. 2707–2742, 2008.
[12] Tzafestas, S.: Concerning Human-Automation Symbiosis in the Society and the Nature. Int'l. J. of Factory Automation, Robotics and Soft Computing, 1(3), pp. 16–24, 2006.
[13] Ferreira, P., Doltsinis, S. and N. Lohse, Symbiotic Assembly Systems – A New Paradigm, Procedia CIRP, vol. 17, pp. 26–31, 2014, ISSN 2212-8271, http://dx.doi.org/10.1016/j.procir.2014.01.066
[14] Krüger J. T., K. Lien, A. Verl. Cooperation of human and machines in assembly lines. CIRP Annals-Manufacturing Technology 58.2, 2009.
[15] Ferreira, P. and N. Lohse. Configuration Model for Evolvable Assembly Systems. in 4th CIRP Conference On Assembly Technologies, 2012.
[16] Brunner, S. G.; Steinmetz, F.; Belder, R. & Dömel, A. RAFCON: A Graphical Tool for Engineering Complex, Robotic Tasks Intelligent Robots and Systems, 2015. IROS 2015. IEEE/RSJ International Conference, 2016.
[17] Design, Execution and Post-Mortem Analysis of Prolonged Autonomous Robot Operations. Accepted for R-AL. Final reference will follow.
[18] J. P. Holdren, "A National strategic plan for advanced manufacturing," Report of the Interagency working group on Advanced Manufacturing IAM to the National Science and Technology Council NSTC Committee on Technology CoT, Executive Office of the President National Science and Technology Council, Washington, D.C., 2012.
[19] http://europa.eu/rapid/press-release_MEMO-14-193_en.htm

[20] Duguay et al. From mass production to flexible/agile production. Int. J.of Operations & Production Management, 17(12), pp. 1183–1195, 1997.
[21] Krüger, J., Lien, T. K., Verl, A., Cooperation of Human and Machines in Assembly Lines, Annals of the CIRP, 58/2, pp. 628–646, 2009.
[22] Sawaragi et al.. Human-Robot collaboration: Technical issues from a viewpoint of human-centred automation. ISARC, pp. 388–393, 2006.
[23] Billings Human-centered Aircfraft Automation Phylosophy, NASA Technical Memorandum 103885, NASA Ames Research Center, 1991.
[24] Eeva Järvenpää, Pasi Luostarinen, Minna Lanz and Reijo Tuokko. Adaptation of Manufacturing Systems in Dynamic Environment Based on Capability Description Method, Manufacturing System, Dr. Faieza Abdul Aziz (Ed.), ISBN: 978-953-51-0530-5, 2012.
[25] Koren, Y., Heisel, U., Jovane, F., Moriwaki, T., Pritschow, G., Ulsoy, G., et al. Reconfigurable manufacturing systems. CIRP Annals: Manufacturing Technology, 48(2), pp. 527–540, 1999.
[26] Occupational Safety and Health Administration, https://osha.europa.eu.
[27] Autor, D. H. "Why are there still so many jobs? The history and future of workplace automation." The Journal of Economic Perspectives 29.3, pp. 3–30, 2015.
[28] European Working Conditions Surveys, http://www.eurofound.europa.eu/surveys

PART II

9

Digital Models for Industrial Automation Platforms

Nikos Kefalakis, Aikaterini Roukounaki and John Soldatos

Kifisias 44 Ave., Marousi, GR15125, Greece
Email: nkef@ait.gr; arou@ait.gr; jsol@ait.gr

This chapter presents the role and uses of digital models in industrial automation applications of the Industry 4.0 era. Accordingly, it reviews a range of standard-based models for digital automation and their suitability for the tasks of plant modelling and configuration. Finally, the chapter introduces the digital models specified and used in the scope of the FAR-EDGE automation platform, towards supporting digital twins and system configuration use cases.

9.1 Introduction

The digital modelling of the physical world is one of the core concepts of the digitization of industry and the fourth industrial revolution (Industry 4.0). It foresees the development of digital representations of physical world objects and processes as a means of executing automation and control operations, based on digital operations functionalities (i.e. at the cyber rather than at the physical world) [1]. The motivation for this stems from the fact that digital world operations can be flexibly altered or even undone at a low cost, while this is impossible in the physical world. Hence, plant operators can experiment with operations over digital models, run what-if scenarios and ultimately derive optimal deployment configurations for automation operations, while also deploying them on the field based on IT applications and tools, such as Industrial Internet of Things (IIoT) tools.

The concept of simulating and experimenting with automation operations in the realm of digital models of the plant is conveniently called "digital twin" and is a key enabler of the digitization of industrial processes. One of the automation platforms developed by the co-authors of this book, namely the FAR-EDGE edge computing platform, takes advantage of this concept based on the integration of digital models that represent the manufacturing shopfloor, as well as other physical and logical components of the automation platform such as edge gateways. In particular, the FAR-EDGE reference architecture and platform design specify a set of digital models as an integral element of the FAR-EDGE automation platform. In line with the Industry 4.0 "digital twin" concept, these digital models serve several complementary and important objectives:

- **Digital Simulation:** FAR-EDGE implements digital twins in order to support digital simulations of the plant, including what-if scenarios. The latter can be evaluated and used to decide optimal configurations of automation elements.
- **Semantic Interoperability:** The FAR-EDGE digital models provide a uniform representation of the concepts and entities that comprise a FAR-EDGE deployment, which boosts semantic interoperability across diverse digital systems and physical devices. The use of common data model provides a uniform vocabulary for describing various entities (e.g. sensors, CPS devices, SCADA Supervisory Control and Data Acquisition systems, production systems) across different applications in the automation, analytics and simulation domains of the platform.
- **Information Exchange:** The digital models in FAR-EDGE provide the means for exchanging information across different FAR-EDGE deployments. This is closely related to the above-listed semantic interoperability objective: By exchanging information in a common agreed format, two or more different FAR-EDGE deployments can become interoperable despite differences in their internal implementation details.
- **System Configuration:** The design and deployment of digital models is a key prerequisite for performing automation and control operations at IT (Information Technology) timescales. As part of the digitization of industrial processes, automation systems (i.e. Operational Technology (OT)) can be configured through IT systems and tools. The latter configure and update digital models, which reflect the status of the physical world. In this way, automation and configuration operations are performed at the level of IT rather than at the level of OT (Operational

Technology). This requires a synchronization between digital models and the status of the physical world, which is challenging to implement.

This chapter provides insights into the digital models that are used to support information exchange, digital simulations, semantic interoperability and digital operations as part of the FAR-EDGE platform. It first analyses the rationale behind the specification and implementation of digital models in the FAR-EDGE platform, along with some of the main requirements that drive the specification of the models. These requirements include standards compliance, extensibility, high performance, as well as support of FAR-EDGE functionalities in the platform's simulation domain. Following the review of these requirements for the FAR-EDGE digital models, we present a number of standards-based models (i.e. digital models and schemas specified as part of Industry 4.0 standards) against their suitability in supporting these requirements. As part of this chapter, we highlight the suitability of AutomationML and the standards-based schemas that it comprises (e.g. CAEX) for the simulation functionalities of the FAR-EDGE platform. Accordingly, we introduce a range of new proprietary models that can represent FAR-EDGE deployment configurations, based on concepts that cover the platform's edge computing model to automation and distributed data analytics. Specifically, we introduce new digital models that reflect concepts specified and used as part of the FAR-EDGE RA and the edge computing infrastructure of the project, such as edge gateways, data channels, measurement devices, as well as live data streams. These concepts can be blended with AutomationML and CAEX concepts as a means of putting plant models (e.g. CAEX instances) in the context of FAR-EDGE edge computing deployments.

Another important part of the chapter is the presentation of the linking between the above-listed models for edge computing configurations with the AutomationML-based models used for digital twins and digital simulations as part of the platform. The presented methodology is based on well-known concepts from the areas of data models linking and interoperability, including the concept of common repositories and registries for data models interoperability.

This chapter is structured as follows:

- Section 9.2 following the chapter's introduction presents the rationale behind the use of digital models in Industry 4.0 in general and in FAR-EDGE in particular;
- Section 9.3 reviews a set of standards-based digital models, which are commonly used for plant modelling and representation;

- Section 9.4 introduces the proprietary FAR-EDGE data models that are used for configuring the distributed data analytics functionalities of the platform;
- Section 9.5 presents a methodology for linking the FAR-EDGE proprietary data models with standards-based data models used for digital twins' representations in the platform's simulation domain.
- Section 9.6 is the final and concluding section of the chapter.

9.2 Scope and Use of Digital Models for Automation

9.2.1 Scope of Digital Models

Industry 4.0 applications are based on Cyber-Physical Systems (CPS). One of their main characteristics is that they implement automation functionalities at the cyber layer of production systems. In this context, they also take advantage of digital models as a pool of schemas and functions that are used for the digital representation of the factory, including the synchronization of their digital properties with the status of the real-world entities that they represent. At a finer level of detail, the functionalities of digital models support the operations and features that are described in the following paragraphs.

9.2.2 Factory and Plant Information Modelling

Primarily, digital models enable modelling of information at the factory and plant levels. In particular, the models provide a digital representation of the factory, which includes information about the elements (e.g. systems, devices and people) that comprise the plant. Automation and analytic applications can access the models in order to obtain information about the configuration of the plant, which they can use for implementing and validating automation processes. For example, in the FAR-EDGE project, digital models provide information about the hierarchical relationships between physical and logical entities in the scope of a FAR-EDGE deployment such as the sensors and devices that are associated with a given station.

In principle, a detailed and exhaustive description of the plant facilitates the implementation of many different processes and applications, including automation and analytics, as well as enterprise processes. However, developing and maintaining a detailed and exhaustive representation of the plant is very challenging. Therefore, FAR-EDGE and other digital automation platforms model only a subset of the plant, according to a "mini-world" that pertains to target automation and analytics use cases. Nevertheless, the digital modelling process can be open and extensible, in order to provide opportunities for supporting a broader set of functionalities and use cases, based on a fair additional effort.

9.2.3 Automation and Analytics Processes Modelling

Beyond a static representation of the structure of a factory and a plant, digital models should be able to represent the more dynamic automation and analytics processes, which form part of the plant's dynamic behaviour. Such processes should be represented based on the elements that are entailed in each processes, including their relationships and their evolution over time. Again, instead of an exhaustive modelling and representation of all possible workflows (e.g. through appropriate state machines), most automation platforms (including the FAR-EDGE automation platform) tend to focus on the processes that comprise a set of target use cases.

9.2.4 Automation and Analytics Platforms Configuration

The modelling of automation and analytics processing provides also a basis for their configuration and reconfiguration, as a means of changing the automation or the analytics logic based on IT functions. For example, using the digital model for an analytics process, it is possible to configure the devices and other data sources entailed in analytics processes, as well as the analytics (e.g. machine learning) algorithms applied on their data. This can provide increased flexibility in configuring and deploying different automation and analytics workflows in a factory. It can also support the implementation of the popular "plug and produce" concept [2].

9.2.5 Cyber and Physical Worlds Synchronization

As already outlined, the digital models can enable the configuration of automation functions and workflows at IT rather than OT times. In particular, automation operations can be configured at the IT layer of a digital automation platform, while being reflected in the physical world. The idea behind this configuration approach is that dealing with IT functions is much easier and more flexible than dealing with OT technology.

In order to provide this IT layer flexibility, there is a need to reflect changes in the IT layer to the OT layer (i.e. to the field) and vice versa. Hence, mechanisms for synchronizing the status of the physical world with its digital representation are needed based on digital models.

The synchronization between the physical and digital worlds can be also used to improve the results of digital simulations based on the so-called digital twins. In particular, it allows digital simulation applications to operate not only based on simulated data, but also with real data stemming from the synchronization of the physical and digital worlds. This can facilitate more accurate and realistic simulations, given that part of them can rely on real data that are seamlessly blended in the simulation application. The development of such realistic simulations is therefore based on dynamic access to plant information, which is illustrated in the following paragraph.

9.2.6 Dynamic Access to Plant Information

Digital models facilitate dynamic access to the status of the plant, at the cyber layer of the digital factory automation systems. Access to such information is needed in order to identify the configuration of automation processes, as well as the status of production processes and KPIs (Key Performance Indicators). Hence, digital models can serve as a vehicle for representing dynamic up-to-date information about the field, in addition to static (or semi-static) metadata of the shopfloor. One prominent use of the dynamic access to plant information involves the use of real-life data in order to boost the performance and accuracy of digital simulations, as outlined in the previous paragraph.

It should be underlined that the digital models specify the schema used to model the structure of the plant information. In the scope of the digital platform's operation, this schema is populated with instance data, which reflect the status of the plant at a given time instant. Hence, dynamic access to plant information is based on querying the instance of the plant database, which will follow the structure of the digital models. The concept of dynamic access to plant information and the importance of the synchronization between the digital models and the actual status of the plant is presented in Figure 9.1.

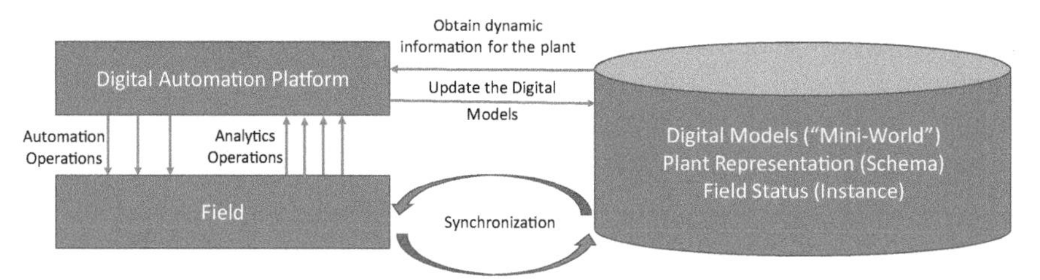

Figure 9.1 Digital Models and Dynamic Access to Plant Information.

9.3 Review of Standards Based Digital Models

In the following paragraphs, we provide a review of representative standards-based schemas that can be used for digital modelling in the scope of digital automation platforms. The review is by no means exhaustive, yet it covers some of the most popular models and schemas. Moreover, it provides some insights in terms of the ability of these standards to support the requirements and functionalities illustrated in the previous paragraph. Readers can also consult similar works on reviewing digital models (e.g. [3]), including works that have performed a comparative evaluation of alternative models [4].

9.3.1 Overview

For over a decade, various industrial standards have been developed, including information models that are used for information modelling in factory automation. Several standards come with a set of semantic definitions, which are typically used for modelling and exchanging data across systems and applications. These standards include, for example, the IEC 62264 standard that complies with the mainstream ISA-95 standard for factory automation. IEC 62264 boosts interoperability and integration across different/heterogeneous enterprise and control systems. Likewise, ISA-88 for batch processes comes with IEC 61512, and IEC 62424 supports exchange of data between process control and productions tools, while IEC 62714 covers engineering data of industrial automation systems [5]. Several of these standards are referenced and/or used by the RAMI 4.0 reference model [6], which is driving the design and development of several digital automation platforms. In the following paragraphs, we briefly describe some of these standards.

9.3.2 IEC 62264

IEC 62264 is a standard for enterprise-control system integration. It is based on the ANSI/ISA-95 hierarchy for automation systems. With reference to this hierarchy, the standard covers the domain of manufacturing operations management (i.e. Level 4) and the interface content and transactions within Level 3 and between Level 3 and Level 4. Hence, the standard is primarily focused on the integration between manufacturing operations and control, rather than on pure control (i.e. Levels 1, 2 and 3) operations only.

In practice, the standard defines activity models, function models and object models in the MOM (Manufacturing Operations Management) domain. The models are hierarchical and describe the MOM domain and its activities, the interface content and associated transactions within MoM level and between MoM and Enterprise level. Examples of entities that are modelled by the standard include materials, equipment, personnel, product definition, process segments, production schedules, product capabilities, production performance and more.

Note that IEC 62264 is among the standards referenced and used in RAMI 4.0. Due to its compliance with RAMI 4.0, IEC 62264 meets several of the requirements listed in the previous paragraph. However, it is focused on Level 3 and Level 4 entities of the ISA-95 standards and hence it is not very appropriate for use cases involving Levels 1, 2 and 3.

9.3.3 IEC 62769 (FDI)

The Information Model that is associated with the IEC 62769 (FDI) standard aims at reflecting the topology of the automation system. It therefore represents the devices of the automation system, as well as the communication networks that connect them. It includes attributes that are appropriate for modelling the main properties, relationships, operations of networks and field devices.

IEC 62769 is appropriate for modelling the field layer of the factory. This makes it appropriate for several of automation use cases, yet it does not provide the means for mapping and modelling some of the edge computing concepts of the FAR-EDGE automation platform (e.g. edge gateways and ledger services).

9.3.4 IEC 62453 (FDT)

IEC 62453 Field Device Tool (FDT) is an open standard for industrial automation integration of networks and devices. It provides standardized software to enable intelligent field devices that can be integrated seamlessly into automation applications, from the commissioning tool to the control system. FDT supports the coupling of software modules, which have been implemented as representatives for field devices and are therefore able to provide and/or exchange information. However, IEC 62453 is limited to the modelling of networks and devices and hence not suitable for plant-wide modelling.

9.3.5 IEC 61512 (Batch Control)

IEC 61512 – Batch control is also referenced by RAMI 4.0. It models batch production records, including information about production of batches or elements of batch production. IEC 61512 focuses on batch manufacturing and production processes.

9.3.6 IEC 61424 (CAEX)

IEC 61424 (CAEX) provides the means for modelling a plant in a hierarchical way and in an XML format (i.e. CAEX is provided as an XML Schema through an XML Schema language (XSD) file). CAEN abstracts a plant by considering it as a set of interconnected modules or components. CAEX models and stores such modules in an object-oriented way and based on object-oriented concepts such as classes, encapsulation, class libraries, instances, instance hierarchies, inheritance, relations, attributes and interfaces.

CAEX separates vendor-independent information (e.g. objects, attributes, interfaces, hierarchies, references, libraries, classes) and application-dependent information such as certain attribute names, specific classes or object catalogues. CAEX is appropriate for storing static metadata, but it is not designed to hold dynamic information. Nevertheless, it can be extended with special classes that could hold dynamic information and behaviour of the various modules.

IEC 61424 provides a sound basis for modelling the meta-data of a plant, which is one of the requirements for the digital models of an automation platform. However, there is also a need for supporting dynamic information as well, which asks for extensions to this model. CAEX is part of AutomationML compliant modelling, and as such, it is used in scope of FAR-EDGE in order to support the digital twins that are used from the simulation functionalities of the platform.

9.3.7 Business to Manufacturing Markup Language (B2MML)

B2MML is an XML implementation of the ANSI/ISA-95, Enterprise-Control System Integration, family of standards (ISA-95). As such, it is closely related to the above-listed IEC 62264 international standard, i.e. it provides a data representation that is fully compliant to the scope and semantics of IEC 62264. In practice, B2MML comprises a series of XML schemas, which are available as XML Schema language (XSD) files. Hence, B2MML supports the modelling of a large number of different entities, which represent MOM objects and transactions, as well as other interfaces between the enterprise and control layers.

B2MML is an excellent choice for supporting integration of business systems (such as Enterprise Resource Planning (ERP) and Supply Chain

Management (SCM) systems), with control systems (e.g. SCADA, DCS) and manufacturing execution systems (MES). This holds not only for B2MML compliant business systems (i.e. systems that support directly the interpretation of B2MML messages), but also for legacy ERP/SCM systems which can be made B2MML-compliant based on the implementation of relevant middleware adapters that transform B2MML to their own semantics and vice versa.

The language can be considered RAMI 4.0-compliant, given that RAMI 4.0 uses ISA-95 concepts and references of relevant standards (such as IEC 62264). It is also important that the B2MML schemas provide support for the entire ISA-95 standard, rather than a subset of it.

B2MML is characterized by compatibility with enterprise systems (e.g. ERP and PLM systems), which makes it appropriate for supporting information modelling for use cases involving enterprise-level entities and concepts. Furthermore, B2MML can boost compatibility with a wide range of available ISA-95-compliant systems, while at the same time adhering to information models referenced in RAMI 4.0. Therefore, B2MML could be exploited in the scope of use cases involving enterprises systems and entities, as soon as it is used in conjunction with additional models supporting concepts and entities for the configuration of an automation platform (e.g. like edge node, edge gateways and edge processes in the scope of an edge computing platform like FAR-EDGE).

9.3.8 AutomationML

AutomationML is an XML-based open standard, which provides the means for describing the components of a complex production environment. It has a hierarchical structure and is commonly used to facilitate consistent exchange and editing of plant layout data across heterogeneous engineering tools. AutomationML takes advantage of existing standards such as PLCopen XML or COLLADA. It provides the means for modelling plant information and automation processes based on objects structured in a hierarchical fashion, including information about geometric, model logic, behaviour sequences and I/O connections. AutomationML comprises different standards that support modelling for various entities and concerns. In particular, it relies on the following standards:

- CAEX (IEC 62424), in order to model topological information.
- COLLADA (ISO/PAS 17506) of the Khronos Group in order to model and implement geometry concepts and 3D information as well as Kinematics (i.e. the geometry of motion). Support for Kinematics ensures

also the modelling of connections and dependencies among objects as part of motion planning.
- PLCopen XML (IEC 61131) in order to model sequences of actions, internal behaviour of objects and I/O connections.

Note that AutomationML and the three above-listed standards are also in the list of Industry 4.0 standards that are directly connected to RAMI 4.0 in order to boost semantic interoperability.

AutomationML satisfies several of the requirements of the digital modelling requirements in FAR-EDGE and is appropriate for supporting digital simulations based on the development of a "digital twin" of the plant. It is therefore the standards-based digital model that supports plant modelling at the FAR-EDGE simulation domain. Moreover, the proprietary digital models that are used in FAR-EDGE can be linked to instances of AutomationML/CAEX digital models, towards ensuring uniqueness of the referenced entities and bridging of the diverse concepts that are captured by the two models. This is further discussed in Section 9.5.

9.4 FAR-EDGE Digital Models Outline

9.4.1 Scope of Digital Modelling in FAR-EDGE

In line with the uses of digital models that are described in Section 9.2, the FAR-EDGE digital automation platform leverages digital modelling for a dual purpose:

- **Data persistence and plant modelling for digital simulation and digital twins**. This is the reason why FAR-EDGE uses digital models for its simulation functionalities. The respective digital models are based on AutomationML, which has been described in the previous section.
- **Configuration of the FAR-EDGE platform, including configuration of its automation and analytics functionalities**. In particular, the FAR-EDGE platform holds digital presentations of the logical and physical configurations of FAR-EDGE components such as data sources, devices and edge gateways. The FAR-EDGE platform makes use of these configurations in order to configure its analytics and automation functionalities, based on functionalities such as the definition and configuration of data sources, association of these data sources to gateways and more. Specifically, the platform offers APIs and tools for manipulating these data models towards configuring the platform. The respective data models are proprietary and complement the use

of AutomationML in the simulation domain. The following paragraphs present briefly the data modelling entities used for the configuring data analytics in FAR-EDGE. The latter models come with an open-source implementation of functionalities for their management and are part of the FAR-EDGE platform. They are outlined in the following paragraphs.

9.4.2 Main Entities of Digital Models for Data Analytics

The proprietary FAR-EDGE data models that are used for configuring distributed data analytics functionalities, model factory data and metadata, along with the analytics functions and workflows that process them.

Factory Data and Metadata

The representation of factory data and metadata is based on the following entities:

- **Data Source Definition (DSD):** This defines the properties of a data source in the shopfloor, such as a data stream from a sensor or an automation device.
- **Data Interface Specification (DI):** The DI is associated with a data source and provides the information need to connect to it and access its data, including details like network protocol, port, network address and more.
- **Data Kind (DK):** This specifies the semantics of the data of the data source, which provides flexibility in modelling different types of data. The DK is an XML specification and hence it can be used to define virtually any type of data in an open and extensible way.
- **Data Source Manifest (DSM):** A DSM specifies a specific instance of a data source in line with its DSD, DI and DK specifications. Multiple manifests (i.e. DSMs) are therefore used to represent the data sources that are available in the factory in the scope of the FAR-EDGE automation platform.
- **Data Consumer Manifest (DCM):** This models an instance of a data consumer, i.e. any application that accesses a data sources.
- **Data Channel Descriptor (DCD):** A DCD models the association between an instance of a consumer and an instance of a data source. It is useful to keep track of the established connections and associations between data sources and data consumers.
- **LiveDataSet:** This entity models and represents the actual dataset that stem from an instance of a data source that is represented through

a DSM. Hence, it references a DSM, which drives the specification of the types of the attributes of the LiveDataSet in line with the DK. A LiveDataSet is associated with a timestamp and keeps track of the location of the data source in case it is associated with a mobile (rather than a stationary) edge node. Hence, it has a location attribute as well. In principle, the data source comprises a set of name–value pairs, which adhere to different data types in line with the DK of the DSM.

- **Edge Gateway:** This entity models an edge gateway of a FAR-EDGE edge computing deployment. In the scope of a FAR-EDGE deployment, data sources are associated with an edge gateway. This usually implies not only a logical association, but also a physical association, i.e. an edge gateway is deployed at a station and manages data sources in close physical proximity to the station.

Based on the above entities, it is possible to represent the different data sources of a digital shopfloor in a modular, dynamic and extensible way. This is based on a repository (i.e. registry) of data sources and their manifests, which keeps track of the various data sources that register to it. The FAR-EDGE platform includes such a registry, which provides dynamicity in creating, registering and using data sources in the industrial plant.

Factory Data Analytics Metadata

In order to facilitate the management and configuration of analytics functions and workflows over the various data sources, the FAR-EDGE digital models specify a number of analytics-related entities. In particular:

- **Analytics Processor Definition (APD):** This specifies a processing function to be applied on one or more data sources. In the scope of FAR-EDGE, three processing functions are defined, including functions that pre-process that data of a data source (i.e. Pre-Processors), functions that store the outcomes of the processing (i.e. Store Processors) and functions that analyse the data from the data sources (i.e. Analytics Processors). These three types of processors can be combined in various configurations over the data sources in order to define different analytics workflows.
- **Analytics Processor Manifest (APM):** This represents an instance of a processor that is defined through the APD. The instance specifies the type of processors and its actual logic through linking to a programming function. In the case of FAR-EDGE, the latter is a class/programme implemented in the Java language.

- **Analytics orchestrator Manifest (AM):** An AM represents an entire analytics workflow. It defines a combination of analytics processor instances (i.e. of APMs) that implements a distributed data analytics task. The latter is likely to span multiple edge gateways and to operate over their data sources.

9.4.3 Hierarchical Structure

The FAR-EDGE Digital Models for distributed data analytics follow a hierarchical structure, which defines the different relationships between the various entities. For example, an edge gateway comprises multiple data source manifests. Each one of the latter is associated with a data source definition. Likewise, LiveDataSets are associated with instances of data sources, i.e. data sources manifests. As an example, Figure 9.2 illustrates a snapshot of the FAR-EDGE digital models structure, which shows the association of each edge gateway with data source manifests and data analytics manifests. A more detailed presentation of the hierarchical structure of our data models is beyond the scope of this chapter. Interested readers can consult directly our XML schemas, which are part of our open source implementation of the

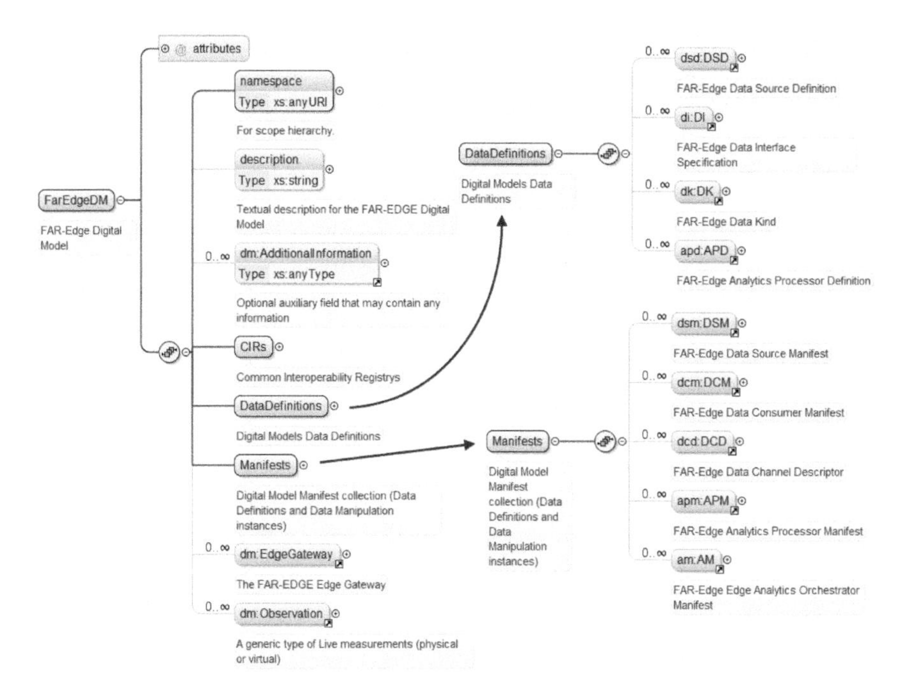

Figure 9.2 Snapshot of the FAR-EDGE Digital Models Structure.

FAR-EDGE digital models repository that is also an integral part of the FAR-EDGE platform.

9.4.4 Model Repository Open Source Implementation

As part of the open source implementation of the FAR-EDGE automation platform, we have implemented a data models repository, which provides support for the entities outlined in the previous paragraphs, including support for managing data kinds, data interfaces, data source definitions and analytics processor definitions. The implementation of the open source repository supports create, update, delete, get and discover functionalities, which are defined as follows:

- **Create:** This operation provides the means of creating an instance of the entity.
- **Update:** This allows updating an existing instance of the entity.
- **Delete:** This permits the deletion of an instance from the repository.
- **Get:** This fetches an instance of an entity based on its unique identifier.
- **Discover:** This helps model users to dynamically discover instances of one or more entities subject to given criteria.

The FAR-EDGE digital models repository implementation is available at the GitHub of the project at: https://github.com/far-edge/DigitalModels. The implementation comprises all schemata (i.e. see far-edge.dm.schemata) along with relevant ("generated") documentation in HTML (HyperText Markup Language) and PDF (Portable Data Format) formats. It also provides access to Java libraries, i.e. annotated libraries according to the JAXB (Java Architecture for XML Binding) framework in a proper Maven project (see far-edge.dm.commons). This open source implementation can provide a basis for researchers and engineers who might opt to implement their own digital models based on a similar approach. At the same time, they provide a means for implementing, using or even extending the FAR-EDGE analytics framework.

9.5 Simulation and Analytics Models Linking and Interoperability

The review of models in Section 9.3 justified the suitability of AutomationML for supporting the FAR-EDGE digital simulation functionalities. Furthermore, in Section 9.4, we introduced a digital model for representing and configuring the analytics functionalities of the FAR-EDGE platform. The use of a dedicated model for each of the two functional domains of the platform (i.e. analytics, simulation) provides flexibility to developers and deployers of analytics and simulation solutions, since they can use the model of their choice. Nevertheless, it could also create consistency and interoperability issues, especially in cases where functionalities and data from the two different domains need to be combined. To alleviate such problems, there is not only a need for linking entities in the two different models, so as to allow developers and deployers to access information for an entity in any of the two models, but also for combining information from the two models when needed. The merits of such linking become evident when considering the following examples:

- A digital simulation that needs to access information stemming from data analytics on real-life shopfloor data. For instance, a digital simulation may need to access maintenance-related parameters of a piece of equipment, following proper data analytics over sensor data (e.g. analytics vibration or ultrasound data for a machine). To this end, the machine representation in the simulation model (e.g. AutomationML) needs to be linked to the corresponding representation in the data model used for the distributed data analytics of the platform.
- Another digital simulation application that needs to analyse data sources using the distributed data analytics engine. In such a case, the simulation application needs to convey to the analytics engine the data sources to be used. To this end, there is a need for linking the representations of devices and data sources in the simulation domain, with the corresponding representations of the very same devices in the analytics domain.

In order to realize this linking, the FAR-EDGE data models include placeholders for data linking entities, i.e. linking of two representations of the same object/entity in different domains. In particular, both DSMs and logical entities in the simulation domain are linked based on a Universally Unique IDentifier (UUID). DSMs are assigned a UUID in an analytics domain whenever they are created and introduced to the system. Likewise, simulation applications assign a UUID to the main entities entailed in the simulation. The linking and harmonization of these UUIDs provide the means for linking the entities of two models.

This linking concept resembles to the concept of a Common Interoperability Registry (CIR), which is very commonly used in O&M (Operations and Maintenance). This registry is destined to provide "Yellow-Pages" lookup for all systems. This facilitates location of an object in any of the systems where it is registered, as soon as it is referenced with its UUID. Hence, different systems and models that have different identifiers for the very same entity or objects are glued together and are able talk "on-line". The main vehicle for this gluing is the specification and use of globally unique identifiers, which are linked to "local" object identifiers, i.e. identifiers pertaining to each one of the models.

9.6 Conclusions

This chapter has analysed the rationale behind the specification and integration of digital models in emerging digital automation platforms, which included a discussion of the main requirements that drive any relevant digital modelling effort. Moreover, it has presented a range of standards-based digital models, notably models that are used for semantic interoperability and information exchange in Industry 4.0 systems and applications. Following this review, it has illustrated why AutomationML is suitable for supporting the digital simulation functionalities of the FAR-EDGE platform.

The chapter has also introduced a proprietary model for representing and configuring the analytics part of the platform. This model provides the means for modelling and representing data sources and analytics workflows based on appropriate manifests. The respective models are implemented and persisted in a models repository, which is provided as a set of schemas and open source libraries as part of the FAR-EDGE digital automation platform. Hence, they can serve as a basis for using the FAR-EDGE digital models in analytics scenarios, as well as for implementing similar digital modelling ideas.

As part of this chapter, we have also outlined how globally unique identifiers can be used to link different models that refer to same entity or object in the factory based on their own local identifiers. The use of such global identifiers permits the association of entities referenced and used in both the AutomationML models of FAR-EDGE simulation and the FAR-EDGE models of the analytics engine. As part of our implementation roadmap, we also plan to implement a Common Interoperability Registry (CIR) that will keep track of all global identifiers and their mapping to local identifiers used by the digital models of the simulation, analytics and automation domains. This will strengthen the generality and versatility of our approach to digital model interoperability.

Overall, this chapter can be a good start for researchers and engineers who wish to start working with digital modelling and digital twins in Industry 4.0, as it presents the different use cases of digital models, along with the specification and implementation of a digital model for distributed data analytics in industrial plants.

Acknowledgements

This work was carried out in the scope of the FAR-EDGE project (H2020-703094). The authors acknowledge help and contributions from all partners of the project.

References

[1] H. Lasi, P. Fettke, H.-G. Kemper, T. Feld, M. Hoffmann, 'Industry 4.0', Business & Information Systems Engineering, vol. 6, no. 4, pp. 239, 2014.

[2] G. Di Orio, A. Rocha, L. Ribeiro, J. Barata, 'The prime semantic language: Plug and produce in standard-based manufacturing production systems', The International Conference on Flexible Automation and Intelligent Manufacturing (FAIM 2015), 23–26 June 2015.

[3] W. Lepuschitz, A. Lobato-Jimenez, E. Axinia, M. Merdan, 'A survey on standards and ontologies for process automation', in Industrial Applications of Holonic and Multi-Agent Systems, Springer, pp. 22–32, 2015

[4] R. S. Peres, M. Parreira-Rocha, A. D. Rocha, J. Barbosa, P. Leitão and J. Barata, 'Selection of a data exchange format for industry 4.0 manufacturing systems,' IECON 2016 - 42nd Annual Conference of the IEEE Industrial Electronics Society, Florence, pp. 5723–5728, doi: 10.1109/IECON.2016.7793750, 2016.

[5] 'IEC 62714 engineering data exchange format for use in industrial automation systems engineering - automation markup language - parts 1 and 2', in International Electrotechnical commission, pp. 2014–2015.

[6] K. Schweichhart, 'Reference Architectural Model Industrie 4.0 - An Introduction', Deutsche Telekom, April 2016 online resource: https://ec.europa.eu/futurium/en/system/files/ged/a2-schweichhart-reference_architectural_model_industrie_4.0_rami_4.0.pdf

10

Open Semantic Meta-model as a Cornerstone for the Design and Simulation of CPS-based Factories

Jan Wehrstedt[1], Diego Rovere[2], Paolo Pedrazzoli[3], Giovanni dal Maso[2], Torben Meyer[4], Veronika Brandstetter[1], Michele Ciavotta[5], Marco Macchi[6] and Elisa Negri[6]

[1]SIEMENS, Germany
[2]TTS srl, Italy
[3]Scuola Universitaria Professionale della Svizzera Italiana (SUPSI), The Institute of Systems and Technologies for Sustainable Production (ISTEPS), Galleria 2, Via Cantonale 2C, CH-6928 Manno, Switzerland
[4]VOLKSWAGEN, Germany
[5]Università degli Studi di Milano-Bicocca, Italy
[6]Politecnico di Milano, Milan, Italy
E-mail: janchristoph.wehrstedt@siemens.com; rovere@ttsnetwork.com; pedrazzoli@ttsnetwork.com; dalmaso@ttsnetwork.com; torben.meyer@volkswagen.de; veronika.brandstetter@siemens.com; michele.ciavotta@unimib.it; marco.macchi@polimi.it; elisa.negri@polimi.it

A key enabler towards the fourth industrial revolution is the ability to maintain the digital information all along the factory life cycle, despite changes in purpose and tools, allowing data to be enriched and used as needed for that specific phase (digital continuity). Indeed, a fundamental issue is the lack of common modelling languages, and rigorous semantics for describing interactions – physical and computational – across heterogeneous tools and systems, towards effective simulation. This chapter describes the definition of a semantic meta-model meant to describe the functional characteristics of a CPS, which are relevant from its design and simulation for its integration and coordination in an industrial production environment.

Actually, digital continuity needs to be empowered by a standardized, open semantic meta-model capable of fully describing the properties and functional characteristics of the CPS simulation models, as a key element to empower multidisciplinary simulation tools. The hereby described meta-model is able to provide a cross-tool representation of the different specific simulation models defining both static information (3D models, kinematics chains, multi-body physics skeletons, etc.) and behavioural information (observable properties, inverse kinematics processors, motion-low computation functions, resource consumption logics, etc.).

10.1 Introduction

In order to empower simulation methodologies and multidisciplinary tools for the design, engineering and management of CPS-based (Cyber Physical Systems) factories, we need to target the implementation of actual digital continuity, defined as the ability to maintain digital information all along the factory life cycle, despite changes in purpose and tools.

A Semantic Data Model for CPS representation is the foundation to achieve digital continuity, because it provides a unified description of the CPS-based simulation models that different simulation tools can rely on to operate.

Cyber Physical Systems are engineered systems that offer close interaction between cyber and physical components. CPS are defined as the systems that offer integrations of computation, networking, and physical processes, or in other words, as the systems where physical and software components are deeply intertwined, each operating on different spatial and temporal scales, exhibiting multiple and distinct behavioural modalities, and interacting with each other in a myriad of ways that change with context [2, 3]. From this definition, it is clear that the number and complexity of features that a CPS data model has to represent are very high, even if limited to the simulation field. Moreover, many of the aspects that concur to define a CPS for simulation (3D models, kinematics structures, dynamic behaviours, etc.) have been already investigated and formalized by many well-established data models that are, or can be considered, to all extents data exchange standards.

For these reasons, the goal of an effective CPS Semantic Data Model is providing a gluing infrastructure that refers existing interoperability standards and integrates them into a single extensible CPS definition. This approach reduces the burden on the simulation software applications to access the new data structures because they mainly add a meta-information level whereas data for specific purposes is still available in standard formats.

AutomationML [1] is a standard technology that is based on this "Integration philosophy" and defines the semantics of many elements of the manufacturing systems so that it is suitable to be adopted as the foundation of our CPS Semantic Data Model.

10.2 Adoption of AutomationML Standard

The meta-data model needs basis on which data is saved and processed. The goal of AutomationML is to interconnect engineering tools in their different disciplines, e.g. mechanical plant engineering, electrical design, process engineering, process control engineering, HMI development, PLC programming, robot programming, etc. It is a standard focused on data exchange in the domain of automation engineering, defined in four whitepapers that focus each on one of the following aspects:

1. Architecture and general requirements;
2. Role class libraries;
3. Geometry and kinematics;
4. Logic.

The data exchange format defined in these documents is the Automation Markup Language (AML), an XML schema-based data format and has been developed in order to support the data exchange in a heterogeneous engineering tools landscape for the production.

Engineering information is stored following the Object-Oriented Paradigm, and physical and logical plant components are modelled as data objects encapsulating different aspects. An object may consist of other sub-objects and may itself be part of a larger composition or aggregation. Typical objects in plant automation comprise information on topology, geometry, kinematics and logic, whereas logic comprises sequencing, behaviour and control. Therefore, an important focus in the data exchange in engineering is the exchange of object-oriented data structures, geometry, kinematics and logic.

AML combines existing industry data formats that are designed for the storage and exchange of different aspects of engineering information. These data formats are used on an "as-is" basis within their own specifications and are not branched for AML needs. The core of AML is the top-level data format CAEX that connects the different data formats (e.g. COLLADA for geometries or PLCOPEN-XML for logic). Therefore, AML has an inherent distributed document architecture. The goals and basic concepts of AutomationML are well aligned with our objectives, and it can be used as the base of the semantic meta data model; nonetheless, it is mainly a specification for data exchange only and it falls short when it comes to describe some operational aspects of simulation. For these reasons, we decided to extend AML aiming at targeting a more integrated connection between real/digital CPS and simulation tools.

10.3 Meta Data Model Reference

This chapter documents the *Meta Data Model* developed. It is organized into eight sections that correspond to the eight main semantic areas in which the data model is organized:

1. *Base Model (§10.3.1)*: documents low-level utility classes that are used for the definition of high-level classes of the other sections.
2. *Assets and Behaviours (§10.3.2)*: documents, classes and concepts related to the possibility of using external data sources to define additional resource models.
3. *Prototypes Model (§10.3.3)*: introduces the concepts of resource prototypes and resource instances that are at the basis of the model reuse paradigm and documents the classes defining the resource model prototypes.
4. *Resources Model (§10.3.4)*: documents all the classes related to representation of intelligent and passive resources constituting the model of a manufacturing plant.
5. *Device Model (§10.3.5)*: documents all the classes related to the representation of the data connection with the physical devices, including the definition of all the relevant I/O signals that are exchanged with the digital counterpart.
6. *Project Model (§10.3.6)*: documents all the classes that represent complex multi-disciplinary simulation projects and that enable simulation tools to share plant models and results.
7. *Product Routing Model (§10.3.7)*: documents all the classes related to the definition of a discrete product, of the manufacturing processes and of the production plans that should be used for plant simulation.
8. *Security Model (§10.3.8)*: documents the classes that are related to the access control and that define the authentication and authorization levels needed to work on a certain resource.

Each section is introduced with a diagram view (based on UML Class Diagram) that contains only the classes composing that specific data model area and their relationships with the main classes belonging to the other data model areas. Therefore, it is possible to find the same class representation (e.g. Property class) in many different diagrams, but each class is documented only once in the proper semantic section.

10.3.1 Base Model

This section documents some low-level and general-purpose classes that are shared by other higher-level models described in the following sections. In particular, the classes related to the possibility of modelling generic, simple and composite properties of plant resources are documented (Figure 10.1).

10.3.1.1 Property

Property is an abstract class derived by IdentifiedElement and represents runtime properties of every resource and prototype. These properties are relevant information that can be dynamically assigned and read by the simulation tools.

10.3.1.2 CompositeProperty

CompositeProperty is a class derived by Property and represents a composition of different properties of every resource and prototype. This composition is modelled to create a list of simple properties of the resource, or even a multilevel structure of CompositeProperty instances. Figure 10.2 shows a possible application of the base model classes to represent properties, meta information and documentation of a sample CPS. A resource (in this case, CPS4) can have many properties instances associated to it and these properties can be simple (as ToolLength, EnergyConsumption and TempCPS4) or composite that allow creating structured properties (CurrProd).

10.3.2 Assets and Behaviours

The goal of the CPS Semantic Data Model is providing a gluing infrastructure that refers existing interoperability standards and integrates them into a single extensible CPS definition. For this reason, the implemented model includes the mechanisms to reference external data sources (Figure 10.3).

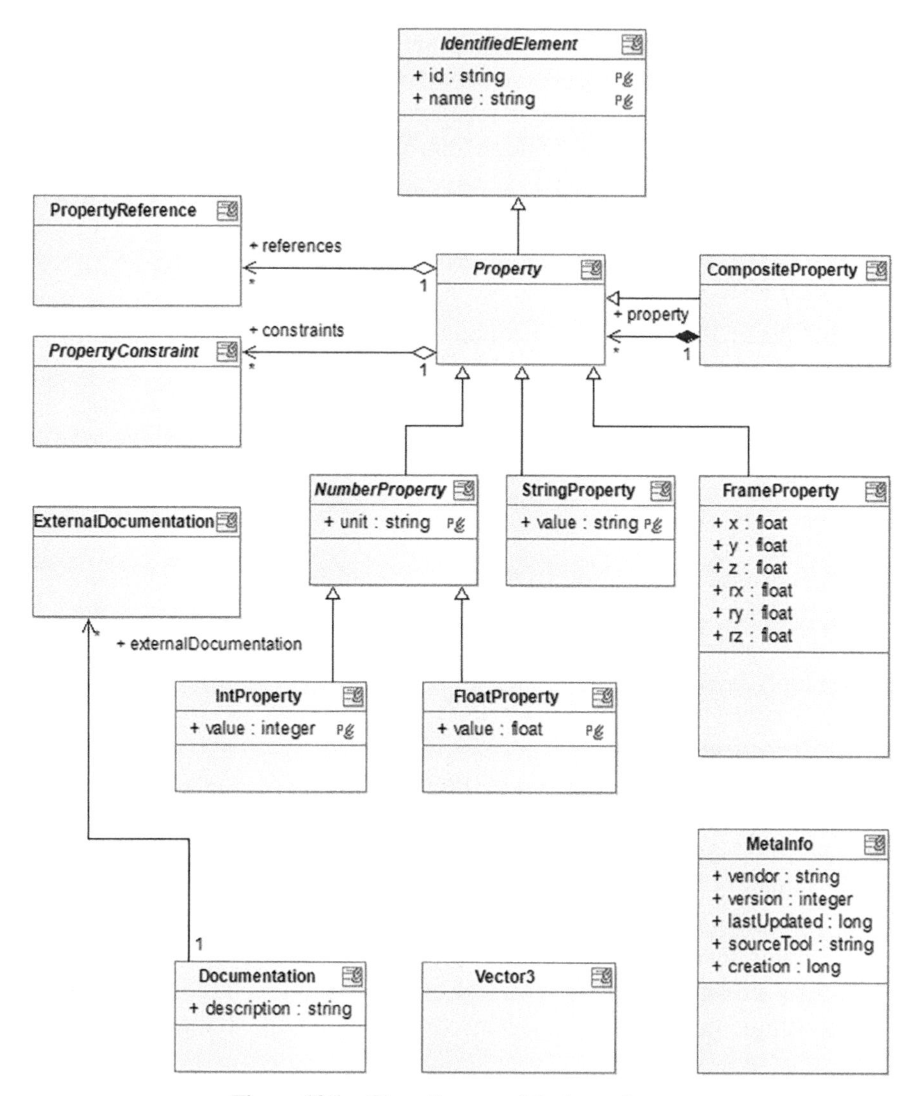

Figure 10.1 Class diagram of the base classes.

10.3.2.1 ExternalReference

ExternalReference is abstract and extends IdentifiedElement. This class represents a generic reference to a data source that is external to the *Meta Data Model* (e.g. a file stored on the Central Support Infrastructure (CSI, see Chapter 13)). The external source can contain any kind of binary

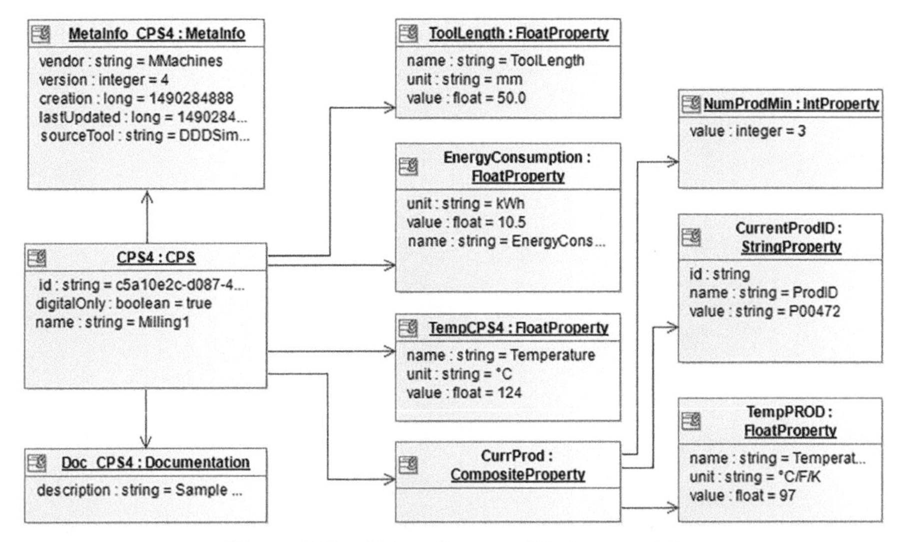

Figure 10.2 Object diagram of the base model.

data in proprietary or interoperable format, depending on the type of resource. Using external references allows avoiding re-defining data models and persistency formats for all the possible technical aspects related to a certain resource. The approach that has been adopted is like AutomationML one, where additional data is stored in external files using already existing standards (e.g. COLLADA for 3D models or PLCopen for PLC code).

10.3.2.2 Asset

Asset is an extension of ExternalResource. This class represents a reference to an external relevant model expressed according to interoperable standard or binary format that behavioural models want to use. An important feature that the CPS data model should support is the possibility to create links between runtime properties and properties defined inside assets and between properties defined by two different assets. Assets can be considered static data of the CPS because they represent self-contained models (e.g. 3D Models) that should be slowly changing.

10.3.2.3 Behaviour

Behaviour is an extension of ExternalResource. This class represents a reference to runnable behavioural models that implement: (i) functionalities and operative logics of the physical systems and (ii) raw data stream aggregation

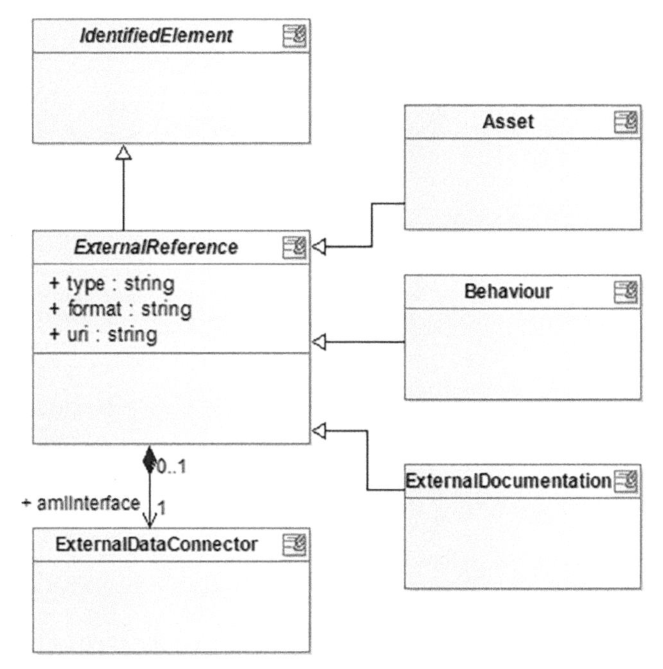

Figure 10.3 Class diagram for assets and behaviours.

and processing functions. Simulation Tools should be able to use directly the former to improve reliability of simulations, whereas the latter should run inside the CSI to update the runtime properties of the CPS model.

10.3.3 Prototypes Model

This section is meant to describe the classes and concepts related to the definition of prototype resources that can be defined once and reused many times to create different plant models.

10.3.3.1 Prototypes and instances

One of the most exploited features of manufacturing plants is the fact that they are mostly composed of standard “off-the-shelf” components (machine tools, robots, etc.) that are composed in a modular way. Thanks to this and with a good organization of modules, in fact, it is possible to speed up the simulation set up, reusing as much as possible already developed models. For this reason, usually simulation software tools adopt a mechanism based on the definition of libraries of models that can be applied to assemble a full plant layout.

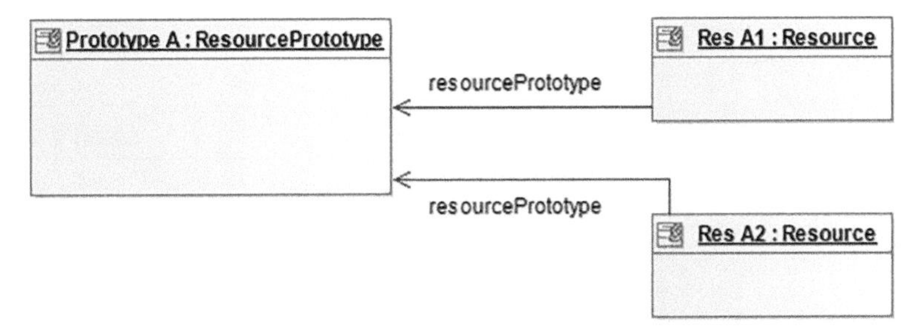

Figure 10.4 Prototype-resource object diagram.

The data model aims at natively supporting the same efficient re-use approach implementing the classes to describe "ready to use" resources, called "prototypes" and "instances" of such elements that are the actual resources composing plants. The relationship that exists between prototypes and instances is the same that in OOP exists between a class and an object (instance) of that class.

A prototype is a Resource model that is complete from a digital point of view, but it is still not applied in any plant model. It contains all the relevant information, assets and behaviours that simulation tools may want to use and, ideally, device manufacturers should directly provide Prototypes of their products ready to be assembled into production line models.

As shown in Figure 10.4, a Resource instance is a ResourcePrototype that has become a well-identified, specific resource of the manufacturing plant. Each instance shares with its originating Prototype the initial definition, but during life cycle, its model can diverge from the initial one because properties and even models change. Therefore, a single ResourcePrototype can be used to instantiate many specific resources that share the same original model.

10.3.3.2 Prototypes and instances aggregation patterns

An important aspect that *Meta Data Model* defines is the one related to the composition of resources into higher-level resources. This concept is at the basis of the creation of a hierarchy of resources within a plant and it is an intrinsic way of organizing the description of a manufacturing system. Nevertheless, depending on each specific discipline, there are many ways resource instances (and therefore CPSs) can be grouped in a hierarchical structure. For example, spatial relationships define the topological hierarchy of a system, but from a safety grouping or electrical perspective, the same resources should be organized into different hierarchies (e.g. in the

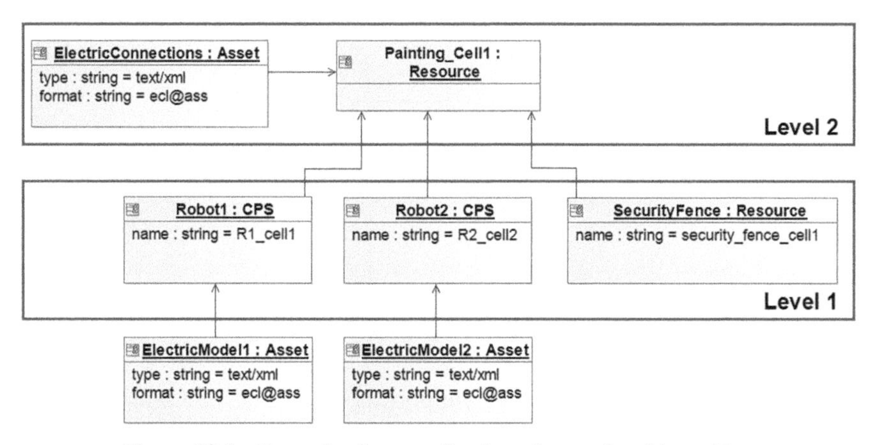

Figure 10.5 Example of usage of main and secondary hierarchies.

automotive, a cell safety group contains the robot and the surrounding fences, but from an electrical point of view, fences are not represented at all).

For this reason, *Meta Data Model* provides an aggregation system that is based on two levels:

- a first main hierarchy structure that is implemented in the two base classes for prototypes and instances, AbstractResourcePrototype and AbstractResource (Figure 10.6);
- a second level, discipline-dependent, that is defined in parallel to the main one and that should be contained inside domain-specific Assets.

The former hierarchy level is meant to provide a reference organization of the plant that enables both simulation tools and the CSI to access resources in a uniform way. In fact, the main hierarchy has the fundamental role of controlling the "visibility level" of resources, setting the lower access boundaries that constrain the resources to which the secondary ("parallel") hierarchies should be associated.

Figure 10.5 shows an example of application of the main resources hierarchy and the secondary, domain-specific one. The main hierarchy organizes the two robots and the surrounding security fence with a natural logical grouping since Robot1, Robot2 and SecurityFence belong physically to the same production cell, Painting_Cell1. Even if this arrangement of the instances is functional from a management point of view, it is not directly corresponding to the relationships defined in the electrical schema of the plant, for which the only meaningful resources are the two robots. Imagining that an electric connection exists between the two robots, a secondary, domain-specific schema (in this case, the domain is the electric design) needs to be defined separately. The Painting_Cell1 resource acts as the aggregator of the two robot CPS; therefore, it has the "visibility" on the two resources of the lower level (Level 1), meaning that they exist and it knows how to reference them. For this reason, the electrical schema that connects Robot1 and Robot2 is defined at Level 2 as the "ElectricConnections" Asset associated to the Painting_Cell1. This asset, if needed, is allowed to make references to each electric schema of the lower-level resources.

10.3.3.3 AbstractResourcePrototype

AbstractResourcePrototype is abstract and extends IdentifiedElement (see Figure 10.6). It represents the base class containing attributes and relationships that are common both to prototypes of intelligent devices and to prototypes of simple passive resources or aggregation of prototypes. The main difference between prototype and instance classes is that the former does not have any reference to a Plant model, because they represent "not-applied" elements.

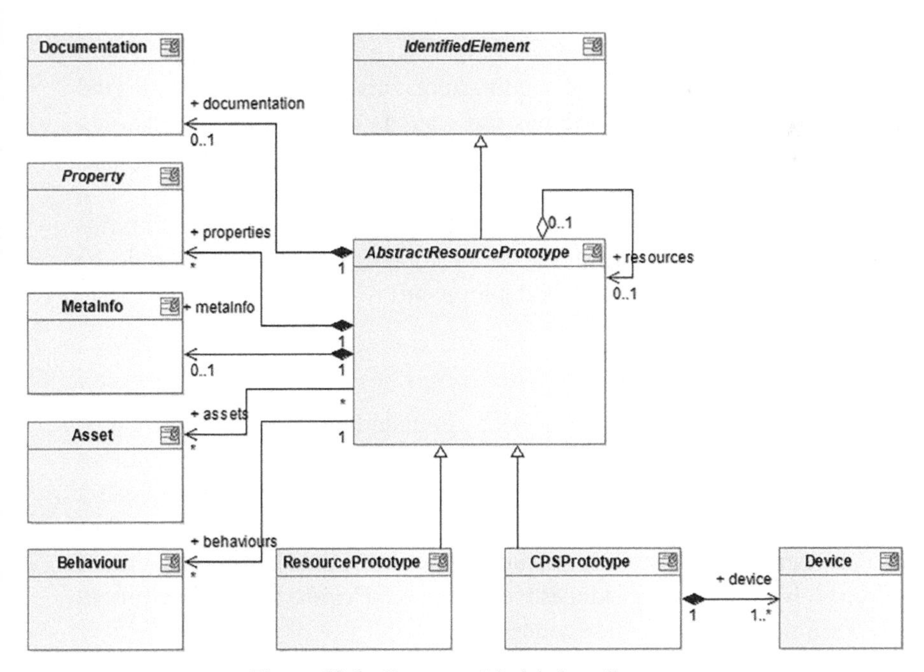

Figure 10.6 Prototype Model class diagram.

Each AbstractResourcePrototype can aggregate other AbstractResource-Prototype (i.e. CPSPrototype and ResourcePrototype instances), and it can use its Assets and Behaviours to create higher-level complex models and functionalities starting from the lower-level ones.

10.3.3.4 ResourcePrototype

ResourcePrototype extends AbstractResourcePrototype. This class represents the prototype of a generic passive resource of the plant that does not have any electronic equipment capable of sending/receiving data to/from its digital counterpart, or an aggregation of multiple resource prototypes. Examples of simple resources are cell protection fences, part positioning fixtures, etc.

Resource class is the direct instance class of a ResourcePrototype.

Since a ResourcePrototype must be identifiable within the libraries of prototypes, its ID attribute should be set to a valid UUID that should be unique within an overall framework deployment.

10.3.3.5 CPSPrototype

CPSPrototype extends AbstractResourcePrototype. This class represents a prototype of an “intelligent” resource that is a resource equipped with an electronic device, capable of sending/receiving data to/from its digital counterpart. A CPSPrototype defines the way its derived instances should connect to the physical devices to maintain synchronization between shop floor and simulation models. CPS class is the direct instance class of a CPSPrototype. Since a CPSPrototype must be identifiable within the libraries of prototypes, its ID attribute should be set to a valid UUID that should be unique within an overall framework deployment.

10.3.4 Resources Model

From *Meta Data Model* perspective, each simulated plant can be represented as a bunch of resources (machine tools, robots, handling systems, passive elements, etc.). Each resource can have a real physics counterpart to which it can be connected or defined from a product life cycle management point of view. This section of the model is meant to document the classes that support the description of resource instances (see §**10.3.4.1 Prototypes and instances** for the definition of the instance concept).

10.3.4.1 AbstractResource

AbstractResource is abstract and extends IdentifiedElement (Figure 10.7). This class represents the generalization of the concept of plant resource. As cited at the beginning of the section, a plant is a composition of intelligent devices (e.g. machines controlled by PLC, IoT ready sensors, etc.) or passive elements (fences, fixtures, etc.). Even if such resources are semantically different, from a simulation point of view, they have a certain number of common properties. This fact justifies, from a class hierarchy perspective, the definition of a base class that CPS and Resource classes extend.

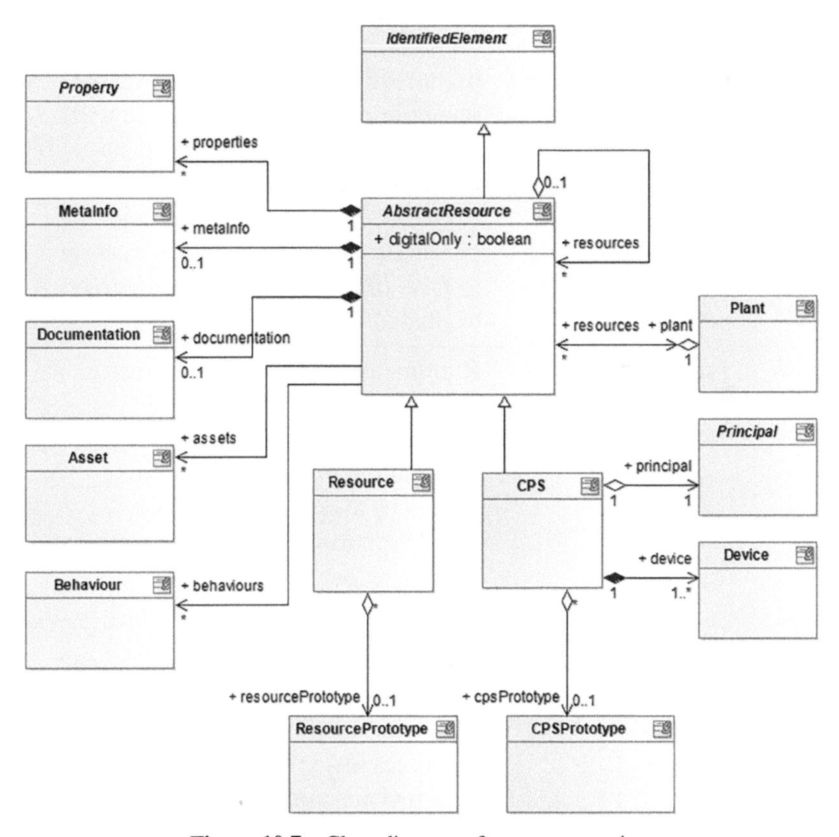

Figure 10.7 Class diagram of resources section.

An AbstractResource is identified by its ID, which must be unique within the same plant.

Field	Type	Description
digitalOnly	Boolean	This flag indicates whether this resource (be it a CPS or a simple resource) has a physical counterpart somewhere in the real plant or if it is purely a virtual element. In design phase of a plant that goes on green field, resources will all have digitalOnly = true, while during the reconfiguration of a plant, there will be a mixed condition with some resources having the flag set to true (the ones existing in the running production lines) and some others set to false (the ones that are going to be evaluated with simulation).
properties	Property[]	Runtime properties of the resource. Each property of the resource represents a relevant piece of information that can be shared (accessed and modified) by the simulation tools and by the functional and behavioural models. The length of the array can be 0 to n.
resources	AbstractResource[]	List of the resources that this instance aggregates. This field implements the hierarchy relationships among resources inside a plant. See §**Prototypes and instances aggregation patterns**. The length of the array can be 0 to n.

10.3.4.2 CPS

CPS extends AbstractResource. This class represents each "intelligent" device belonging to the plant equipped with an electronic device capable of sending/receiving data to/from its digital counterpart. A CPS can be connected with the physical device to maintain synchronization between shopfloor and simulation models. A CPS can be an aggregation of other CPSs and simple Resources, using its Assets and Behaviours to aggregate lower-level models and functionalities.

Each CPS must be identified by a string ID that must be unique within the plant.

Field	Type	Description
cps-Prototype	CPS-Prototype	Each CPS can be an instantiation of a prototype CPS that has been defined in a library of models (usually stored in the CSI) that simulation tools can access and use. See §**10.3.4.1 Prototypes and instances**. This field can be null if the CPS does not derive from the instantiation of a prototype.
device	Device	Represents the description of the device that ensures the data connection between the physical and digital contexts. This object characterizes all the I/Os that can be received and sent from and to the real equipment. This field cannot be null, while it is possible that the device, even if fully defined, is not connected to real electronic equipment.
principal	Principal	Each CPS has a related access level that is defined in compliance with the security data model described in section "Security Data Model" and implemented by the CSI.

10.3.5 Device Model

This section contains the documentation of the classes needed to model the electronic equipment of the intelligent resources. This equipment is described in terms of the interfaces that can be used by the digital tools to open data streams with the real devices (Figure 10.8).

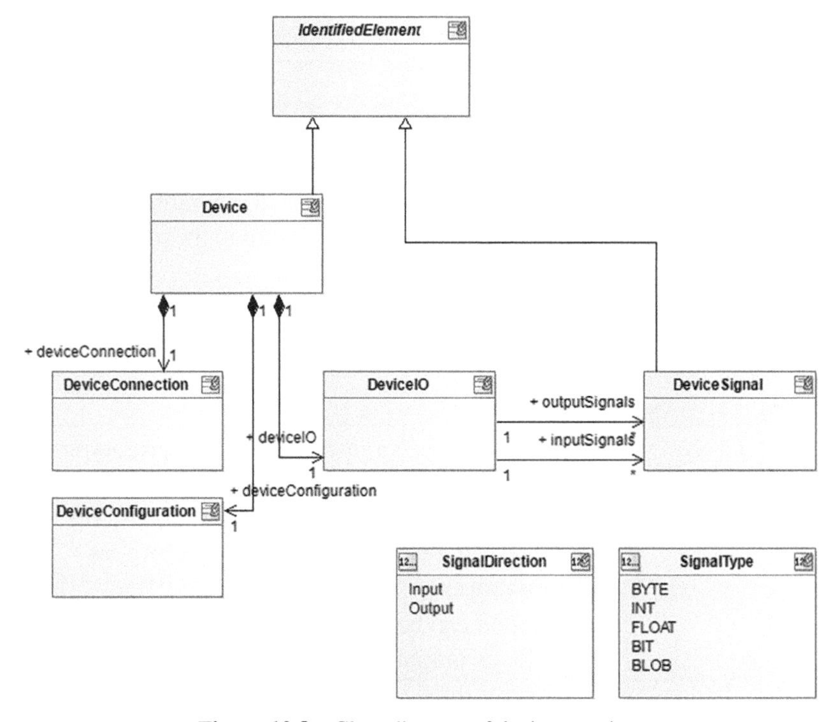

Figure 10.8 Class diagram of devices section.

10.3.5.1 Device

Device is an IdentifiedElement and represents an electronic equipment of physical layer that can be connected to the digital counterpart to send/receive data.

Field	Type	Description
device-Connection	Device-Connection	It contains all the details to open data streams with the physical device. E.g. for Ethernet-based connections, it contains IP address as well as information on ports, protocols and possibly the security parameters to apply to receive access rights to the specific resource. The field can be null.
device-Configuration	Device-Configuration	It contains details on the device hardware and software configuration (e.g. version of the running PLC code). This object can be updated dynamically based on data read from the physical

Field	Type	Description
		device to reflect the actual working condition of the device. The field can be null.
deviceIO	DeviceIO	It contains the map of Input/Output data signals that can be exchanged with the physical device. The field cannot be null. If no signal can be exchanged with the device, the DeviceIO map is present but empty. Normally, this should not happen (except during the drafting phase) because if a device does not allow any data exchange with its digital counterpart, then it should be treated as a passive resource.

10.3.5.2 DeviceIO

DeviceIO represents a map of input and output signals that can be exchanged with a specific device. Moreover, the DeviceIO represents the communication between CPS on IO-Level.

Field	Type	Description
input-Signals	Device-Signal[]	Array of DeviceSignal describing input signals. Signal direction is seen by the device; therefore, this is the list of data that can be sent TO the device. The field cannot be null. Length of the array can be 0 to n All DeviceSignal instances belonging to this collection must have *direction* attribute set to *SignalDirection.Input.*
output-Signals	Device-Signal[]	Array of DeviceSignal describing output signals. Signal direction is seen by the device; therefore, this is the list of data that can be received FROM the device. The field cannot be null. Length of the array can be 0 to n All DeviceSignal instances belonging to this collection must have *direction* attribute set to *SignalDirection.Output.*

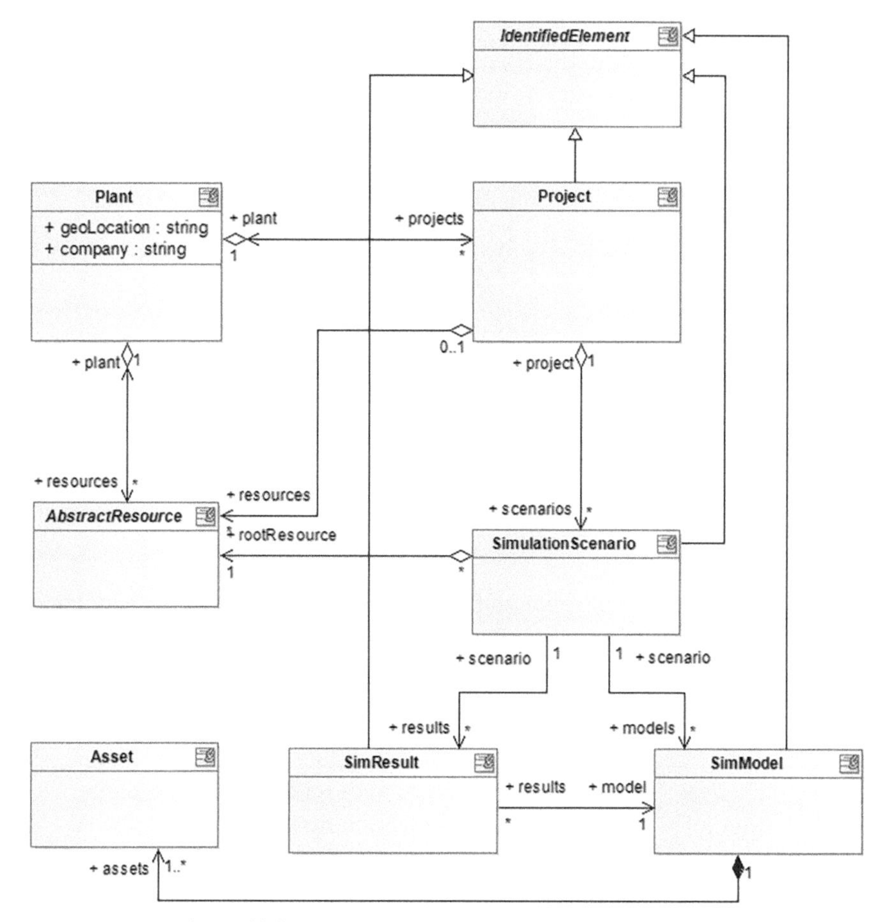

Figure 10.9 Class diagram of the Project Model section.

10.3.6 Project Model

This section describes the classes related to the management of projects, scenarios and results of simulations for a certain plant that are produced and consumed by simulation tools (Figure 10.9).

10.3.6.1 Project

A project is an IdentifiedElement. It can be considered mainly as a utility container of different simulation scenarios that have been grouped together because they are related to the same part of the plant (e.g. different scenarios for the same painting cell of the production line).

A project could identify a design or a reconfiguration of a part of the plant for which each SimulationScenario represents a hypothesis of layout of different resources.

10.3.6.2 Plant

Plant is an extension of IdentifiedElement and represents an aggregation of projects and resources. A plant instance could be considered as an entry point for simulation tools that want to access models stored on the CSI. It contains references to all the resource instances that are subject of SimulationScenarios. In this way, it is possible to have different simulation scenarios, even with simulation of different types, bound to a single resource instance.

Note: the fact that different simulations of different nature can be set up for the same resource (be it a cell, a line, etc.) is not related to the concept of multi-disciplinary simulation that is, instead, implemented by the Simulation Framework and refers to the possibility of running concurrent, interdependent simulations of different types.

The ID of the Plant must be unique within the overall framework deployment.

10.3.6.3 SimulationScenario

SimulationScenario is an extension of IdentifiedElement and represents the run of a SimModel producing some SimResults. A simulation scenario refers to a root resource that is not necessarily the root resource instance of the whole plant, because a simulation scenario can be bound to just a small part of the full plant. A simulation scenario can set up a multi-disciplinary simulation, defining different simulation models for the same resource instance to be run concurrently by the Simulation Framework.

10.3.6.4 SimModel

SimModel is an IdentifiedElement and represents a simulation model within a particular SimulationScenario. Each model can assemble different behavioural models of the root resource into a specific simulation model, creating scenario-specific relationships that are stored inside simulation assets that can be expressed both in an interoperable format (e.g. AutomationML) when there is need for data exchange among different tools and in proprietary formats.

The ID of a SimModel instance must be unique within a Simulation Scenario.

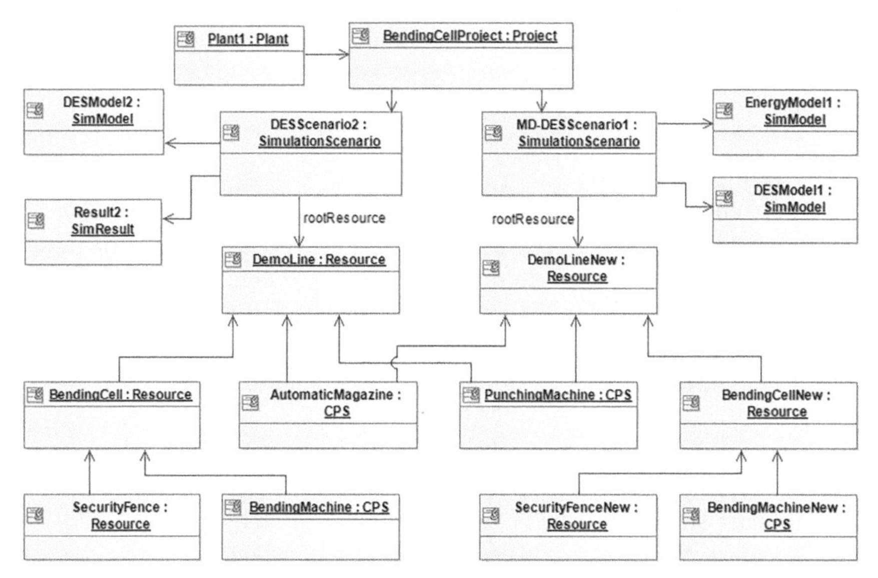

Figure 10.10 Object diagram of the Project Model.

The object diagram shown below (Figure 10.10) shows a possible application of the Project Model: a set of simple resources and CPS is organized into two hierarchies: one representing the actual demo line and a second hierarchy modelling a hypothesis of redesign of the demo plant. All the Resource and CPS instances belong to the plant model Plant1 (relationships in this case have not been reported to keep the diagram tidy). The user wants to perform two different simulations, one for each root resource. For this reason, he/she sets up two SimulationScenario instances: MD-DESScenario1 and DESScenario2. Each one refers to a different root resource. The former is a multi-disciplinary scenario of the DemoPlantNew that will use a combination of a DES model and an Energy Consumption model, while the latter represents a simple DES-only scenario of the original DemoPlant. These scenarios are aggregated in a Project instance (BendingCellProject) that belongs to the Plant1 project and that is meant to compare the performance of the plant using two different setups of the bending cell. For DESScenario2, there are already simulation results Result2.

10.3.7 Product Routing Model

In this paragraph, a description of the product routing section of the meta data model is given. Structural choices as well as requirements consideration are reported, with a particular focus on the validation points that have been reviewed by experts. In order to describe this part of the model, each class is treated separately and clusters of functional areas have been created for simplicity. All attributes, cardinality indications and relationships are described with respect to the single entity and in the general data model perspective.

10.3.7.1 Relationship between product routing model and ISO 14649-10 standard

The product routing section of the data model has been developed according to the ISO 14649-10 standard, "Industrial automation systems and integration – Physical device control – Data model for computerized numerical controllers – Part 10: General process data", which was deeply analysed and chosen as best-fitting standard for the product feature – operation coupling part. Its characteristics and focus areas are suitable from the functional point of view, as it tackles some aspects that the model needs to cover in exactly the same application environment. In fact, it supports the communication between CAD and CNC. ISO 14649-10 specifies the process data that is generally needed for NC programming in any of the possible machining technologies. These data elements describe the interface between a computerized numerical controller and the programming system (i.e. CAM system or shopfloor programming system). On the programming system, the programme for the numerical controller is created. This programme includes geometric and technological information. It can be described using this part of ISO 14649 together with the technology-specific parts (ISO 14649-11, etc.). This part of ISO 14649 provides the control structures for the sequence of programme execution, mainly the sequence of working steps and associated machine functions.[1] The standard ISO 14649-10 gives a set of terms and a certain hierarchy among them, though without specifying the type of relations. Being focused on process data for CNC (Computerized Numerical Control), the terminology is deeply technical in describing all different types of manufacturing features, mechanical parameters and measures. The relationship between workpiece features, operations and sequencing is of relevance for the purpose of this work, so a number of entities have been selected. Only after that, the distinction between classes and attributes was made, together with the definition of the types of relationships and references among the classes.

[1] ISO 14649. http://www.iso.org/iso/catalogue_detail?csnumber=34743

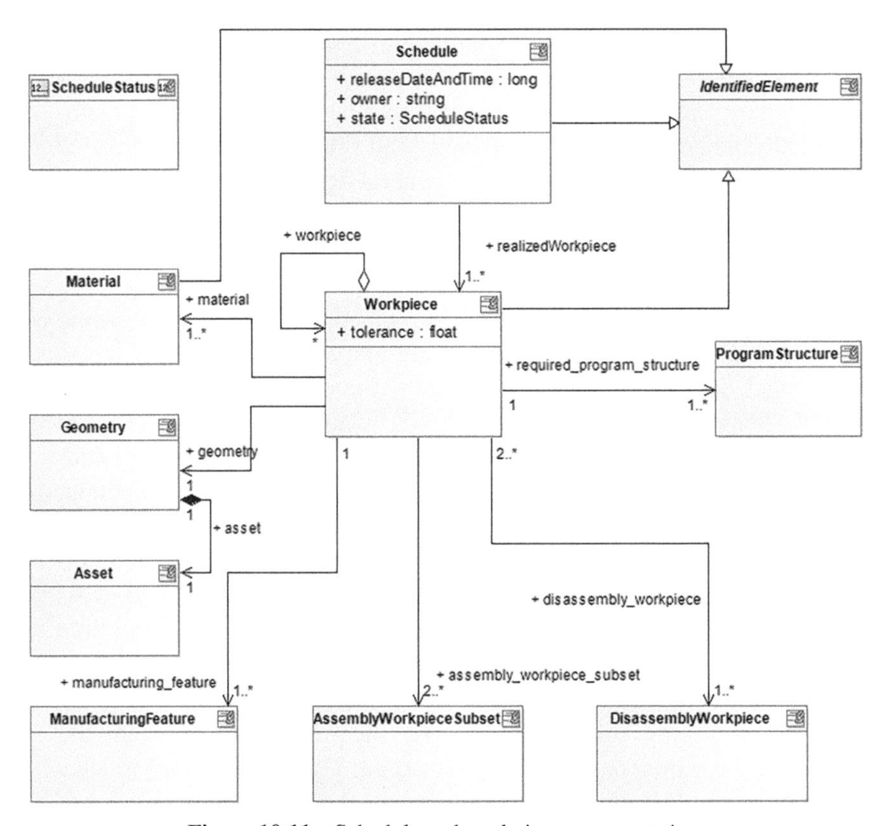

Figure 10.11 Schedule and workpiece representation.

10.3.7.2 Workpiece

Workpiece class (Figure 10.11) represents the part or product that needs to be machined, assembled or disassembled. Each schedule realizes at least one workpiece, but it may also realize different product variants, with various features. Each product variant is a different instantiation of the class "Workpiece" and extends the IdentifiedElement class. Being a central entity for the data model, the workpiece has a further development side that concerns the production scheduling and product routing. Manufacturing methods and instructions are not contained in the workpiece information but are determined by the operations themselves.

10.3.7.3 ProgramStructure

ProgramStructure determines how the different operations are executed for a specific work piece, i.e. in series or parallel (see also Figure 10.12).

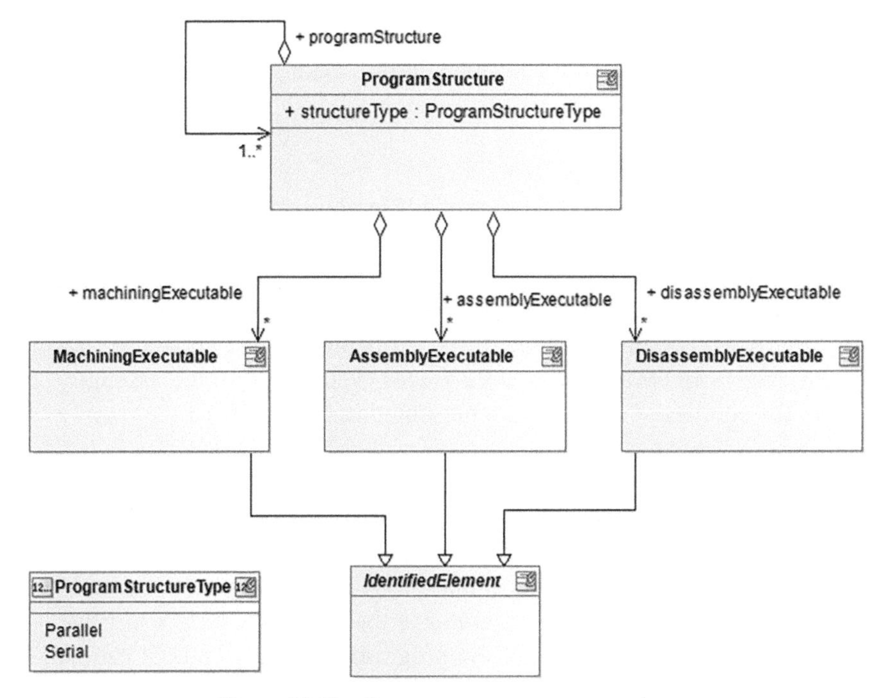

Figure 10.12 Program structure representation.

A program structure, at low level, is composed of single, ordered steps, called "Executables". Depending on the type of program structure, the executables are realized in series or parallel. The program structure thus defines how the different steps are executed and at the same time gives some flexibility in the choice, by taking into account data from the system.

10.3.7.4 ProgramStructureType

Enumeration representing the allowed types of a ProgramStructure instance (Figure 10.12).

10.3.7.5 MachiningExecutable

Machining executables initiate actions on a machine and need to be arranged in a defined order. They define all those tasks that cause a physical transformation of the workpiece. MachiningExecutable class extends the IdentifiedElements class and is a generalization of machining working steps and machining NC functions, since both of these are special types of machining executables. Hierarchically, it is also a sub-class of program structures, being their basic units, as it constitutes the steps needed for the execution of the program structure. Starting from the machining executable, the connected classes are represented in Figure 10.12.

10.3.7.6 AssemblyExecutable

AssemblyExecutable also extends IdentifiedElement class. AssemblyExecutable are a specialization of program structures and generalizations of working steps or NC functions. As in the case of machining executables, they initiate actions on a machine and need to be arranged in a defined order: assembly executables include all those operations that allow creating a single product from two or more work pieces. Starting from the assembly executable, the connected classes are represented in Figure 10.12.

10.3.7.7 DisassemblyExecutable

DisassemblyExecutable is derived from IdentifiedElement. DisassemblyExecutables are generalizations of working steps or NC functions. As in the case of machining and assembly executables, they are also a specialization of program structures, being their basic units, as these three classes constitute the steps needed for the execution of the program structure. Thus, it can be imagined that one or more machining executables, one or more assembly executables and one or more disassembly executable compose program structure. Disassembly executables also initiate actions on a machine and need to be arranged in a defined order: disassembly executables perform an opposite activity with respect to assembly, which means that from a single part it extrapolates more than one part. Starting from the disassembly executable, the connected classes are represented in Figure 10.12.

10.3.7.8 MachiningNcFunction

MachiningNcFunction is an IdentifiedElement and a specialization of MachiningExecutable (Figure 10.13) that differentiates from the machining working step for the fact that it is a technology-independent action, such as a handling or picking operation or rapid movements. It has a specific purpose and given parameters. If needed, other parameters regarding speed or other technological requirements can be added as attributes.

10.3.7.9 MachiningWorkingStep

MachiningWorkingStep is an IdentifiedElement that is also a specialization of MachiningExecutable, the most important one for the purpose of this work. It is the machining process for a certain area of the workpiece, and as such,

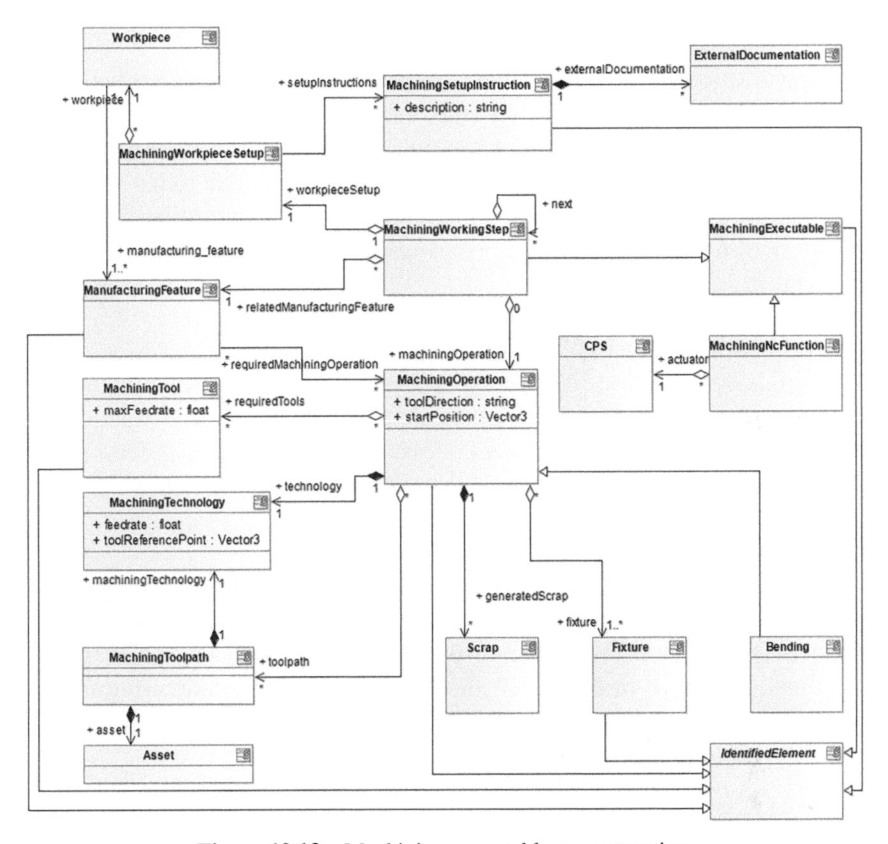

Figure 10.13 Machining executable representation.

it is related to a technology like milling, drilling or bending. It cannot exist independent of a feature, but rather specifies the association between a distinct feature and an operation to be performed on the feature. It creates an unambiguous specification, which can be executed by the machine. An operation can be replicated for different features, while a working step is unique in each part program as it spans for a defined period of time and relates to a specific workpiece and a specific manufacturing feature. Each working step thus defines the conditions under which the relative operation has to be performed. This means also that the operation related to the machining working step must be in the list of possible operations related to a certain manufacturing feature (Figure 10.13).

10.3.7.10 MachiningWorkpieceSetup

MachiningWorkpieceSetup has a direct reference to the workpiece and is defined for each machining working step, since it defines its position for machining. In fact, it may change according to the position of the single machining feature on the workpiece. In fact, also the reference to the manufacturing feature for which it is defined is unique: a single workpiece setup, in fact, refers to only one machining working step that is meant to realize a defined feature.

10.3.7.11 MachiningSetupInstructions

For each single operation in time and space, precise setup instructions may be specified, connected to the workpiece setup, such as operator instructions and external material in the forms of tables, documents and guidelines. MachiningSetupInstructions class extends the IdentifiedElement class.

10.3.7.12 ManufacturingFeature

ManufacturingFeature is an IdentifiedElement that is a characteristic of the workpiece, which requires specific operations. For 3D simulation and Computer Aided Design, it is fundamental to have the physical characteristics specifications: as shown in Figure 10.13, the workpiece manufacturing features are a relevant piece of information for modelling and simulation, as they determine the required operations.

10.3.7.13 MachiningOperation

MachiningOperation is an IdentifiedElement that specifies the contents of a machining working step and is connected to the tool to be used and a set of technological parameters for the operation. The tool choice depends on the specific working step conditions (Figure 10.13). The more information is specified for tool and fixture, the more limited the list of possible matches is. Therefore, only the relevant, necessary values should be specified.

10.3.7.14 MachiningTechnology

MachiningTechnology collects a set of parameters, such as feed rate or tool reference point. The addition of new attributes would expand the possibilities of technological specifications.

10.3.7.15 FixtureFixture

Fixture class is an IdentifiedElement that represents the fixtures required by machining operations, if any. Given that the same operation may be performed under different conditions, the choice of a fitting fixture is done for the single working step.

10.3.7.16 Assembly and disassembly

In Figures 10.14 and 10.15, assembly-executable and disassembly-executable branches are examined, even though their development is very similar to the machining executable branch. In fact, they differ only for a low number of details and specifications. These differences are presented in the following subsections.

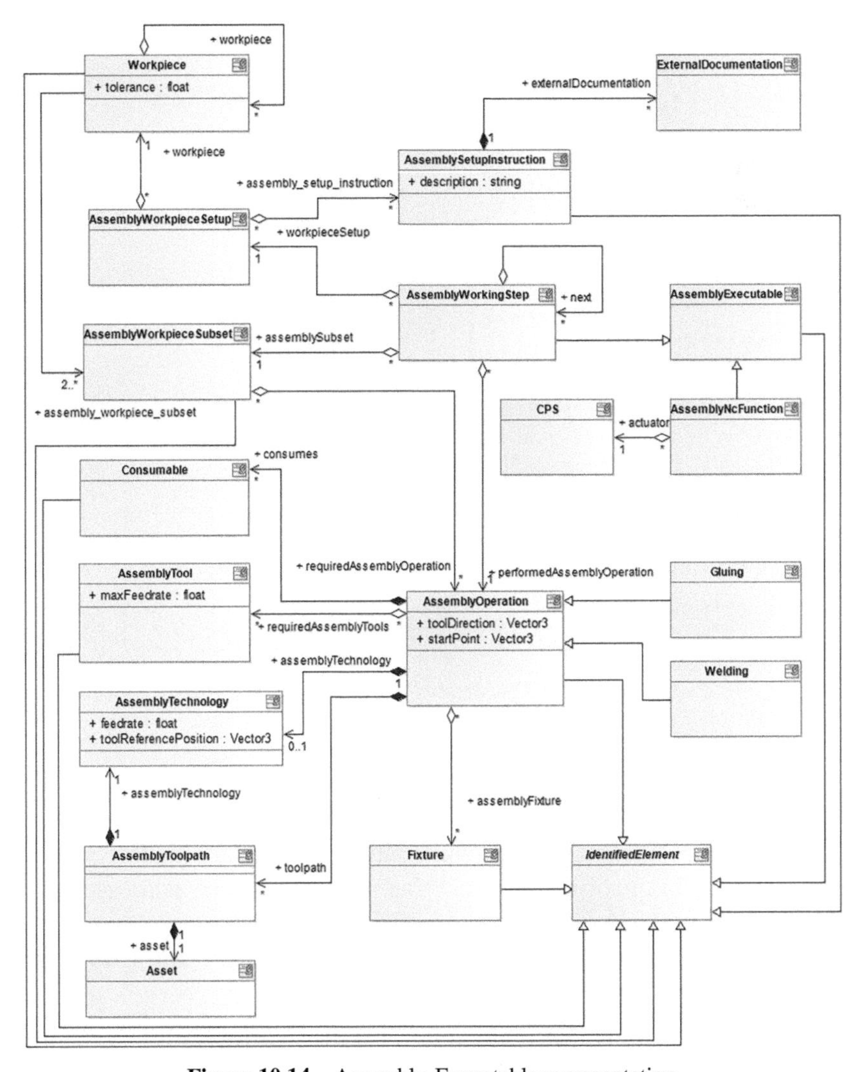

Figure 10.14 Assembly-Executable representation.

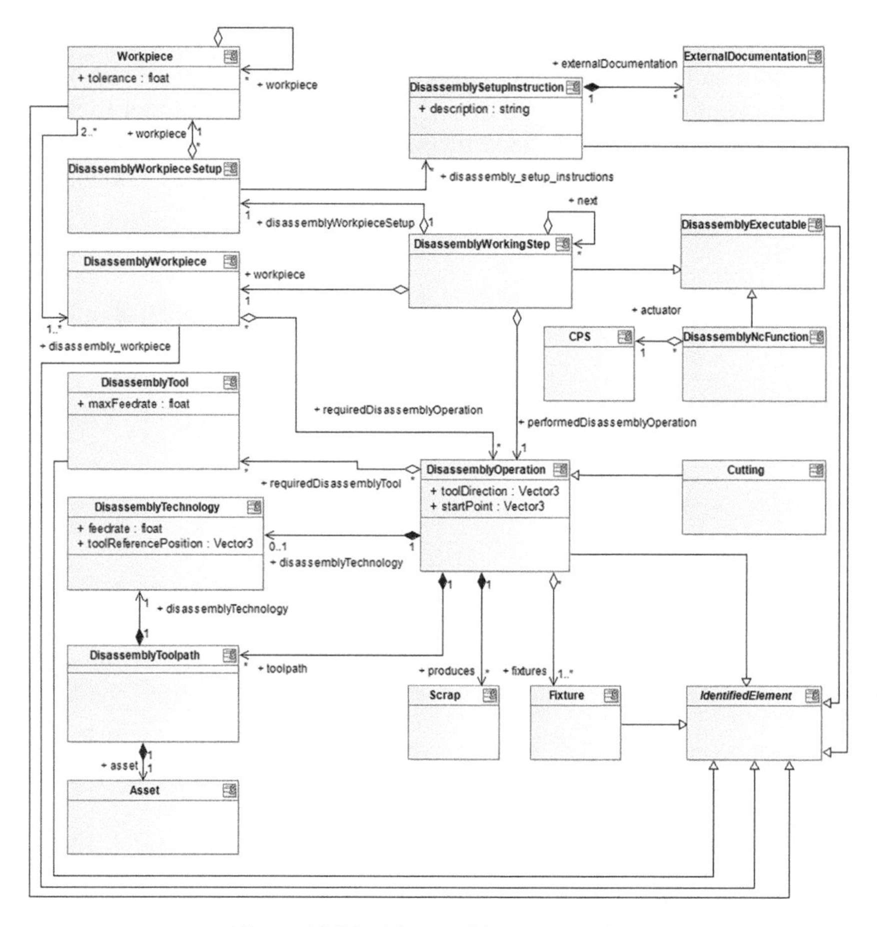

Figure 10.15 Disassembly representation.

10.3.8 Security Model

The phases of requirement gathering and analysis highlighted that security and privacy are two of the principal issues that must be properly addressed in a simulation platform.

Here, security and privacy will be enforced focusing mainly on the following aspects:

- The implementation of suitable **authentication/authorization** mechanisms
- **Securing communication and data storage** via encryption

These aspects fall under the so-called Privacy-Enhancing Technologies (PETs).

More in detail, authentication is the process of confirming the identity of an external actor in order to avoid possible malicious accesses to the system resources and services. Authentication, however, is only one side of the coin, it is in fact tightly coupled with the concept of authorization, which can be defined as the set of actions a software system has to implement in order to grant (authenticated) users the permission to execute an operation on one or more resources. Authentication and authorization are concepts related to both security (unwanted possible catastrophic access to inner resources) and privacy and data protection issues (malicious access to other users' data).

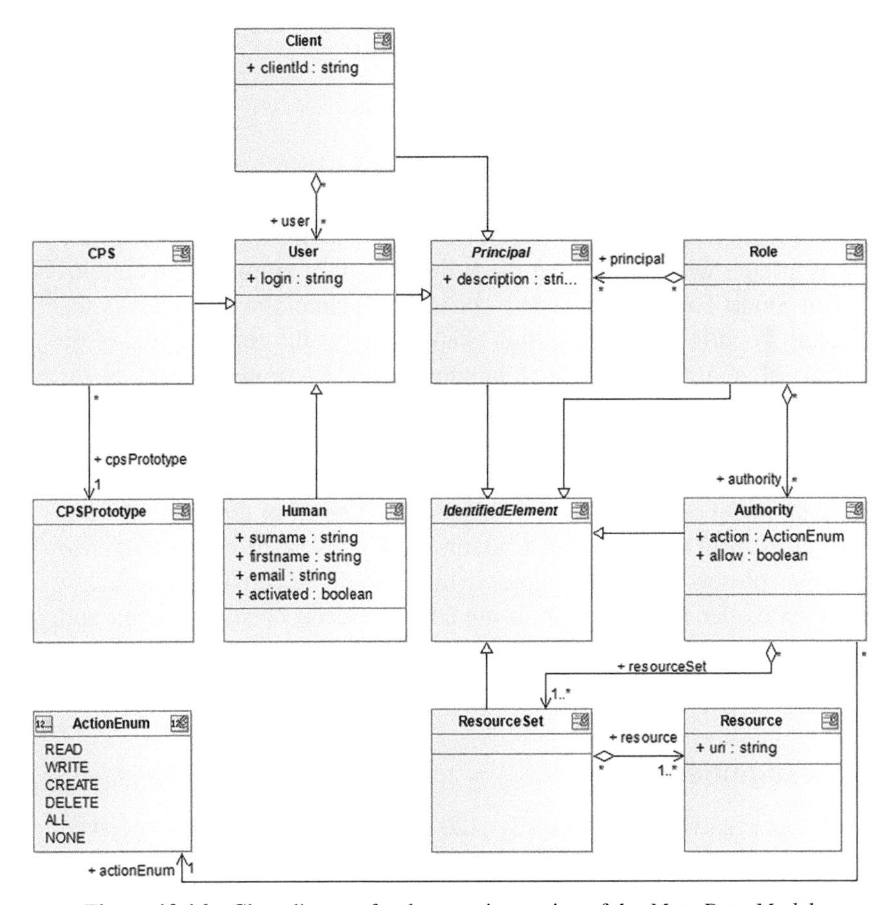

Figure 10.16 Class diagram for the security section of the *Meta Data Model.*

Securing communication is the third piece of this security and privacy puzzle, and it is as necessary as authentication and authorization. As a matter of fact, most physical devices (e.g. wireless networks) show very few privacy guaranties, and in many cases, it is practically impossible to secure wide networks against eavesdroppers. Nonetheless, confidentiality and privacy are fundamental rights (acknowledged by the European Convention on Human Rights) and must be enforced over often unsecure (communication and storage) infrastructures. For this reason, the simulation platform is committed to employ state-of-the-art encryption mechanisms (e.g. SSL and TLS) on both data storage and transport.

In the following sections of the document, the part of *Meta Data Model* devoted to security/access control management is reported and discussed. The elements of the meta model that play a role in security-related scenarios are depicted in Figure 10.16.

10.4 Conclusions

Multidisciplinary simulation is increasingly important with regard to the design, deployment and management of CPS-based factories. There are many challenges arising when exploiting the full potential of simulation technologies within Smart Factories, where a consistent technological barrier is the lack of digital continuity. Indeed, this chapter targets the fundamental issue of the lack of common modelling languages and rigorous semantics for describing interactions – physical and digital – across heterogeneous tools and systems towards effective simulation applicable along the whole factory life cycle.

The data model described in this chapter is the result of the joint effort of different actors from the European academia and industry. From the reference specifications presented in this chapter, which should be considered as a first release of a broader collaboration, a model has indeed been developed and has subsequently been validated within both an automotive industry use case and a steel carpentry scenario.

Acknowledgements

This work was achieved within the EU-H2020 project MAYA, which received funding from the European Union's Horizon 2020 research and innovation programme, under grant agreement No. 678556.

References

[1] www.automationml.org, accessed on March 24, 2017.
[2] Weyer, Stephan, et al.: Towards Industry 4.0-Standardization as the crucial challenge for highly modular, multi-vendor production systems. IFAC-PapersOnLine, 48. Jg., Nr. 3, S. 579–584, 2015.
[3] Baudisch, Thomas and Brandstetter, Veronika and Wehrstedt, Jan Christoph and Wei{\ss}, Mario and Meyer, Torben: Ein zentrales, multiperspektivisches Datenmodell fur die automatische Generierung von Simulationsmodellen fur die Virtuelle Inbetriebnahme. Tagungsband Automation 2017.

11

A Centralized Support Infrastructure (CSI) to Manage CPS Digital Twin, towards the Synchronization between CPS Deployed on the Shopfloor and Their Digital Representation

Diego Rovere[1], Paolo Pedrazzoli[2], Giovanni dal Maso[1], Marino Alge[2] and Michele Ciavotta[3]

[1]TTS srl, Italy
[2]Scuola Universitaria Professionale della Svizzera Italiana (SUPSI), The Institute of Systems and Technologies for Sustainable Production (ISTEPS), Galleria 2, Via Cantonale 2C, CH-6928 Manno, Switzerland
[3]Università degli Studi di Milano-Bicocca, Italy
E-mail: rovere@ttsnetwork.com; pedrazzoli@ttsnetwork.com; dalmaso@ttsnetwork.com; marino.alge@supsi.ch; michele.ciavotta@unimib.it

In order to support effective multi-disciplinary simulation tools in all phases of the factory life cycle, it is mandatory to ensure that the Digital Twin mirrors constantly and faithfully the state of the CPS. CPS nameplate values change over time due to situation and strain. Thereupon, this chapter describes the future CPS as equipped with special assets named Functional Models to be uploaded to CSI for synchronization and data analysis. Functional Models are essentially software routines that are run against data sent by the CPS. Such routines can regularly update CPS reference values, estimate indirect metrics, or train predictive models. Functional Models are fully managed (registered, executed, and monitored) by the CSI middleware.

11.1 Introduction

The main purpose of the CSI is to manage CPS Digital Twins (DTs) allowing the synchronization between CPS deployed on the shopfloor and their digital representation. In particular, during the whole factory life cycle, the CSI will provide services (via suitable API endpoints) to analyze the data streams coming from the shopfloor and to share simulation models and results among simulators.

In this chapter, we present the implementation of a distributed middleware developed within the frame of MAYA European project, tailored to enable scalable interoperability between enterprise applications and CPS with especial attention paid to simulation tools. The proposed platform strives for being the first solution based on both Microservices [1, 2] and Big Data [3] paradigms to empower shopfloor CPS along the whole plant life cycle and realize real-digital synchronization ensuring at the same time security and confidentiality of sensible factory data.

11.2 Terminology

Shopfloor CPS – With the expression "Shop-floor CPS" we refer to Digital-Mechatronic systems deployed at shopfloor level. They are physical entities that intervene in various ways in the manufacture of a certain product. For the scope of this chapter, Shopfloor CPS (referred to as Real CPS or simply CPS) can communicate to each other and with the CSI.

CPS Digital Twin (or just Digital Twin) – In the smart factory, each shopfloor CPS is mirrored by its virtual alter ego, called Digital Twin (DT). The Digital Twin is the semantic, functional, and simulation-ready representation of a CPS; it gathers together heterogeneous pieces of information. In particular, it can define, among other things, Shopfloor CPS performance specifications, Behavioral (simulation) Models, and Functional Models.

Digital Twin is a composite concept that is specified as follows:

CPS Prototype (or just Prototype) – Chapter 12 proposes a meta-model that paves the way to a semantic definition of CPS within the CSI. Following the Object-Oriented Programming (OOP) approach, we distinguish between a Prototype (or class) and its derived instances. A CPS prototype is a model that defines the structure and the associate semantic for a certain class of CPS.

A prototype defines fields representing both general characteristics of the represented CPS class and the state of a specific Shopfloor CPS.

CPS Instance – Once a shopfloor CPS is connected to the CSI platform, a set of processes are run to instantiate, starting from a CPS prototype, the Digital Twin. The Digital Twin is an instance of a specific CPS prototype. Therefore, a CPS instance can be defined as the computer-based representation (live object in memory or stored in a database) of its Digital Twin, which can be considered a more abstract concept even independent of this implementation within the CSI.

Behavioral Models – These are simulation models, linked to the semantic representation of a CPS (prototype and instance) and stored within the CSI. Each Digital Twin can feature behavioral models of different nature to enable the multi-disciplinary approach to simulation.

Functional Models – In layman's terms, functional models are pieces of software to be run on a compliant platform created to analyze data coming from the shopfloor. Data can enter a platform in the form of streams or imported from other sources (text files, excel, databases, etc.). The results of the analysis are used to enrich the Digital Twin implementing the real-to-digital synchronization. They can be used, for instance, to update license plate data of Digital Twins or to enable predictive maintenance specific on the considered CPS.

11.3 CSI Architecture

The overall CSI component diagram is shown in Figure 11.1: a relevant part of the platform consists of a microservice-based infrastructure devoted to administrative tasks related to Digital Twins and a Big Data deployment accountable for processing shopfloor data. Since the two portions of our middleware have different requirements, being also grounded on different technological solutions, in what follows, they are presented and discussed separately.

11.3.1 Microservice Platform

In a nutshell, the microservice architecture is the evolution of the classical Service Oriented Architecture (SOA) [4] in which the application is seen as a suite of small services, each devoted to a single activity. Within the CSI,

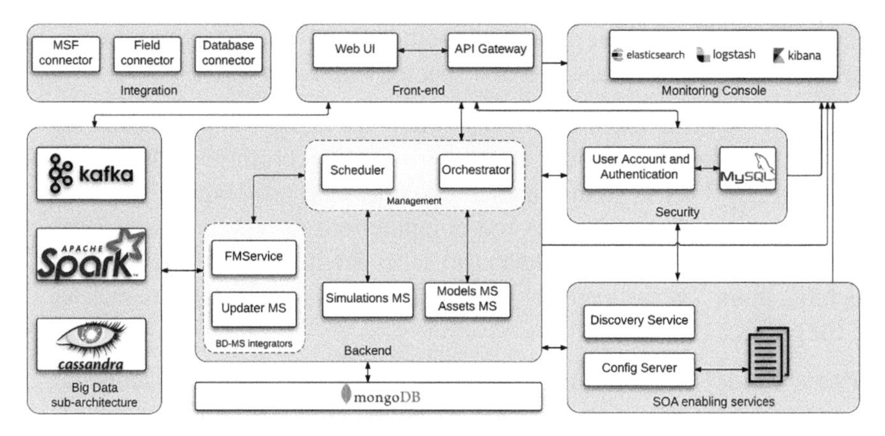

Figure 11.1 CSI Component Diagram.

each microservice exposes a small set of functionalities and runs in its own process, communicating with other services mainly via HTTP resource API or messages. Four groups of services can be identified and addressed in what follows.

11.3.1.1 Front-end services

Front-end services are designed to provide the CSI with a single and secure interface to the outer world. As a consequence, any other service can be accessed only through the front-end and only by trusted entities. The main services in this group are:

Web-based UI

The Web-based UI is a Web application for human–machine interaction; it provides a user friendly interface to register new CPS or to execute queries. Administration tools such as security management and platform monitoring are available as well.

API Gateway

The API Gateway, instead, is a service designed to provide dynamic and secure API routing, acting as a front door for the requests coming from authorized players, namely users via the Web UI and devices/CPS executing REST/WebSocket calls. In layman's terms, all the other platform services are accessible only through the gateway and only by trusted entities.

The gateway is based on Netflix Zuul[1] for dynamic routing, monitoring, and security, and Ribbon[2], a multi-protocol inter-process communication library that, in collaboration with Service Registry (see SOA enabling services), dispatches incoming requests applying load-balance policy. The API gateway, finally, offers an implementation of the Circuit Breaker[3] pattern impeding the system to get stuck in case the target back-end service fails to answer within a certain time.

11.3.1.2 Security and privacy

Security policies are enforced by the User Account and Authentication (UAA) service, which is in charge of the authentication and authorization tasks:

UAA Service

In a nutshell the main task of this service is to check users' (human operators, CPS or microservices) credentials to verify the identity and issuing a time-limited OAuth2 [13] token to authorize a subset of possible actions that depends on the particular role the user has been assigned to. Users' data, roles and permission are stored in a relational database: currently, MySQL[4] database is used to this end.

It is worth to notice that authentication and authorization is required not only for human users and CPS but also to establish a trustful collaboration between microservices avoiding malevolent and tampering actions.

11.3.1.3 SOA enabling services

SOA enabling services: this group of services has the task to support the microservice paradigm; it features:

Service Registry

This service provides a REST endpoint for service discovering. This service is designed to allow transparent and agnostic service communication and load balancing. Based on Netflix Eureka[5], it exposes APIs for service registration and for service querying, allowing the services to communicate without referring to specific IP addresses. This is especially important in the scenario in which services are replicated in order to handle a high workload.

[1]https://github.com/Netflix/zuul/wiki

[2]https://github.com/Netflix/ribbon

[3]https://martinfowler.com/bliki/CircuitBreaker.html

[4]www.mysql.com

[5]https://github.com/Netflix/eureka/wiki

Configuration Server

The main task of this service is to store properties files in a centralized way for all the micro-services involved in the CSI. This is a task of paramount importance in many scenarios involving the overall life cycle of the platform. Among the benefits of having a configuration server, we mention here the ability to change the service runtime behavior in order to, for example, perform debugging and monitoring.

Monitoring Console

This macro-component with three services implements the so-called ELK stack (i.e., Elasticsearch, Logstash, and Kibana) to achieve log collection, analyzing, and monitoring services. In other words, logs from every microservice are collected, stored, processed, and presented in a graphical form to the CSI administrator. A query language is also provided to enable the administrator to interactively analyze the information coming from the platform.

11.3.1.4 Backend services

To this group belong those services that implement the Chapter 12 meta-data model and manage the creation, update, deletion, storage, retrieval, and query of CPS Digital Twins as well as simulation-related information. In particular, the CSI features the following services:

Orchestrator

The Orchestrator microservice coordinates and organizes other services' execution to create high-level composite business processes.

Scheduler

Service for the orchestration of recurring action. Example of those jobs are: importing data from external sources at regular intervals, updating CPS Prototypes and instances, removing from internal databases stale data, and sending emails enclosing a report on the system's healthy to administrators.

Models MS/Assets MS

Models and Assets microservices handle the persistence of Digital Twin information (their representation and assets, respectively) providing endpoints for CRUD operations. In the current version of the CSI, these two components are merged into a single service in order to streamline the access to MongoDB and avoid synchronization issues.

FMService
This service is able to communicate with the Big Data platform; its main task is to submit the Functional Models to Apache Spark, to monitor the execution, cancel, and list them.

Updater MS
This service is designed to interact with the Big Data platform (in particular with Apache Cassandra) to retrieve data generated by the Functional Models.

Simulations MS
This service is appointed to managing the persistence of simulation-related data within a suitable database.

11.3.2 Big Data Sub-Architecture

Big Data technologies are becoming innovation drivers in industry [5]. The CSI is required to handle unprecedented volumes of data generated by the digital representation of the factory in order to keep updated the CPS nameplate information. To this end, a data processing platform, specifically a Lambda architecture [6], has been implemented according to the best practices of the field. The Lambda Architecture was introduced as a generic, linearly scalable, and fault-tolerant data processing architecture. In particular, both data in rest and data in motion patterns are enforced by the platform, making it suitable for both stream and batch processing.

The Lambda Architecture encompasses three layers, namely batch, speed, and serving layers. The batch layer is appointed to the analysis of large quantities (but still finite) of data. A typical scenario is that wherefore the data ingested by the system are inserted in NoSQL Databases. Pre-computation is applied periodically on batches of data. The purpose is to offer the data a suitable aggregated form for different batch views. Note that the batch layer has a high processing latency because it is intended for historical data.

The speed layer is in charge of processing infinite streams of information. It is the purpose of the Speed Layer to offer a low latency, real-time data processing. The speed layer processes the input data as they are streamed in and it feeds the real-time views defined in the serving layer.

The Serving Layer has the main responsibility to offer a view on the results of the analysis. The layer responds to queries coming from external systems; in this particular case, the serving layer provides an interface that integrates with the rest of the CSI.

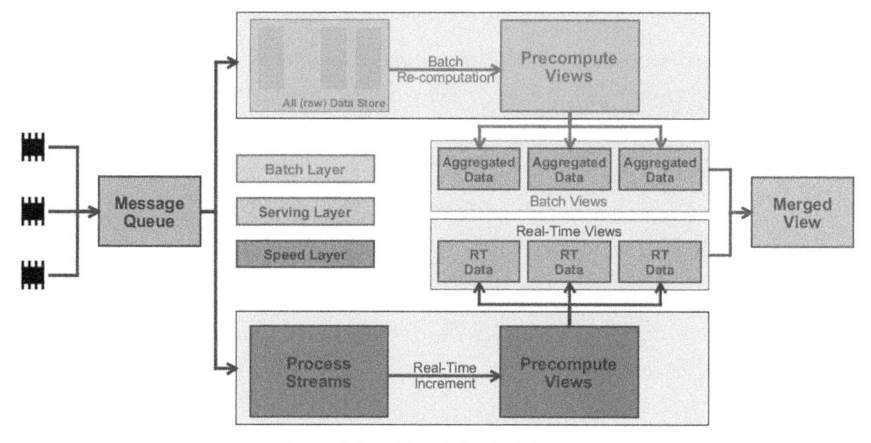

Figure 11.2 Lambda Architecture.

Designing and setting up a Big Data environment, here in the form of the Lambda Architecture (Figure 11.2), is a complex task that starts with doing some structural decisions. In what follows, some high-level considerations about the technological choices made are presented:

11.3.2.1 Batch layer

The field of Big Data is bursting with literally hundreds of tools and frameworks, each with specific characteristics; however, recently, some new solutions have appeared on the market that natively extend MapReduce [7] paradigm reduce, and, among other things, provide a more flexible and complete programming paradigm paving the way to the realization of new and more complex algorithms.

The solution selected to implement this layer, Apache Spark [8], claims to be up to 100× faster than Hadoop on memory and up to 10× faster on disk. This is mainly due to a particular distributed, in memory data structure called Resilient Distributed Datasets (RDD). Shortly, Apache Spark attracted the interest of important players and gathered a vast community of contributors, only to mention a few: Intel, IBM, Yahoo!, Databricks, Cloudera, Netflix, Alibaba, and UC Berkely. Moreover, Spark implements both map-reduce and streaming paradigm, features out-of-the-box an SQL-like language for automatic generation of jobs, and supports several programming languages (Java, Scala, Python, and R).

11.3.2.2 Stream processing engine

If the batch processing engine enables the analysis of large historical data (often referred to as Data at Rest), then the stream processing engine is the component of the Lambda Architecture that is in charge of continuously manipulating the incoming data in quasi real-time fashion (i.e., the Data in Motion scenario). Recently, stream processing has increased in popularity. Only within the Apache Foundation, we identified several tools supporting different flavors of stream processing. Among them is Spark Streaming [9], the tool used to implement this layer.

Spark Streaming relies on Spark core to implement micro-batching stream processing. This means that the elements of the incoming streams are grouped together in small batches and then manipulated. As a consequence, Spark shows a higher latency (about 1 second). Spark Streaming is a valid alternative owing to the rich API, the large set of libraries, and its stability.

Spark can work in standalone mode featuring on its own resource manager or it can rely on external resource managers (as YARN). Other resource managers exist (e.g. Apache Mesos), but they are related more to cluster management than on Big Data. Nonetheless, Spark can be executed over both YARN and Mesos.

11.3.2.3 All data store

A central role in the Lambda Architecture is played by the All Data Store, which is the service in charge of storing and retrieving the historical data to be analyzed. Depending on the type of data entering the system, this element of the platform can be realized in different ways. In MAYA, we decided to implement it through a NoSQL database particularly suitable for fast updates, Apache Cassandra [10]. It is the most representative champion of the column-oriented group. It is a distributed, linear scalable solution capable of ensuring high volumes of data. Cassandra is widely adopted (it is the most used column-oriented database) and features an SQL-like query language named CQL (Cassandra Query Language) along with a Thrift[6] interface. As far as stream views are concerned, Cassandra has been successfully used to handle time series for IoT and Big Data.

11.3.2.4 Message queueing system

In a typical Big Data scenario, data flows coming from different sources continuously enter the system; the most used integration paradigm to handle

[6]https://thrift.apache.org/

data flows consists in setting up a proper message queue. A message queue is a middleware implementing a publisher/subscriber pattern to decouple producers and consumers by means of an asynchronous communications protocol. Message queues can be analyzed under several points of view, in particular policies regarding Durability, Security, Filtering, Purging, Routing, and Acknowledgment, and message protocols (as AMQP, STOMP, MQTT) must be carefully considered.

Message queue systems are not a novelty and many proprietary as well as open-source solutions have appeared on the market in the last years. Among the open-source ones, there is Apache Kafka [11]. A preliminary analysis seems to demonstrate that Kafka is the most widely used in big players' production environments as, for instance, in LinkedIn, Yahoo!, Twitter, Netflix, Spotify, Uber, Pinterest, PayPal, Cisco, and Coursera among the others. Kafka is written in Java and originally developed at LinkedIn; it provides a distributed and persistent message passing system with a variety of policies. It relies on Apache Zookeeper [12] to maintain the state across the cluster. Kafka has been tested to provide close to 200,000 messages/second for writes and 3 million messages/second for reads, which is an order of magnitude more that its alternatives.

11.3.2.5 Serving layer

This layer provides a low-latency storage system for both batch and speed layers. The goal of this layer is to provide an engine able to ingest different types of workloads and query them showing a unified view of data. The rationale is that the outcomes of the different computations must be suitably handled to later be further processed. In particular, batch views will contain steady, structured, and versioned data, whereas stream views will contain time-related data. Within the CSI, we have adopted the following flexible approach: in case Batch activities are required, the serving layer is implemented by means of Apache Cassandra NoSQL database, otherwise Apache Kafka is exploited. Notice that it is not uncommon to use a persistent and distributed message system as serving layer as, for example, in ORYX2[7], where precisely Kafka is used.

11.3.3 Integration Services

Technically, these services do not belong to the CSI at the moment, but we envision their development in the following phases of the project with the aim

[7]http://oryx.io/

of streamlining the interaction processes with external tools and databases; in particular, at the moment of writing we foresee the following services:

MSF Connector
This component passes the CPS id, the simulation model in AutomationML format, and the simulation types requested by the user. The MSF sends in return the simulation results per each simulation type requested.

Field Connector
This service serves to bridge the gap between the communication layer and the field in case of CPS non-compliant with the CSI. In particular, it will create suitable WebSocket channels for data streams coming from the field and root those data to the Big Data platform inside the CSI.

DB Importer
Database Importers will be in charge of importing valuable data from external databases to enable the execution of Functional Models on those data.

11.4 Real-to-Digital Synchronization Scenario

Several usage scenarios are possible to be executed within the CSI. Nonetheless, we propose the following as a reference use case, as it involves a good part of CSI components and functionalities. The objective is to use it as a reading key to better understand the relationships among the CSI and how they are reflected into the architecture. The considered scenario concerns the automated processing of data streams coming from CPS and can be described as follows:

1. A human operator registers a new CPS. This action can be performed via the graphical UI or by means of available REST [13] endpoints;
2. The CPS logs in on the CSI, its digital identity is verified and the Digital Twin is activated;
3. The Functional Model featured by the Digital Twin (if any) is set up, scheduled, and executed;
4. WebSocket channel is established between the CPS and CSI. The CPS starts sending data to the platform;
5. The Functional Model periodically generates updates for a subset of attributes of the corresponding Digital Twin;
6. The CPS disconnects from the CSI and consequently the related Functional Models is halted and dismissed.

Figure 11.3 describes in UML the main actions carried out by the CPS and by the CSI in the scenario at hand. In particular, the CPS connects by logging in on the platform, at that point it is associated to a WebSocket endpoints and it can start sending data up to the CSI. The CSI, on the other hand, launches the execution of the Functional Model associated with the CPS.

A deeper insight is gained by means of Figure 11.4; in it, the interactions among services within the CSI are highlighted. It is clear, in fact, that the CPS connects with the CSI via the API ***Gateway***. In the current version

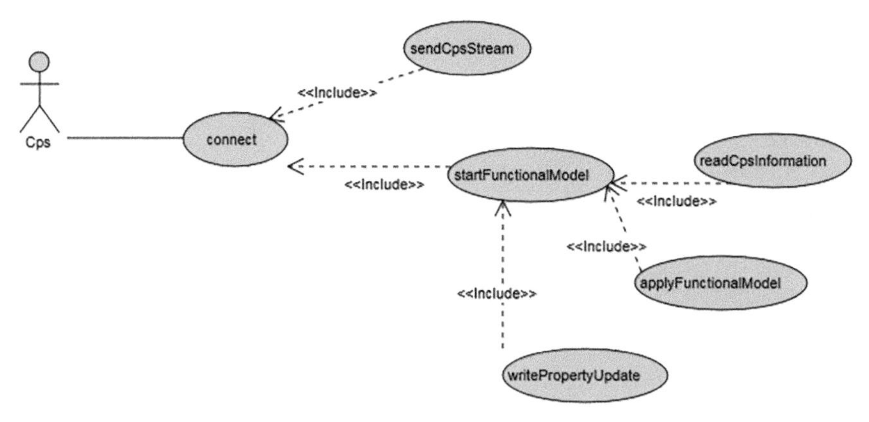

Figure 11.3 CPS connection.

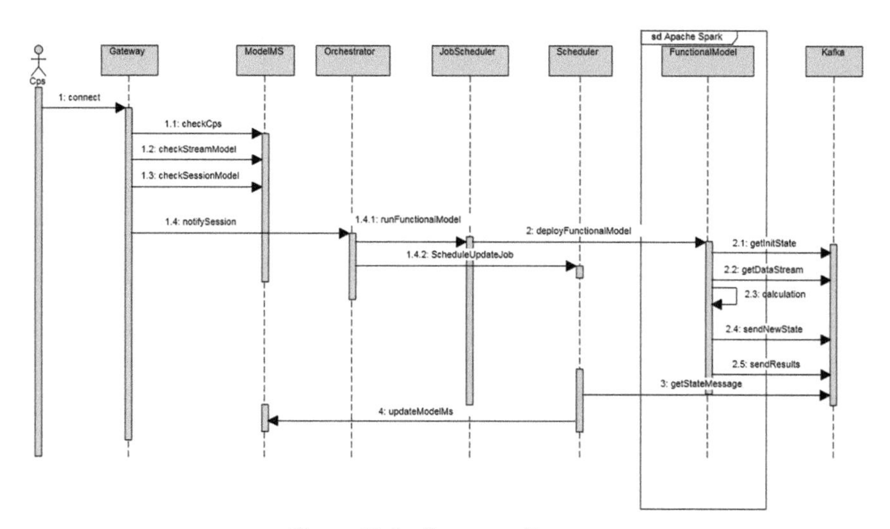

Figure 11.4 Sequence diagram.

of the CSI, the Gateway is in charge of checking whether the CPS asking for being attended is legit (it must have been created within the platform beforehand). To do this, the Gateway interrogates the ***Models MS*** service. The Gateway then creates a WebSocket endpoint for the CPS, redirects the incoming workload to Kafka, and notifies the ***Orchestrator***. This, in turn, is in charge of running the Functional model(s) associated with the CPS. The Functional models are executed within the Big Data platform (in Apache Kafka cluster) and in particular they use Kafka not only as source of data but also as the endpoint where to post the results of the computation. Meanwhile the Orchestrator has scheduled a recurrent job on the ***Scheduler*** that picks up the updated from the output Kafka topic and uses them to update the nameplated values of the CPS Digital Twin.

During the whole process, the Security is present in the form of SSL connection, CPS log in via OAuth2, and service-to-service authorization and authentication. We outlined the real-to-digital synchronization in Figure 11.5, wherein the reader can spot the presence of all the players present in the sequence diagram plus the UAA Service in charge of the authentication and authorization tasks. The actions performed by this service are pervasive and would have made the sequence diagram unintelligible.

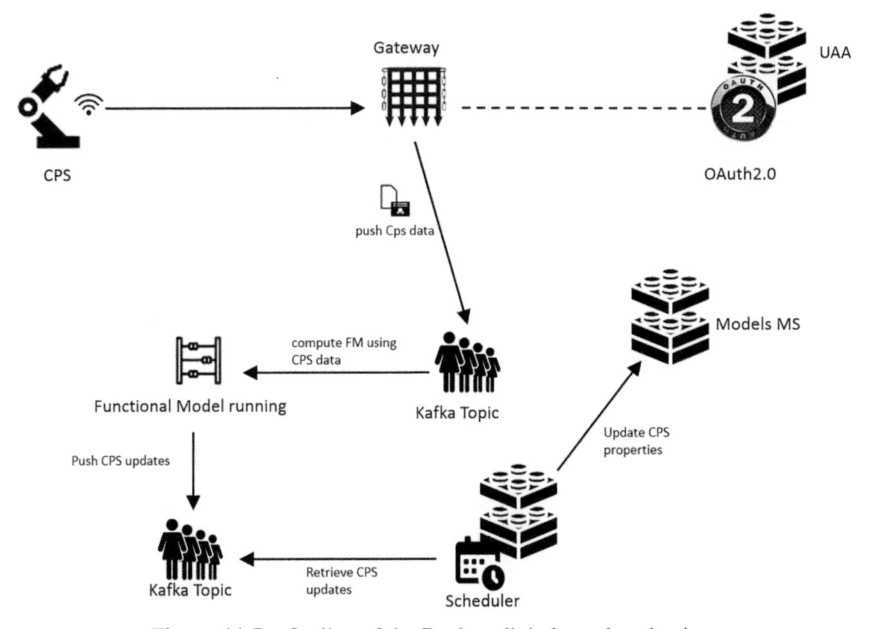

Figure 11.5 Outline of the Real-to-digital synchronization.

11.5 Enabling Technologies

CSI aims at being the first reference middleware for smart factories based on a composite Microservices/Big Data approach paying particular attention to security concerns. In the following paragraphs, we examine the reasons behind the technical choices made.

11.5.1 Microservices

The Microservices approach proposes to have numerous small code bases managed by small teams instead of having a giant code base that eventually every developer touch with the result of making more complex, slow, and painful the process of delivering a new version of the system.

In a nutshell, the microservice architecture is the evolution of the classical Service-Oriented Architecture (SOA), in which the application is seen as a suite of small services, each devoted to as single activity. Each microservice exposes an atomic functionality of the system and runs in its own process, communicating with other services via HTTP resource API (REST) or messages.

The adoption of the microservice paradigm provides several benefits, as well as presents inconveniences and new challenges. Among the benefits of this architectural style, the following must be enumerated:

Agility – Microservices fit into the Agile/DevOps development methodology [2], enabling business to start small and innovate fast by iterating on their core products without affording substantial downtimes. A minimal version of an application, in fact, can be created in shorter time reducing time-to-market and up-front investment costs, and providing an advantage with respect to competitors. Future versions of the application can be realized by seamlessly adding new microservices.

Isolation and Resilience – Resiliency is the ability of self-recovery after a failure. A failure in a monolithic application can be a catastrophic event, as the whole platform must recover completely. In a microservice platform, instead, each service can fail and heal independently with a possibly reduced impact on the overall platform's functionalities. Resilience is strongly dependent on compartmentalization and containment of failure, namely Isolation. Microservices can be easily containerized and deployed as single process, reducing thus the probability of cascade-fail of the overall application. Isolation, moreover, enables reactive service scaling and independent monitoring, debugging, and testing.

Elasticity – A platform can be subject to variable workloads especially on seasonal basis. Elasticity is the ability to respond to workload changes provisioning or dismissing computational power. This is usually translated into scaling up and down services. This process can be particularly painful and costly in case of on premise software; easier and automated in case of cloud-based applications. Nonetheless, microservices allows for a finer grain approach, in which services in distress (e.g., that are not meeting their Quality of Service) can be identified and singularly scaled taking full advantage of cloud computing since it requires the provisioning of just the right amount of resources. This approach can lead to substantial savings in the cloud that usually implements pay-per-use provisioning policies.

As far as the challenges and drawbacks derived by the choice of adopting microservices are concerned, we mention here:

Management of Distributed Data – As each microservice might have its private database, it is difficult to implement business transactions that maintain data consistency across multiple databases.

Higher Complexity of the Resulting System – Proliferation of small services could translate into a tangle Web of relationships among them. Experienced teams must be put together to deal in the best possible way with microservice platforms.

11.5.2 Cloud Ready Architecture: The Choice of Docker

Containerization services (among which the most known is definitely Docker [14]) and microservice are two closely related yet different aspects of the same phenomenon; although containerization is not essential to realize microservice architectures, it is certainly true that it enables microservices to fully realize their potential; Docker's *agility*, *isolation*, and *portability*, in fact, powered the rise and success of the microservice pattern while the latter gathered an ever-increasing interest around containers. It can be safely said that there are now two faces of the same coin and have made the fortune of each other.

At this point, it is important to answer to the simple question: what is a containerization system? A containerization system (hereinafter, we will use Docker and containerization system interchangeably) is a para-virtualization platform that exploits isolation features of Linux kernel, as namespaces and *cgroups* (recently also Windows' ones), to create a secure and isolate environment for the execution of a process. Each process running in a container has access to its own file system and libraries, but it shares with other containers the underpinning kernel.

This approach is defined para-virtualization because, unlike virtualization systems that emulate hardware to execute whole virtual machines to run atop, there is no need to emulate anything. Moreover, Docker do not depend on specific virtualization technologies and, therefore, it can run wherever a Linux kernel available. The overall approach results to be lightweight with respect to more traditional hypervisor-based virtualization platform allowing for a better exploitation of the available resources and for the creation of faster and more reactive applications. In the light of these considerations, it should be clear how Docker fits perfectly for microservices, as it isolates containers to one process and makes it simple and fast to handle the full life cycle of these services.

The current version of the CSI is provided with a set of scripts for automatic creation of Docker images for each of the services involved in the platform. Deployment scripts, which rely on a tool called Docker-compose, are provided as well to streamline the deployment on a local testbed. Nonetheless, a similar approach can be used to execute the platform on the most important Clouds (e.g. Amazon ECS, Azure Container Service).

11.5.3 Lambda Architecture

A very important subset of CSI functionalities consists in the capability to handle unprecedented volume generated by the digital representation of the factory. To this end, a Big Data platform has been integrated with the microservice one. The phrase Big Data usually refers to a large research area that encompasses several facets. In this work, in particular, we refer to Big Data architectures. The following benefits deserve to be enumerates:

Simple but Reliable – The CSI Big Data platform has been implemented employing a reduced number of tools; all of them are considered state of the art, are used in production by hundreds of companies worldwide, and are backed by large communities and big Information and Communications Technologies players.

Multi-paradigm and General Purpose – Batch and Stream processing as well as ad-hoc queries are supported and can run concurrently. Moreover, the unified execution model, coupled with a large set of libraries, permits the execution of complex and heterogeneous tasks (as machine learning, data filtering, ETL, etc.).

Robust and Fault Tolerant – In case of failure, the data processing is automatically rescheduled and restarted on the remaining resources.

Multi-tenant and Scalable – In MAYA, this means that several Functional Models can run in parallel sharing computational resources. Furthermore, in case more resources are provisioned and the platform will start to exploit them without downtimes.

The downside of this approach is that it is fundamentally and technologically different for the rest of the platform and required quite an integration work. For this reason, the main elements of the CSI Big Data architecture had to be interfaced with expressly created microservices (as FMserver and Updates MS, see Section 4.1.4 for more details). Finally, Big Data solutions generally require steep learning curves to be fully exploited being moreover really resource eager.

11.5.4 Security and Privacy

Security and privacy issues assume paramount importance in Industrial IoT. Here, we enforce those aspects since the earliest stages of the design, focusing on suitable Privacy-Enhancing Technologies (PETs) that encompass Authentication, Authorization, and Encryption mechanisms.

More in detail, authentication is the process of confirming the identity of an actor in order to avoid possibly malicious accesses to the system resources and services. Authentication can be defined as the set of actions a software system has to implement in order to grant the actor the permissions to execute an operation on one or more resources.

Specifically, seeking for more flexibility, we implemented a role-based access control model that permits the authentication process to depend on the actor's role. Suitable authentication/authorization mechanisms (based on the Oauth2 protocol) have been developed for human operators, and services and CPS.

Securing communication is the third piece of this security and privacy puzzle, as no trustworthy authentication and authorization mechanism can be built without the previous establishment of a secure channel. For this reason, the CSI committed to employ modern encryption mechanisms (e.g. SSL and TLS) for the communication and data storage as well.

11.6 Conclusions

This document presented the Centralized Support Infrastructure built within the H2020 MAYA project: an IoT middleware designed to support simulation in smart factories. To the best of our knowledge, it represents the first example of Microservice platform for manufacturing. Since security and privacy are sensitive subjects for the industry, special attention has been paid on their enforcement from the earliest phases of the project. The proposed platform has been here described in detail in connection with CPS and simulators. Lastly, the overall architecture has been discussed along with benefits and challenges.

Acknowledgements

The work hereby described has been achieved within the EU-H2020 project MAYA, which has received funding from the European Union's Horizon 2020 research and innovation program, under grant agreement No. 678556.

References

[1] N. Dragoni et al., "Microservices: yesterday, today, and tomorrow," in *Present and Ulterior Software Engineering*, Springer Berlin Heidelberg, 2017.

[2] S. Newman, *Building microservices.* " O'Reilly Media, Inc.," 2015.

[3] J. Manyika et al., "Big data: The next frontier for innovation, competition, and productivity," 2011.

[4] S. Newman, *Building microservices.* " O'Reilly Media, Inc.," 2015.

[5] C. Yang, W. Shen, and X. Wang, "Applications of Internet of Things in manufacturing," in *Proceedings of the 2016 IEEE 20th International Conference on Computer Supported Cooperative Work in Design, CSCWD 2016*, pp. 670–675, 2016.

[6] R. Drath, A. Luder, J. Peschke, and L. Hundt, "AutomationML-the glue for seamless automation engineering," in *Emerging Technologies and Factory Automation, 2008. ETFA 2008. IEEE International Conference on*, pp. 616–623, 2008.

[7] J. Dean and S. Ghemawat, "MapReduce: Simplified Data Processing on Large Clusters," *Proc. OSDI - Symp. Oper. Syst. Des. Implement.*, pp. 137–149, 2004.

[8] M. Zaharia, M. Chowdhury, M. J. Franklin, S. Shenker, and I. Stoica, "Spark?: Cluster Computing with Working Sets," *HotCloud'10 Proc. 2nd USENIX Conf. Hot Top. cloud Comput.*, p. 10, 2010.

[9] M. Zaharia, T. Das, H. Li, T. Hunter, S. Shenker, and I. Stoica, "Discretized Streams: Fault-Tolerant Streaming Computation at Scale," *Sosp*, no. 1, pp. 423–438, 2013.

[10] A. Lakshman and P. Malik, "Cassandra," *ACM SIGOPS Oper. Syst. Rev.*, vol. 44, no. 2, p. 35, 2010.

[11] J. Kreps and L. Corp, "Kafka: a Distributed Messaging System for Log Processing," *ACM SIGMOD Work. Netw. Meets Databases*, p. 6, 2011.

[12] P. Hunt, M. Konar, F. P. Junqueira, and B. Reed, "ZooKeeper: Wait-free Coordination for Internet-scale Systems," in *USENIX Annual Technical Conference*, vol. 8, p. 11, 2010.

[13] R. T. Fielding and R. N. Taylor, "Principled Design of the Modern Web Architecture," *ACM Transactions on Internet Technology*, vol. 2, no. 2. pp. 407–416, 2002.

[14] D. Jaramillo, D. V. Nguyen, and R. Smart, "Leveraging microservices architecture by using Docker technology," in *Conference Proceedings - IEEE SOUTHEASTCON*, 2016, July 2016.

PART III

12

Building an Automation Software Ecosystem on the Top of IEC 61499

Andrea Barni, Elias Montini, Giuseppe Landolfi, Marzio Sorlini and Silvia Menato

Scuola Universitaria Professionale della Svizzera Italiana (SUPSI), The Institute of Systems and Technologies for Sustainable Production (ISTEPS), Galleria 2, Via Cantonale 2C, CH-6928 Manno, Switzerland
Email: andrea.barni@supsi.ch; elias.montini@supsi.ch; giuseppe.landolfi@supsi.ch; marzio.sorlini@supsi.ch; silvia.menato@supsi.ch

The adoption of Cyber Physical System (CPS) technologies at European level is constrained by a still emerging value chain and by the challenging transformation of manufacturing processes and business ecosystems that their deployment requires. This issue becomes even more challenging when the concept of CPS is exploited to propose cyber-physical machines and manufacturing systems, where the complexity of the controlling intelligence and of the digital counterpart explodes. As a matter of fact, the market behind CPS has a potential that is still scarcely supported by methodologies and tools able to foster the rise of a solid ecosystem required for a relevant market uptake. Multi-sided platforms (MSPs) have demonstrated to play the pivotal role of providing the environments and the technological infrastructures able to match make the needs of manifold user insisting on them. The manufacturing sector did not remain untouched by this trend and moves its first step towards the integration of platform logics across value networks: the CPS business ecosystem is one of those.

In this chapter, beyond an analysis of the current state of the automation value network, the design and implementation of a multi-sided platform for CPS deployment within the automation sector are described. The proposed platform can provide the infrastructure to incentivize CPS adoption, creating the technological and value drivers supporting the transition towards new paradigms for the development of the software components of a mechatronic system. Developing an infrastructure on the top of which the CPS value network can be instantiated and orchestrated, the proposed platform provides the technical means to incentivize the creation of an ecosystem able to support especially SMEs in their transition towards Industry 4.0.

12.1 Introduction

Technological innovation is the main engine behind economic development that aims at supporting companies in adapting to the rhythm of the market dictated by globalization [1, 2]. According to Stal [3], innovation is the development of new methods, devices or machines that could change the way in which things happen. The fourth industrial revolution, pursuing the extensive adoption of innovative technologies and systems, increasingly impact almost every industry. According to Bharadwaj et al. [4], "digital technologies (viewed as combinations of information, computing, communication, and connectivity technologies) are fundamentally transforming business strategies, business processes, firm capabilities, products and services, and key interfirm relationships in extended business networks".

The automation industry has historically a leading role in experimenting and pushing this transformation, with technological and process-related innovation being assimilated all-inclusively across the whole value network [5]. However, the characteristics that the automation market acquired in the last decades, where complex value networks support standard-based technologies relying on legacy systems, make purely technological advancements no more enough to satisfy the need of innovation expressed by the market. As proposed in the Technology-Organization-Environment Framework [6], the propensity of companies towards the adoption of innovations is indeed not only dependent on the technology per se, but it is influenced by the technological context, the organizational context, and the environmental context. The technological context includes the internal and external technologies that are relevant to the firm. The organizational context refers to the characteristics and resources of the firm. The environmental context includes the size and structure of the industry, the firm's competitors, the macroeconomic context, and the regulatory environment [6–8]. These three elements present both constraints and opportunities for technological innovation and influence the way a firm sees the need for, searches for, and adopts new technology [9].

In this context, the European initiative Daedalus supports companies in facing opportunities and challenges of Industry 4.0 starting from the overcoming of the rigid hierarchical levels of the automation pyramid. This is done by supporting CPS orchestration in real time through the IEC-61499 automation language, to achieve complex and optimized behaviors impossible with other current technologies. To do so, it proposes a methodology and the related supporting technologies that, integrated within an "Industry platform[1]" and brought to the market by means of a Digital Marketplace, are meant to foster the evolution of the automation value network. The desired evolution is expected not only to impact on how companies manage their production systems, providing extended functionalities and greater flexibility across the automation pyramid, but also to broadly impact the automation ecosystem, creating and/or improving connections, relationships and value drivers among the automation stakeholders.

In the following sections, the principal characteristics of the current automation domain are analysed by focussing on the stakeholders (hereinafter complementors) that populate the ecosystem and on the structure of the relationships among them. The Daedalus platform is therefore presented by highlighting, beyond technological aspects described in Chapter 5, the value exchanges managed by the digital marketplace. The impact that the creation of such ecosystem has on the complementors is eventually discussed through an analysis of the to-be business networks.

12.2 An Outlook of the Automation Value Network

The automation industry is an interdisciplinary field, which involves a wide variety of tasks, product portfolios (machinery, control, equipment, small elements, etc.), technologies (robotics, software, etc.), standards and services, serving different sectors, with distinct requirements and needs. This environment is populated by many stakeholders, which, through complex and articulated value chains, collaborate to develop complete automation solutions. This section aims to provide an overview of this complex domain,

[1]An industry platform is defined as: foundation technologies or services that are essential for a broader, interdependent ecosystem of businesses [17, 18].

describing its characteristics, its players and the relation that they have established over time.

12.2.1 Characteristics of the Automation Ecosystem Stakeholders

The automation environment is currently populated by several different players, which can be grouped into five macro-categories:

- Component suppliers (CSs);
- Automation solutions providers (ASPs);
- Equipment and machines builders (E&MBs);
- System integrators (SIs);
- Plant owners (POs).

These macro-categories are the most relevant entities concurring in the design and development of industrial automation solutions in which different hardware & software elements, characterized by a high granularity of functionalities and complexity are integrated into the building of complex mechatronic systems. In Figure 12.1, connections, flows, and relationships among those of the automation domain have been thus summarized by providing a general

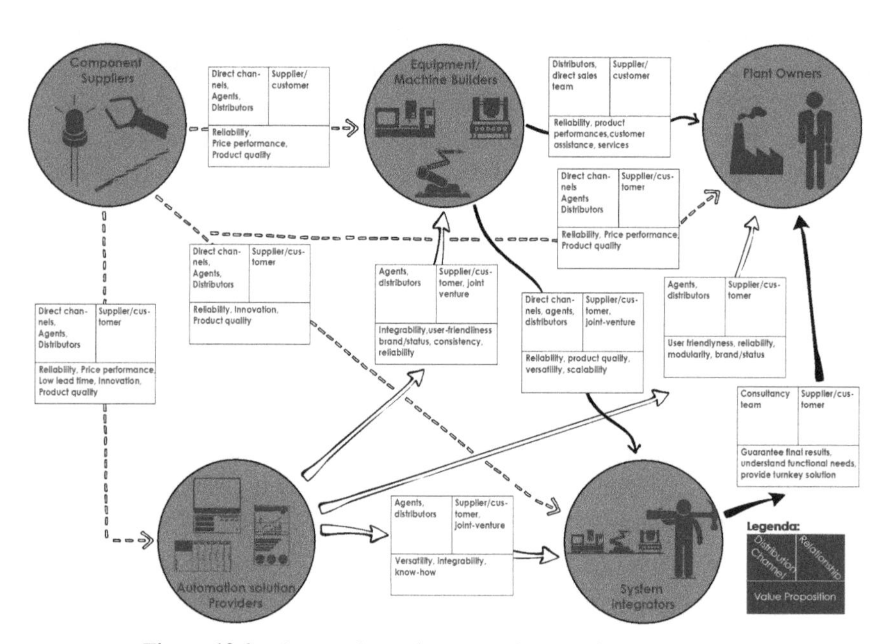

Figure 12.1 Automation value network general representation.

overview of the current automation value chains with a particular focus on: (i) Type of relation, (ii) Distribution Channels and (iii) Value-proposition. These elements drawn in boxes among complementors are intended to be univocal: for example, the value proposition of one player can vary a lot in accordance to the customer he serves.

The main interactions that can exist among the automation players are here presented with the aim of not covering all the possible interactions (biggest players frequently group under their umbrella more than one of the proposed stakeholders' functions; similarly, it is frequent that companies establish partnerships exposing a unique contact point with the customer), but describing the most common ones. The resulting schema highlights the linearity of the current automation ecosystem, where automation solutions, i.e. manufacturing lines, are the result of a "chronological" (even if very complex) interaction among players that goes from the granularity of low intelligence components, to the high integration and desired smartness of full manufacturing lines.

In the existing value chains, the automation solution to be purchased is still selected merely considering its hardware elements. Despite the great commitment exerted in software development to create integrated and versatile automation solutions, resulting in high impacts the software has in terms of costs and implementation efforts, but still it is not a primary decision-making parameter. The decision between a solution or another one depends first on the hardware (the component, the control system, the machine, the equipment, the production line) and, only in second instance, the software to integrate, coordinate, and/or use the entire system is selected. To this end, in the schema, it is not underlined in the relevance of the software, being considered a player in the background.

12.2.1.1 Automation solution providers

The automation solution provider (ASP) produces controllers for automation, such as PLC's, servo-drivers, HIM, safety devices and a wide variety of products. Companies such as Siemens, Allen-Bradley, Turk, Omron, Phoenix Contact, Rexroth, Mitsubishi and Schneider/Modic provide the hardware components of the control solution, the software to develop the programs and the standards on which the logic controls are based.

The choice between the different automation solution brands is based on several parameters including integrability, flexibility, scalability and reliability.

Controllers have a relevant impact on the realization of an automation solution. Developing a plant totally based on one single brand guarantees reliability and easy integration. Nevertheless, this decision involves a strong dependency, which can bring disadvantages in terms of costs, flexibility and change opportunities.

ASPs have usually a strict relation with machine builders. Depending on the adopted business model, it may happen that the automation element is directly provided by the equipment & machine builder (E&MB), which is an ASP itself (e.g. ABB and Yaskawa). In other cases, E&MBs develop some kind of drivers within their products, allowing to work with different automation solutions. For example, they may create a special driver for communication with profibus for Siemens, or Etherner/IP for Allen Bradley or just leave an open port of communication like MODBUS, to work with any other device. In many cases, an E&MB proposes to its customer automation solutions that are compliant with its machines. The customer decides which one to implement.

Customers have a relevant role for the ASP's business and their relationship can be resumed in two categories:

- Self-service: customers have a limited interaction with automation solution providers' employees. The main relation channel is often the website, where self-help resources such as white papers, case studies, videos and answers to frequently asked questions are available. There is often the possibility to use also a personal assistance in the form of phone and e-mail support (e.g. ABB).
- Consulting: direct sales forces consult the costumer to ensure that all the needs are met. The main objective is to establish a long-term relationship. Technical support is provided through personal and on-site assistance, but also through phone or online resources.

ASPs can offer consultancy services either directly or through the support of system integrators. Some system integrators (SIs) prefer to remain completely independent from ASPs, while others choose to form alliances, which take the form of membership in an ASP's program. This provides to its program members a wide number of benefits including training, advertising, marketing assistance, beta product trials, free product samples and more [5].

12.2.1.2 Components suppliers

Components supplier (CS) produces devices not executing, on their own, complex functionalities (i.e. influencing alone a whole production process).

Their main customers are E&MBs, and sensors, drives, panels, I/O clamps, etc. are typical "deliverables". Production is usually oriented towards a make-to-stock approach in large numbers, aiming at a wide application scope. Their business model mainly focuses on the premium segment and/or on the customization, where it is possible to obtain the largest profits, with a strong emphasis on their home country [10]. CSs usually try to grow through joint ventures and cooperation, exploring adjacent businesses also with horizontal integration.

12.2.1.3 Equipment and machines builders

In the current automation environment, the main business of an equipment & machine builder (E&MB) is the design and production of equipment and machines, through the assembly of different simple components, including low-level controllers and their logic control, in order to obtain more complex and functional systems. The integration and configuration of HMI, PLC, CNC, other accessories and tools depend on the business model of the player. In some cases, these activities are developed in-house, and in others, they are developed by SIs or directly by customers. E&MB is usually characterized by a strong level of internationalization that it intensifies through local value creation and shorter time to market. E&MBs are, with ASPs, the most advanced player from the technological point of view.

Considering their supply chain, the dependency of an E&MB on external suppliers is heavy in the case of high-value added elements such as numerical controls, drives, linear guides, spindles, clamps or specific/custom automation components. Some of these are bought on the market from multinational companies (typical examples are NC, drives and PLC), others are produced by specialized companies working closely with E&MBs (e.g. clamps and tooling).

In some cases, machines and the equipment are sold without the integration of the automation control. It is directly the customer or an SI that implements the automation. Who produces machines usually provides a list of compliant automation solutions, without expressing specific recommendations. It is the ASP that has to promote its products and be able to influence the buyer to install it. In the same way, the ASP does not suggest any E&MB. E&MB's business model has a particular impact on the relation with customers. Sometimes, the E&MBs rely on distributors partners, which sell and implement the basic configuration of their products based on customer needs. In other cases, in particular when the E&MBs is a big company, there is a direct relation, where consultants or agents interact with the customers.

It is necessary to consider also that customers can be both plant owners, who directly purchase the machines and equipment, and the SIs, who purchase it for a third party. Another element, which is influenced by E&MBs, is the machines integration. Some E&MBs provide this service, while others provide only the product and leave the integration to a third-party actor or to the customer. Depending on its needs and the type of equipment and machine, builders can produce basic, highly standardized and high-volume machines or customized ones, involving the customer in the development with a strict collaboration between customer and supplier.

12.2.1.4 System integrators

The main business of system integrators (SIs) focuses on assembly and integration of combined machinery systems, lines and equipment. SI has usually vertical competences in a specific sector (e.g. food and beverage, wood, textile, packing etc.), due to the need of having a deep knowledge of available technologies.

The SI starts from customer's functional needs to design, propose and implement turnkey solutions. These are developed through the combination of existent and new resources. In many cases, SI develops codes in different languages, providing low-level SW (application, libraries, algorithms) to connect, integrate or add basic functionalities to machines, lines and plants. SI can provide support also for request-for-proposal and after-sale maintenance.

Usually, every SI has its main, trusty and reliable suppliers they select among on the basis of specific customer requirements. If it has not specific brand needs, the system integrator appeals on well-known or partner suppliers. SI usually purchases components from different vendors, depending on which integration it is working on and on the customer requirements. The relationship with components suppliers, which is often intermediated by the distributors, is driven by different elements, such as personal relationship and past relationships/experience. In many cases, the system integrator prefers known suppliers, with a long time, inter-personal relationship, which guarantee a service that goes beyond a simple buyer–supplier relationship (e.g. delivery outside the working time). Price is also a relevant aspect, especially when the customer is directly involved in the choice. If the customer has no specific requirements about the automation controllers, HMI, software, etc., the SI adopts the technology he knows better. If the employee knows specific languages and software and there is not the need to change them, then the relative automation solution will be adopted.

12.2.2 Beyond Business Interactions: Limitations and Complexities of the Existing Automation Market

Adopting and using automation solutions requires the involvement of experienced and skilled employees, whose competences are developed during time and cover all the value and supply chain steps. Being able to maintain a sustainable value chain, where all the members have the proper revenue seems to be therefore the winning factor fostering continuity, customer loyalty and product familiarity. For this reason, in particular from the final user point of view, consistency, time continuity and familiarity of the supplier/solution are often more relevant than innovation itself. Also, price and services are relevant aspects to be considered for an automation solution, but if they are aligned between competitors, the personal relationships and the known modus operandi have a higher impact.

Actually, the automation environment is very conservative and slow to make changes: evolutions and innovations are often seen more as concerns than as opportunities of improvement. Automation solution providers, whose main products are PLCs and control systems, are the main rulers. Current PLC technologies, which impact the deployment of industrial automation applications, are a legacy of the 1980s, unsuited for sustaining complex systems composed of intelligent devices. The current control systems, which have at the base the IEC-61131 standard, are outdated from a software engineering point of view. Moreover, they have been implemented by each vendor through specific "dialects" that prevent real interoperability and they are strictly dependent on the hardware on which they run. Therefore, the automation solution brand is considered by the customer as a relevant aspect. If a company wants to access a specific market, it has to adapt its product to that context specificities. For example, in Germany, if a machine does not have Siemens PLC, probably it will not be sold. The low inclination to changes of this sector, due to the strong dependency on reliability and on well-established know-how, does not push the actors towards innovative changes (not even ones pulled by the Industry 4.0 principles).

In addition to the previously mentioned domain's issues, there are different technological limits that should be included to obtain a wide representation. Obsolescence of automation systems has a relevant impact on the all automation solution life cycle. In order to be compliant to the 4th industrial revolution principles, the access to data related to an equipment, a machine, a line, a plant and a factory should be available at any level of the supervisory and management hierarchical architecture. On the contrary, constraints and limits imposed by HMI or SCADA systems, designed and implemented to fulfill the requirements identified at the design stage restrict the access to data. Moreover, should additional information requirements emerge not included/considered in the initial design (e.g. for monitoring performance improvement purposes), existing legacy systems require a modification of the PLCs and a reconfiguration of the SCADA system (and/or HMIs). This upgrading of the system becomes expensive and risky, in particular, if applied to many controllers. In addition, flexibility and optimization of the manufacturing plants do not merely ask to access the raw data available on controllers (like status variables and/or sensors data), but also to computation and/or smart functionalities offered by the increasingly embedded intelligence. The software tools composing the ICT infrastructure of a factory could take advantage of the equipment/machine-embedded computation capabilities with the application of the appropriate functionalities. In classical automation systems, all these kinds of interactions, data elaborations and data delivery are defined at the design stage of the automation software, considering the requirements available in that step. When changes of those specifications should be considered after the automation system is implemented, there could be the need to modify the automation software on many controllers and this requires to be aware of the details about how the systems were implemented. These kinds of modifications are rarely applicable in real productive scenarios and this affects the upgrading and revamping of legacy systems, actually dampening innovation.

12.3 A Digital Marketplace to Support Value Networks Reconfiguration in the Automation Domain

As highlighted in the previous chapter, a classic value chain, characterized by processes linked together in support of a value network is not typical of the automation industry. In fact, the influence of the upstream companies is relevant on the final product, be it a machine, an entire line or a plant. For this reason, value creation in the interdisciplinary automation industry cannot be represented as a chain: it is a value network where the same company can act both as a supplier and as a consumer of products and solutions. In this value network, services along the process steps are becoming more and more important, in particular when offered in connection with a physical product [5] (the so-called "product-related services").

Digital platforms have been widely adopted in the last decade as instruments to support the diffusion of product-related services, reducing transaction costs and facilitating exchanges that otherwise would not have occurred [11]. The main value that the platforms create is the reduction of the barriers of use for their customers and suppliers. A platform encourages producers and suppliers to provide contents, removing hurdles and constraints that are part of the traditional businesses. As for suppliers and producers, platforms create significant value also for consumers, providing ways to access to products and services that they have not even been imagined before. Platforms allow users to trust in strangers, allowing them to enter in their rooms (Airbnb), renting their cars (Uber) and using their applications (Phone and PC marketplaces). Platforms provide and guarantee for users' reliability and quality. New sources of supply can cause undesirable contents, if not filtered, while thanks to the reliability and quality mechanisms that platforms integrate, this issue becomes not relevant.

The platform developed within the Daedalus project follows this trend by extending platform logics to the automation domain. This is done exploiting as a foundation the new evolution of the IEC-61499 standard that envisages the technology on the top of which additional value drivers for the automation complementors can be set up. The Standard allows: (i) the design and modelling of distributed control systems and application execution on distributed resources (not necessarily PLC), (ii) the creation of portable and interchangeable data and models and the re-utilization of the code, (iii) the seamless management of the communication between the different function blocks of an application (independently from the hardware resource they run on) and (iv) the decoupling of the elements of the CPS (its behavioral models) from the physical devices and reside (designed, used and updated) in-Cloud, within the "cyber-world", where all the other software tools of the virtualized automation pyramid can access them and exploit their functionalities.

Among the others, code modularity, reusability and reconfigurability of systems are the main features that are advertised as practical benefits of applying this Standard [12]. The final result is the ability of designing more flexible and competitive automation systems by providing the functionality to combine hardware components and software tools of different vendors within one system as well as the reuse of code [13, 14].

The Daedalus platform is therefore meant to bring together automation complementors and give them the infrastructural support to technologies, services and skills essential for systems improvement through CPS integration and orchestration [15]. This is done by opportunely adapting the functional model already implemented successfully within the IT world for mobile applications developing a digital place (i.e. a marketplace), where automation-related applications and services will be shared among platform users. The digital infrastructure will allow also machine, equipment and components manufacturers to exploit a common platform where to share updates and extended functionalities of their systems.

12.3.1 Architectural Characteristics of the Digital Marketplace

The Digital Marketplace represents the reference interface to be adopted by developers/manufacturers of IEC 61499-compliant CPS(s), interested to sell their products via a multi-sided marketplace and able to match-make product offer with plant owners, equipment manufacturers or system integrators needing their solutions.

The proposed digital infrastructure takes advantage of the faceted nature of a CPS (aggregation of hardware, software and digital twin), to make the decoupling between the mechatronic system, the control application and the digital twin the lever to support the integration of third-party developers and service providers. Thanks to the nature of the IEC 61499, the platform relies on CPS(s), and mechatronic systems may be indeed controlled by different intelligences, potentially made by different developers, which represents a big opportunity for developers who want to create their own control application. Therefore, the Digital Marketplace is not only a repository of CPS, but it provides a set of services enabling developers and manufacturers to test and validate their own products. The Digital Twin is used in this case as the instrument to simulate the mechatronic system in a well-known and certified simulation environment, providing a digital way to validate the cyber part of a CPS.

The Digital Marketplace is a Web-based application that exposes a set of Web services that allow external components, such as portals, IDE, applications, etc. to be connected with the Digital Marketplace and exploit its provided functionalities. At the architectural level, the marketplace is composed of the software components, exposed interfaces and interaction flows proposed in 0, whose main elements are (Figure 12.2):

- The *Persistency Layer:* it is the fundamental layer on which the rest of the architecture is based, and is divided into two main components: the *Repository* and the *Persistency Manager*. The first one represents the knowledge base of the Marketplace and it is composed of the hosted CPS(s), the *Economic Data Model*, meant to describe all economic aspects of the products (pricing strategies, fees, etc.) and the *Semantic Data Model* meant to characterize the submitted product in order

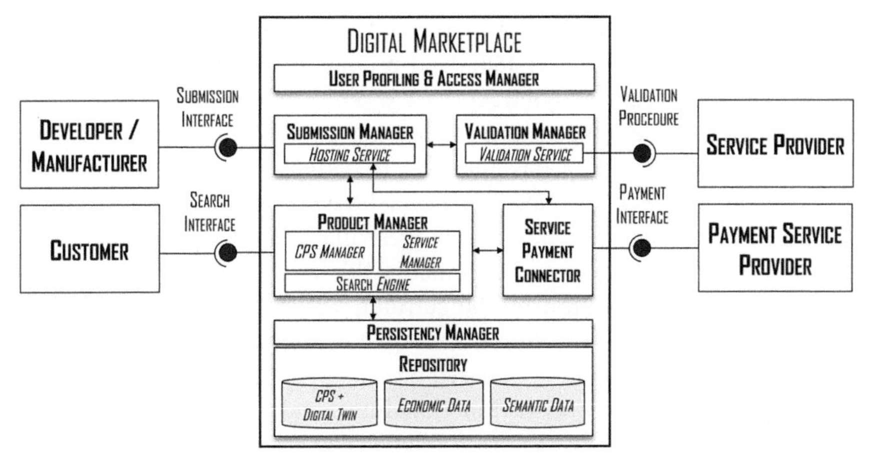

Figure 12.2 Digital Marketplace Architecture.

to be accurately searched and identified. The *Semantic Data Model* is managed by the *Product Manager* component which is also in charge of providing discovery functionalities of the hosted products. In particular, once the CPS has been successfully submitted and validated, it becomes available to customers who want to buy and use it. For this purpose, the Digital Marketplace provides a search engine mechanism based on a set of algorithms meant to result the best possible answer to a search query. The aim of this semantic search is to improve search accuracy by understanding the customer's intent and the contextual meaning of terms as they appear in the searchable dataspace, within the system, to generate more relevant results. Semantic search systems consider various points including context of search, intent, variation of words, synonyms, meaning, generalized and specialized queries, concept matching and natural language queries to provide relevant search results. The semantic search will produce a result containing the list of suggested products, whose characteristics answer the customer's needs.

- The *Submission Manager* is the software component in charge of regulating the CPS submissions process starting from the request, passing through the validation, to the payments. Both the payment service connector and the validation manager are directly connected with the product manager.
  The submission manager exposes a submission interface, which allows to describe the submitted product in terms of: general description of the

product, set of functionalities provided by the CPS, set of compatibilities with existing mechatronic systems and pricing strategies by which the marketplace will manage contracts of the products' usage between the marketplace itself and the customers.
- The *Validation Manager* has the aim to validate all the submitted products in terms of provided functionalities. This component is in charge of managing the validation process that requires to simulate or test the Digital Twin of the submitted CPS into a simulation/testing environment, properly built by the certifier, where both the context of execution and the CPS' operations are replicated. The validation process follows a well-defined protocol based on the objective criteria, aimed to verify if the CPS specifications/functionalities, under certain conditions, are satisfied or not. In order to guarantee the tests repeatability, the CPS tester must publish, into the Digital Marketplace, the testing results accompanied by the applied testing protocol.
- The high-level component belonging to the Digital Marketplace is the User Profiling Manager, which is in charge of managing the user profiling in terms of roles, authentication and authorization.

In order to transform the described Architecture in a functional marketplace, an overall data model, encompassing both the digital integration of all technological elements of the project and the definition of revenue creation mechanisms has been therefore developed. The basic idea of this model is to provide a set of technical functional specifications (by using UML diagrams) aimed to cover the design of all needed mechanisms meant to guarantee the economic and technical aspects behind the Digital Marketplace.

The *Digital Marketplace data model* (Figure 12.3) aimed to cover, on the one hand, all the economic aspects of the products in terms of prices, contracts, etc. and, on the other hand, a detailed description of the hosted CPS.

The *Digital Marketplace data model* not only aims to describe the automation application in terms of "what a certain automation application does" but also how it does something. The design of the data model has the aim to fully characterize a product of the ecosystem in order to be accurately searched and identified. For that reason, the creation of such data model considered aspects like exposed functionalities, compatibilities, specifications, meaning of the application I/O, application extensibilities, description of the logics that the automation application wants to provide and the openness of its source code.

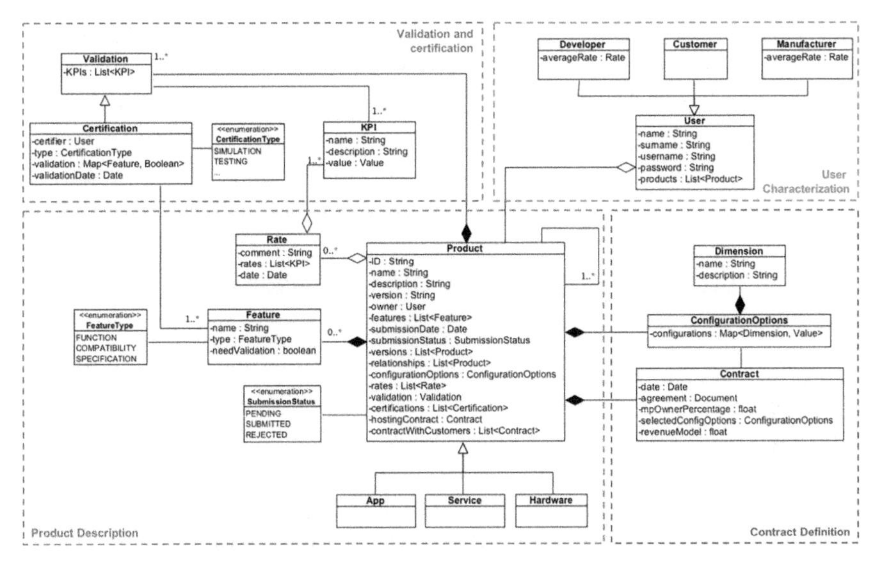

Figure 12.3 Digital Marketplace Data Model.

The designed data model has been divided into five sections, each respective to one of the five packages of the structure presented in the figure above:

1. ***User Characterization package*****:** it contains all data entities related to the user description and characterization. This part of the data model deals with the representation of the User, being it a developer (Developer class) or a manufacturer (Manufacturer class) or a customer (Customer class), and all the related information.
2. ***Product Description package*****:** it contains the data objects needed to describe the products (hardware, application and services) hosted by the Marketplace. This package groups the set of entities needed to formalize the data structure for describing the hosted products in terms of features, possible relationships with other products, product contract configurability, product validation and certification.
3. ***Contract Definition package*****:** this package contains all entities needed to formalize all possible configuration options of the contract that regulate the economic aspects between the Marketplace and the customers about the use of the products.
4. ***Validation and Certification package*****:** this part of the data model is dedicated to formalize those entities meant to support the validation of the submitted product and the optional product certification.

The validation phase has to be intended as part of the product submission process, where, according to the terms of the contract between the developer/manufacturer and the Marketplace, the submitted product undergoes a validation of the provided features. In particular, a validation has to be considered as a more specific validation service provided by the Marketplace and released by a validation service provider.

12.3.2 Value Exchange between the Digital Platform and Its Complementors

If the marketplace described in the previous chapter is the technological infrastructure supporting the exchange of value (products, money, services) among automation stakeholders, the data model underpinning it provides the logical constructs enabling to deploy the rules running ecosystems exchanges. The business model characterizing the Digital Marketplace instantiation is eventually the description of how these rules are managed and how the cost/revenue structure of the marketplace is arranged [16]. The entity of the cost/revenue structure behind value exchange among platform complementors is strictly dependent on the specific implementation scenario that the marketplace will assume (type of platform owner, network of existent suppliers involved, approach to system integrators, etc.). Therefore, the economic dimensions required to assure the profitability of the whole ecosystem have to be defined on a case-by-case basis, in accordance with the specific implementation scenario and the specific business case. The marketplace dynamics driving exchange of value among complementors and more in general, the type of transactions that it can enable can be generalized and discussed without referring to a specific business case.

Four sources of value are at the base of the marketplace dynamics:

1. *Money and credits:* this is the most common form of value that is exchanged by customers and suppliers in return for goods and services delivered. As normally happens, on these transactions, the Marketplace builds its profitability.
2. *Goods and services:* as anticipated in §0, the Marketplace supports the trading of IEC 61499-compliant CPS (aggregation of hardware, software and digital twin), exploiting the independent nature of each CPS component to extend the number of elements that can be traded. Goods and services are therefore hardware components, the related software control part (developed by the company or by application developers), software applications that can support the integration of hardware components

across lines and/or the integration of CPS with IT components of higher levels of the automation pyramid and services provided by third parties related to the deployment of CPS (e.g. integration of simulation services supporting software applications testing and validation).

3. *Information:* the Marketplace is expected to host supporting material for companies/system integrators intended to integrate IEC 61499 technologies in their business and for developers approaching IEC 61499 programming, together with the related software development kit (SDK) supporting software development.
4. *Intangible value*: in order to support customers in selecting hardware and/or software components and services to be deployed, the marketplace supports the delivery of intangible value across each transaction in the form of evaluations of delivered products/services. The customer can therefore rely on a set of credentials of the supplier represented by the evaluation of its work provided by other customers.

In Figure 12.4, for each complementor, the main exchanges supported by the marketplace have been therefore highlighted by representing through arrows the forms of value described above. The direction of the arrow shows whether the value is taken form the platform, delivered to it or both. The diagram also summarizes the impact on the value proposition that the platform supports (further discussed in §12.3.3).

To describe the main interactions occurring among marketplace and Complementors, the complementors have been grouped into four categories of stakeholders (Table 12.1): Customers, Application developers, Service providers and Hardware developers. In the following table, the mapping among the proposed categories and the overall Marketplace complementors is presented. Some of them can play the role of both providers of product/services delivered by the marketplace as well as customers.

12.3.2.1 Customers

The main relationships that customers have with the marketplace are: the purchase of product/services, agreements with product/service providers mediated by the marketplace and rating of the delivered product services. To this end, customers are meant to start the interaction with the marketplace by performing a registration that allows them to store and retrieve data related to their buying experience. By browsing the hardware and software catalogues, customers can select the product/service they are interested in and visualize the software/hardware products or services associated to the selected product.

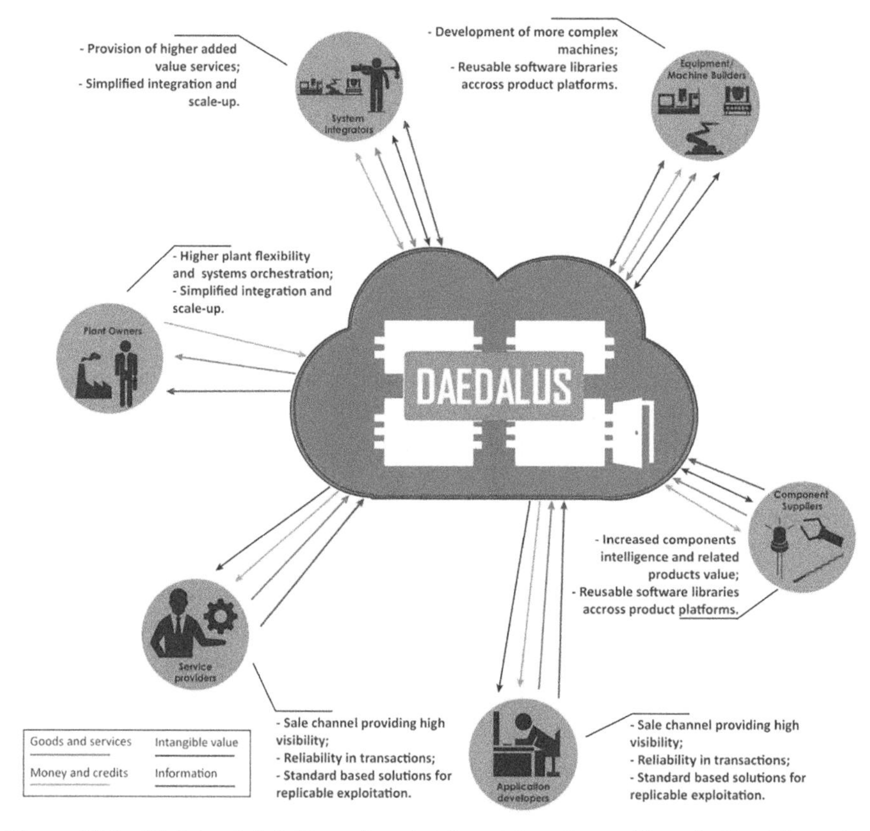

Figure 12.4 High-level definition of marketplace interactions with main Daedalus stakeholders.

Table 12.1 Mapping of stakeholders on Marketplace ecosystem

Type of Stakeholder	Mapping on Marketplace Ecosystem
Customer	Plant owners; system integrators; equipment/machine developers; component suppliers
Application developer	Application developers; system integrators; equipment machine developers
Service providers	Service providers
Hardware developers	Equipment/machine developers; component suppliers

Each product is indeed connected with specific software/hardware/services that can operate together (i.e. if browsing a sensor, then the marketplace suggests the applications supported by the hardware itself and the services of integration supported).

The selection of one product enables the customer to access the contractual area, where the contract among the customer and the marketplace is agreed, and recall the payment service. In its interactions with the marketplace, the customer is not charged for the services provided: it is always the product/service provider that pays a percentage fee.

Once completed the purchase, the customer can exit the marketplace. He will then receive the products/services according to the modalities agreed within the contract. Customer will receive notifications with respect to software updates in order to improve the customer experience and support the maintenance of updated hardware/software functionalities.

12.3.2.2 Hardware developers

Hardware developers are a category of marketplace end-users very important to its deployment; indeed, it is expected that in the first stages of marketplace life, hardware developers will provide both hardware and software applications to run on it. To this end, they will be the first category of stakeholders providing contents of the marketplace.

If maintaining the perspective on the sole hardware, then the marketplace will give hardware manufacturers the possibility to store product catalogues, giving the facilities to define characteristics, specification and costs of their products. As for customers, the first access will require a registration giving them the access to a dedicated page where they will be able to set up the characteristics of their account. In parallel, the marketplace will also provide the infrastructure for the definition of the contracts with customers, leaving manufacturers the freedom, among certain constraints set by the marketplace, to configure the contracts setting the relationship with the customer (i.e. cost of product, type of business model, purchase/product as a service, etc.).

The hardware manufacturer will be billed by the marketplace at two levels: on a first tier, paying a fixed cost for the hosting of the products catalogue and, on a second level, with a percentage fee on each transaction with customers. The economical dimensions of both the fixed cost and the transaction fee will be decided according with the specificity of instantiation of the marketplace.

12.3.2.3 Application developers

Application developers will find on the marketplace the infrastructure to host their applications and sell them. Similarly, to hardware developers, the marketplace will give them the facilities to define characteristics, specification and costs of their products. Moreover, considering the type of product sold, the marketplace will also provide specific contracts templates supporting characteristics of an application sale (in-app purchase, period-based licence, etc.).

The marketplace will be also the place where developers will be able to find, accessing dedicated spaces, the quality procedures and SDK required to develop applications compliant with the ecosystem. These services are provided without additional costs to the developers.

12.3.2.4 Service providers

Service providers are meant to benefit from the marketplace by increasing the visibility of the provided services. Similarly, to other stakeholders, the marketplace gives them the facilities to describe and host their services and set up contracts related to service provision. In exchange, the marketplace charges them a percentage fee for the services sold. The marketplace also make revenues by giving priority advertise for those service providers paying an additional fee.

12.3.3 Opportunities of Exploitation of an Automation Platform

As already mentioned in the previous sections, the creation of a platform-based automation ecosystem is expected to have a relevant impact on the way that automation complementors manage their business. A platform relying on IEC 61499, to support transactions in a complex ecosystem as that of automation, should guarantee to be completely open and hardware-independent, enabling full interoperability and much-deeper portability and reusability of application developments. The specific deployment of technologies and ecosystem should be first targeted to the most innovative and pioneering SMEs and large enterprises in Europe, already oriented to accepting a decentralized approach to automation. These will be the first players that can adapt their business model, in order to be successful in a platform-based automation ecosystem. In the transition toward such ecosystem, opportunities and challenges are clearly generated for all the automation stakeholders. For each of them, a brief description of the expected challenges is provided hereinafter.

12.3.3.1 Opportunities for system integrators

Among the automation players, System Integrators (SIs) are one of the stakeholders closest to the customers. Considering their active role in understanding customers' functional requirements to propose ready-to-use solution, they have a direct vision on their main needs. In this context, SIs are realizing, more than any other automation player, that customers are requiring more flexible and reconfigurable solutions, capable of increasing production performance and providing more advanced functionalities.

On the other hand, in the current automation environment, SIs have a marginal role in adding value for customers and have low technological competences. They usually integrate different components, equipment and machines to provide functional and ready-to-use solutions. They mainly perform the operative part, which does not only allow to cover customer needs, but only to satisfy their functional requirements.

Adhering to a platform-based ecosystem, SIs will no longer be a simple assembler, but they will have the opportunity to add relevant value to the automation solution. This could be done by developing SW for their customers and proposing dedicated solutions, which add functionalities, improve performances and manage orchestration and distributed architecture between the different factory levels. SIs have therefore the opportunity not only to deliver functional solutions meeting customer requirements but also to add functionalities to the systems, increasing the value of the overall proposed solution. Moreover, thanks to reduced hardware dependence, code re-usability and modularity achieved through the adoption of IEC 61499 logics, SW use could be extended in different contexts, for different customers application.

The first opportunity for SIs will be the update of existing legacy automation systems. For the first adoption of platform principles, CPS-izers (systems that are meant to act as an adapter among legacy and IEC 61499 technologies) develop a fundamental role, allowing SIs to transform solutions tied to old legacies to compliant ones. The higher integrability of components, equipment and machines will allow SIs to reduce the effort to provide ready-to-use solutions and to ease the integration of new functionalities by developing dedicated SW. This becomes a relevant activity that is expected to be mainly internalized by SIs. Thanks to the platform and the related marketplace, they will have the opportunity to re-use libraries and algorithms developed by third-party developers to improve or speed up the development of their SW solutions.

All these elements are meant to increase the value proposed to customers, allowing to extend solutions' functionalities and enabling to dedicate more resources to the development of high-level applications and SW, while reducing efforts and resources for components, equipment and machines integration and basic functions programming.

It is necessary to consider that SIs are the players that can achieve the highest benefits from platform-based automation ecosystem, but to which are also required the main transition efforts. In this kind of domain, SI becomes a more advanced player, to which are required more technological competences. It is no more a simple consultant, but it also a product (SW) developer. It is expected that SIs expand their know-how and competences from low operative level to higher, with the objective to provide more added value to its customers, not only through integration, but also through the improvement of performance and functionalities, maintaining them during the whole solution's life cycle.

12.3.3.2 Opportunities for equipment and machines builders

E&MBs, adhering at the ecosystem and adopting the related technologies, have the opportunity to release more advanced products, able to work in flexible and orchestrated production systems. E&MBs can produce complex manufacturing systems as aggregation of CPS, focusing their effort on the assembling and orchestration of the automation tasks of these composing elements. The adoption of platform technologies can allow an E&MB to develop products that can take advantage of all the components (control software, applications and services) IEC-61499 compliant.

For E&MBs, the platform will become a relevant resource, being one of the structural technologies on which its products will be designed and produced. The management of this resource should be performed with particular attention, in order to spread out all the possible benefits and to maximize products' performances.

12.3.3.3 Opportunities for components suppliers

The platform-based ecosystem generates opportunities also to CSs. They have to become capable of releasing more functional, intelligent and independent components. Components can be designed and developed as more complex elements (such as CPS), already equipped with on-board distributed intelligence. A CS should not be focused only on reliability, quality, price and lead time. It should innovate its products adding functionalities. Therefore, CSs will have the opportunity to provide not only hardware, but also SW, adding value to their solutions and increasing the revenue opportunities, creating a closer relation with their customers.

12.3.3.4 Opportunities for automation solutions providers

Thanks to the extended functionalities that it brings by, the IEC 61499 standard can have the potentiality to affirm as a competing standard to the

IEC 61131, currently largely adopted by Programmable Logic Controllers. If this situation actually happens, ASPs are expected to have two behaviors: (i) they can adopt the IEC-61499 standard, implementing their own "dialect" and tools, to create their own IEC-61499 automation ecosystem and (ii) they can try to stop its adoption, taking advantages of their position of strength which ties customer to their legacy solutions.

12.3.3.5 Opportunities for new players

The platform-based ecosystem and in particular the marketplace create the opportunities to all those ICT companies and software developers that aim to make business in the automation market. Application developers (AD) will be a new player of this environment that arises through thanks to the transition to platform-based business model.

These players will have the opportunity to develop compliant software for general-purpose usage scenarios, customizable by CSs, E&MBs, SIs and/or customers for their specific projects. Through the distributed intelligence, software will acquire a more relevant role, through which customers can increase functionalities and performances of equipment, machines, lines and plants, obtain data and/or perform analysis. Added value is provided by guaranteeing special functionalities based on specific competence, quality of implementation and performance achieved.

12.3.3.6 Service providers

SP provides services and support to POs and SIs. Exploiting the IEC-61499 benefits the possibility to develop an extended amount of new services with the aim of creating a digital representation of the system, perform simulation, analysis, test application, and/or store data. The described platform can become the environment where these services are made visible and brought to the market. In this sense, their business model is similar to AD's, but instead of providing SW, SP provides services to be integrated in manufacturing lines design and deployment.

12.4 Conclusions

In the last decades, the automation domain has been characterized by an ecosystem ruled by legacy technologies, where the dominant role of the chosen hardware solutions strongly constrains reusability, upgradability and orchestration of manufacturing systems. This situation led to the rise of important barriers to the shift towards competing or optimized solutions, limiting the potentialities of upgrade and flexibility of the systems.

In this context, the digital platform developed within the DAEDALUS project, relying on the extended functionalities provided by the upgrade and deployment of IEC 61499 in the CPS domain, stands out as a ground-breaking platform able to revolutionize the whole approach to how automation systems are conceived, designed and set up. The infrastructure developed is therefore the first step to achieve the challenge of developing a platform able to foster the creation and deployment of more efficient, flexible and orchestrated production systems, easy to be integrated, monitored and updated. The proposed platform is able to widely manage CPS in their multifaceted sense (HW, SW, Digital Twin), reaching different (even complementary) customers and offering new opportunities to developers in terms of possibility to create own(ed) control applications and of exploiting validation services thanks to the hosted digital twin. As a consequence, the platform drives a reconfiguration of the automation value network, with the aim of releasing the main issues currently faced by the sector and extending the value drivers that characterize their interactions.

Next steps to be carried out in order to create a digital platform meeting the needs of the current industrial markets (customers) are envisaged in (i) the creation of specific mechanisms and procedures, software interfaces, and incentivizing system, all supporting the large adoption of the platform, (ii) further elaborating methodologies and outcomes of processes and services supporting CPS validation, (iii) integrating in the platform value added services for customers (e.g. performances assessment of the machines, management of manufacturing systems, manufacturing data elaboration for predictive maintenance forecasting) and (iv) implementing a business development strategy intended to actually deploy in the market the logics proposed by the Digital Marketplace.

Acknowledgements

The work hereby described has been achieved within the EU-H2020 project DAEDALUS, that has received funding from the European Union's Horizon 2020 research and innovation programme, under grant agreement No. 723248.

References

[1] W. B. Arthur, *The Nature of Technology - What It Is and How It Evolves,* 2011.

[2] K. C. Mussi, F.B., Canuto, "Percepção dos usuários sobre os atributos de uma inovação," *REGE Rev. Gestão,* vol. 15, pp. 17–30, 2008.

[3] R. da S. Pereira, I. D. Franco, I. C. dos Santos, and A. M. Vieira, "Ensino de inovação na formação do administrador brasileiro: contribuições para gestores de curso," *Adm. Ensino e Pesqui.,* vol. 16, no. 1, p. 101, March 2015.

[4] A. Bharadwaj, O. A. El Sawy, P. A. Pavlou, and N. Venkatraman, "Digital Business Strategy: Toward a Next generation of insights," vol. 37, no. 2, pp. 471–482, 2013.

[5] M. Müller-Klier, "Value Chains in the Automation Industry."

[6] R. Depietro, E. Wiarda, and M. Fleischer, "The context for change: Organization, technology and environment," in *The processes of technological innovation,* Lexington, Mass, pp. 151–175, 1990.

[7] J. Tidd, "Innovation management in context: environment, organization and performance," *Int. J. Manag. Rev.,* vol. 3, no. 3, pp. 169–183, September 2001.

[8] J. Tidd, J. Bessant, and K. Pavitt, *Integrating Technological, Market and Organizational Change.* John Wiley & Sons Ltd, 1997.

[9] Z. Arifin and Frmanzah, "The Effect of Dynamic Capability to Technology Adoption and its Determinant Factors for Improving Firm's Performance; Toward a Conceptual Model," *Procedia - Soc. Behav. Sci.,* vol. 207, pp. 786–796, 2015.

[10] Mckinsey&Company, "How to succeed: Strategic options for European Machinery," 2016.

[11] P. Muñoz and B. Cohen, "Mapping out the sharing economy: A configurational approach to sharing business modeling," *Technol. Forecast. Soc. Change,* 2017.

[12] V. Vyatkin, "IEC 61499 as Enabler of Distributed and Intelligent Automation: State-of-the-Art Review," *IEEE Trans. Ind. Informatics,* vol. 7, no. 4, pp. 768–781, November 2011.

[13] M. Wenger, R. Hametner, and A. Zoitl, "IEC 61131-3 control applications vs. control applications transformed in IEC 61499," *IFAC Proc. Vol.,* vol. 43, no. 4, pp. 30–35, 2010.

[14] T. Bangemann, M. Riedl, M. Thron, and C. Diedrich, "Integration of Classical Components Into Industrial Cyber–Physical Systems," *Proc. IEEE,* vol. 104, no. 5, pp. 947–959, May 2016.

[15] G. Landolfi, A. Barni, S. Menato, F. A. Cavadini, D. Rovere, and G. Dal Maso, "Design of a multi-sided platform supporting CPS deployment in the automation market," in *2018 IEEE Industrial Cyber-Physical Systems (ICPS),* pp. 684–689, 2018.

[16] A. Barni, E. Montini, S. Menato, and M. Sorlini, "Integrating agent based simulation in the design of multi-sided platform business model?: a methodological approach," in *2018 IEEE International Conference on Engineering, Technology and Innovation/International Technology Management Conference (ICE/ITMC),* 2018.

[17] A. Gawer, "Platform Dynamics and Strategies: From Products to Services," in *Platforms, Markets and Innovation,* Edward Elgar Publishing.

[18] A. Gawer and M. Cusumano, "Industry Platform and Ecosystem Innovation," *J. Prod. Innov. Manag.,* vol. 31, no. 3, pp. 417–433, 2013.

13

Migration Strategies towards the Digital Manufacturing Automation

Ambra Calà[1], Filippo Boschi[2], Paola Maria Fantini[2], Arndt Lüder[3], Marco Taisch[2] and Jürgen Elger[1]

[1]Siemens AG Corporate Technology, Erlangen, Germany
[2]Politecnico di Milano, Milan, Italy
[3]Otto-von-Guericke University Magdeburg, Magdeburg, Germany
E-mail: ambra.cala@siemens.com; filippo.boschi@polimi.it; paola.fantini@polimi.it; arndt.lueder@ovgu.de; marco.taisch@polimi.it; juergen.elger@siemens.com

Today, industries are facing new market demand and customer requirements for higher product personalization, without jeopardizing the low level of production costs achieved through mass production. The joint pursuit of these objectives of personalization and competitiveness on costs is quite difficult for manufacturers that have traditional production systems based on centralized automation architectures. Centralized control structures, in fact, do not guarantee the system adaptability and flexibility required to achieve increasing product variety at shorter time-to-market. In order to avoid business failure, industries need to quickly adapt their production systems and migrate towards novel production systems characterized by digitalization and robotization.

The objective of this chapter is to illustrate a methodological approach to migration that supports decision makers in addressing the transformation. The approach encompasses the initial assessment of the current level of manufacturing digital maturity, the analysis of priorities based on the business strategy, and the development of a migration strategy. Specifically, this chapter presents an innovative holistic approach to develop a migration strategy towards the digital automation paradigm with the support of a set of best practices and tools. The application of the approach is illustrated through an industrial case.

13.1 Introduction

In recent years, lot of research has been devoted to the improvement of control automation architectures for production systems. Latest advances in manufacturing technologies collaborate under the Industry 4.0 paradigm in order to transform and readapt the traditional manufacturing process in terms of automation concepts and architectures towards the fourth industrial revolution [1]. The increasing frequency of new product introduction and new technological development leads to more competitive, efficient and productive industries in order to meet the volatile market demands and customer requirements.

The Industry 4.0 initiative promotes the digitalization of manufacturing in order to enable a prompt reaction to continuously changing requirements [2]. The envisioned digitalization is supported by innovative information and communication technologies (ICT), Cyber-Physical Systems (CPS), Internet of Things (IoT), Cloud and Edge Computing (EC), and intelligent robots. The control architecture is a key factor for the final performance of these application systems [3]. Therefore, new automation architectures are required to enhance flexibility and scalability, enabling the integration of modern IT technologies and, consequently, increasing efficiency and production performance.

For this purpose, within the last years, a lot of decentralized control architectures have been developed in different research projects highlighting the benefit of decentralized automation in terms of flexibility and reconfigurability of heterogeneous devices [4]. However, after years of research, the reality today shows the dominance of production system based on the traditional approach, i.e. the automation pyramid based on the ISA-95 standard, characterized by a hierarchical and centralized control structure.

The difficulty in adopting new architectural solutions can be summarized in two main problems:

- Enterprises that are reluctant to make the decision to change;
- Projects that fail during the implementation or take-up.

Manufacturers are reluctant to adopt decentralized manufacturing technologies due to their past large investments on their current production facilities, whose current lifetime is long and, therefore, the required changes are sporadic and limited. In addition, methods and guidelines on how to integrate, customize, and maintain the new technologies into the existing ICT infrastructure are unclear and often incomplete. Nevertheless, with the advent of future technologies and with current market requirements, changes during the whole life cycle of the devices and services are necessary.

These changes lead to the transformation of the existing production systems and their migration towards the digital manufacturing of the Industry 4.0 paradigm. The term "migration" refers to the changing process from an existing condition of a system towards the desired one. Here, specifically, the migration is considered as a progressive transformation that moves and the existing production system towards digitalization. Migration strategies are thus essential to support the implementation of digital technologies in the manufacturing sector and the decentralization of the automation pyramid, in order to achieve a flexible manufacturing environment based on rapid and seamless processes as response to new operational and business demands.

Aligned to this vision, the aim of the EU funded project FAR-EDGE (Factory Automation Edge Computing Operating System Reference Implementation) [5] is twofold: it intends not only to virtualize the conventional automation pyramid, by combining EC, CPS and IoT technologies, but also to mitigate manufacturers' conservatism in adopting these new technologies in their existing infrastructures. To this end, it aims at providing them with roadmaps and strategies to guarantee a smooth and low-risk transition towards the decentralized automation control architecture based on FAR-EDGE solutions. Indeed, migration strategies are expected to play an essential role to the success of the envisioned virtualized automation infrastructure. To this end, FAR-EDGE is studying and providing smooth migration path options from legacy-centralized architectures to the emerging FAR-EDGE-based ones.

This chapter aims at describing the migration approach developed within the FAR-EDGE project. After this brief introduction, the state-of-the-art migration processes, change management approaches and maturity models are presented in Section 13.2, providing the founding principles of the FAR-EDGE migration approach presented in Section 13.3. An industrial use case application scenario is presented in Section 13.4, which is assessed and analyzed in Section 13.5, providing an example of migration path alternatives. Finally, Section 13.6 gives an outlook and presents the main conclusions.

13.2 Review of the State-of-the Art Approaches

13.2.1 Migration Processes to Distributed Architectures

There are several other migration processes that have been developed in other projects that allow for a smooth migration between different systems. The work developed in the IMC-AESOP project [6] focused mainly on the implementation of Service Oriented Architecture (SOA) to change the existing systems into distributed and interoperable systems. The migration of systems towards SOA has four major steps, such as Initiation, Configuration, Data Processing, and Control Execution. This migration process makes use of the mediator technology to communicate with the legacy systems, i.e. the old systems. The four steps aim at maintaining the perception of conformity between the several systems' interfaces.

Similarly, the SOAMIG project [7] developed a migration process towards SOA, which is developed as an iterative process and is represented by four phases: Preparation, Conceptualization, Migration and Transition. This migration process aims at a single specific target solution, which is derived step-by-step.

The SMART project [8] performed the analysis of the legacy systems by determining if they can be "linked" to SOA. SMART is an iterative process of six steps: Establish migration Context, Define Candidate Services, Describe Existing Capability, Describe Target SOA Environment, Analyze the Gap, and Develop Migration Strategy. This migration process is mostly used for migrating legacy Information Technology (IT) to SOA.

The MASHUP [9] is another technique for migrating legacy systems into service oriented computing. This migration process proposes a six steps process: Model, Analyze, Map and Identify, Design, Define, and Implement and Deploy. This technique is mainly used to overcome some SOA difficulties, such as the Quality of Service.

The Cloudstep [10] is a step-by-step decision process that supports the migration of legacy application to the cloud, identifying and analyzing the factors that can influence the selection of the cloud solution and also the migration tasks. It comprehends nine activities: Define Organization Profile, Evaluate Organizational Constraints, Define Application Profile, Define Cloud Provider Profile, Evaluate Technical and/or Financial Constraints, Address Application Constraints, Change Cloud Provider, Define Migration Strategy, and Perform Migration.

The XIRUP [11] process aims at the modernization of component-based systems, in an iterative approach. This method comprehends four stages: Preliminary Evaluation, Understanding, Building, and Migration. The ultimate goal of the XIRUP process is to provide cost-effective solutions and tools for modernization.

The different migration processes found in the literature present some similarities, regardless of the domain and target of migration. Generally, following a stepwise approach, first the legacy system and the target system are analyzed and the requirements defined, and then the target system is developed and finally the migration is defined and performed. Processes like SOAMIG and IMC-AESOP focus mainly on the technical constraints and characteristics of the migration, while SMART, MASHUP, and XIRUP pay attention also to business requirements and involved stakeholders, and Cloud-step includes legal, administrative and organizational constraints. In addition, most of the described processes analyze the migration iteratively, but only the XIRUP process considers the integration of the possible new features after the successful validation of the migrated components.

The existing migration processes or methods are all target based, taking only in consideration a specific goal, e.g. service-oriented architectures. While the described processes try to migrate and transform only technologies, now it is fundamental to start considering changing business paradigms. For the implementation of a new business paradigm, in this case Industry 4.0, it is necessary to have a migration process that allows for holistic and continuous improvement. A process that supports the lean approach for continuous improvement, adaptation to change and system's innovation is the migration process proposed by Calà et al. [12] within the PERFoRM project, which constitutes the baseline for the migration strategy towards the digital manufacturing automation presented in this chapter.

13.2.2 Organizational Change Management

Architectures and information systems represent the backbone of enterprises, and their transformation is a part of the comprehensive process of an organizational change. There is a rich management literature addressing the theme of how to introduce, implement, and support changes that impact the role and work of people in the organizations. In his seminal work, Lewin has highlighted how social groups operate in a sort of equilibrium among contrasting interests and that any attempt to force a change may stimulate an increase in opposing forces [13]. Changes have implications on the employees, who, in most cases, show reactions such as concern, anxiety and uncertainty, which may develop into resistance [14]. In order to prevent and overcome resistance, Lewin proposed a three steps process: (i) unfreezing, (ii) moving, and (iii) freezing. The first step aims at destabilizing the equilibrium correspondent to the status-quo, so that current behaviours become uncomfortable and can be discarded, i.e. unlearnt, opening up for new behaviours. In practice, unfreezing can be achieved by provoking some emotional feeling, such as anxiety about the survival of the business; the second step consists in a process of searching for more acceptable behaviours, in which individuals and groups progress in learning; the third steps aim at consolidating the conditions of a new quasi-stationary equilibrium [15].

Lewin's work, by providing insight about the mechanisms that rule human groups and operate within the organizations, and by delivering guidance about change management strategies, has opened the way to following studies. In the last decades, several frameworks and approaches have been defined in order to successfully undertake transformation processes and overcome possible resistance. Starting from the analysis of why change effort fails, Kotter [16] has identified a sequence of eight steps for enacting changes in organizations: (i) creating a sense of urgency, e.g., by attracting the attention on potential downturn in performances or competitive advantage and discussing the dramatic implications of such crisis and timely opportunities to be grasped; (ii) building a powerful guiding coalition, i.e., forming a team of people with enough power, interest and capability to work together for leading the change effort; (iii) creating a vision, i.e., building a future scenario to direct the transformation; (iv) communicating the vision, including teaching by the example of the new behaviours of the guiding coalition; (v) empowering others to behave differently, also by changing the systems and the architectures; (vi) planning actions with short term returns, limited changes that bring visible increases in performances and, through acknowledgment and rewarding practices, can be used as examples; (vii) consolidating improvements, developing policies and practices that reinforce the new behaviours; and (viii) institutionalizing new approaches, by structuring and sustaining the new behaviours. Another quite famous framework for managing changes is the Prosci ADKAR Model [17], which suggests to pursue changes through a sequence of five steps corresponding to the initial letters of ADKAR, i.e. (i) awareness about the need for change; (ii) desire to support the change; (iii) knowledge about how to change; (iv) ability to demonstrate new behaviours and competencies; and (v) reinforcement to stabilize the change.

The focus of some researchers and practitioners has shifted from an episodic to a continuous change.

This type of approach includes the continuous improvement of Kaizen [18], with its three principles: (i) process-orientation, as opposed to result-orientation: (ii) improving and maintaining standards, as opposed to innovations that do not impact on all the practices and are not sustainable; and (iii) people orientation as opposed to an involvement of the employees limited to the higher levels of management.

The concept of a learning organization, capable to build, capture, and mobilize knowledge to adapt to a changing environment has been introduced by Senge in 1990 [19]. The basis for the development of a learning organization consists of five disciplines: (i) mental models, (ii) personal mastery, (iii) systems thinking, (iv) team learning, and (v) building shared vision [20]. Other recent literature supports the theory of an organization that continuously changes through engaging and learning.

The case discussed in this chapter, the migration from conventional centralized automation (e.g., ISA-95) to distributed architectures for the digital shopfloor, concerns a major transformation in which the enterprise information systems play a crucial role for realizing the business vision and converting the strategy into change [21]. The theories and strategies of change management can thus provide some guidance about the path to be followed and the mistakes to avoid for the migration. However, organizations participate in a process of continuous change through engagement and learning [22], which involve the continuous transformation and integration of Enterprise Information Systems [21]. Therefore, rather than targeting the final state of a successfully deployed digital automation model, the migration roadmap should aim at incorporating further continuous transformation of distributed automation architectures in the continuous learning and improvement of the organization, in a never-ending process.

13.2.3 Maturity Models

In order to understand what maturity models are, the basics concepts of maturity models are given. To this aim, it is appropriate to provide some definitions, since the notion of maturity concepts might not be one and the same [23].

Maturity can be defined as "*the state of being complete, perfect or ready*". Adding to this definition, there is another point of view of maturity concept given by Maier et al. in 2012 [23], who believe maturity implies an evolutionary progress from an initial to a desired or normally occurring end stage [24]. This last consideration, which stresses the process toward maturity, introduces another important concept, which is the one of *stages of growth* or *maturity levels.*

The concept of stages of growth started to appear in literature for the first time around the 1970s. In particular, the authors who used these concepts for the first time are Nolan and Crosby in 1979 [25, 26]. The first one published an article where maturity model is seen as a tool to assess in which stage of growth the organization is, assuming it evolves automatically over the time, passing all the stages due to improvements and learning effects [25]. Simultaneously, Crosby [26] proposed a maturity grid for quality management process, as a tool which can be used to understand what is necessary to achieve a higher maturity level, if desired.

From this consideration, it is possible to state that in the same year, two concepts of maturity model have been proposed. On the one hand, Nolan proposed an '*evolutionary model*' that sees the stages of maturity as steps through which every company will improve, and on the other hand, Crosby introduced the "*evolutionist models*" that consider the maturity as a series of steps towards progressively more complex or perfect version of the current status of a company.

Therefore, it has been noticed that, in literature, there is not a general and clear classification of maturity models because of the different interpretation of the maturity concept, of the different approach with which the models (evolutionist/evolutionary) were conceived and according to the different sectors in which they are applied. Nevertheless, Fraser et al. [27] presented a first clear classification per typology of maturity models. In their paper, they distinguish three typologies of maturity models that are, respectively, Maturity grids, Likert-like questionnaires, CMM-like models.

The maturity grids typically illustrate maturity levels in a simple and textual manner, structured in a matrix or a grid. As mentioned before, the first type of maturity grid was the one of Crosby [26], and its main characteristic is that it is not specified what a particular process should look like. Maturity grids only identify some characteristics that any process and every enterprise should have in order to reach high-performance processes [23].

The Likert-like questionnaires are constructed by "questions", which are no more than statements of good practice. The responder to the questionnaire has to score the related performance on a scale from 1 to n. They can be defined as hybrid models, since they combine the questionnaires approach with the definition of maturity. Usually, they have only a description of each level, without specifying the different activities that have to be performed to achieve a precise maturity level.

Finally, there is the Capability Maturity Model (CMM). Its architecture is more formal and complex compared to the first two. They are composed of process areas organized by common features, which specify a number of key practices to address a series of goals. Typically, the CMMs exploit Likert questionnaires to assess the maturity. These models have been improved successively by the Capability Maturity Model Integration (CMMI) [28].

Although Nolan and Crosby have been the pioneers of the maturity assessment tools, as stated by Wendler [29], the maturity models field is clearly dominated by the CMM(I)'s inspired models. For this reason, FAR-EDGE approach is based on this model and, therefore, its relevant features will be described in this chapter.

The CMM was developed at the end of the 1980s by Watts Humphrey and his team from the Software Engineering Institute (SEI) in Carnegie Mellon University. It was used as a tool for objectively assessing the ability of government contractors' processes to perform a contracted software project. Although the focus of the first version of the CMM lies on the software development processes, successively, it has been applied in other process areas [30]. CMM decomposes each maturity level (shown in the Figure 13.1 [38]) into basic parts with the exception of level 1, which is the initial one. These levels define a scale for measuring process maturity and evaluating process capability. Each level is composed by several *key process areas*. Each key process area is organized into five sections called *common features*, which in turn specify *key practices*.

The key process areas specify where an organization should focus on improving processes. In other words, they identify a cluster of related activities, which, if performed collectively, achieve a set of goals considered important for improving process capability.

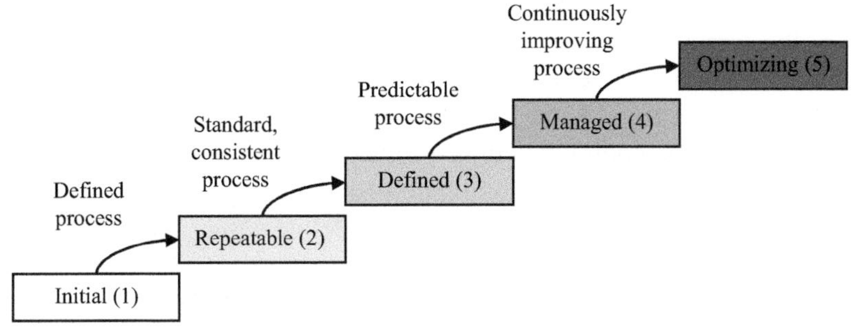

Figure 13.1 CMM's five maturity levels (from [38]).

The practices that describe the key process areas are organized by common features. These are attributes that indicate whether the implementation of a key process area is effective, repeatable and lasting.

Finally, each process area is described in terms of *key practices.* They define the activities and infrastructure for an effective implementation and institutionalization of the key process area. In other words, they describe what to do, but not how to do it.

In 2002, the CMMI was proposed [28]. It is considered as an improvement of the CMM model, but in contrast to this model that was built for software development, the purpose of CMMI has been to provide guidance for improving organizations' processes and their ability to manage the development, acquisition, and maintenance of products or services in general [28]. Furthermore, the focus of this model lies on the representation of the current maturity situation of the organization/process (coherently with the evolutionary model) and on giving indications on how a higher maturity level can be achieved (as proposed by evolutionist model). For these reasons, considering also the FAR_EDGE purposes, the CMMI can be considered as the most appropriate to be taken as a reference model to implement a blueprint migration strategy.

13.3 The FAR-EDGE Approach

The envisioned cyber-physical production and automation systems are characterized by complex smart and digital technology solutions that cannot be implemented in an existing production system in one step without considering their impact on the legacy systems and processes. Only a smooth migration strategy, which applies the future technologies in the existing infrastructures with legacy systems through incremental migration steps, could lower risks and deliver immediate benefits [4]. Indeed, a stepwise approach can mitigate risks at different dimensions of the factory by breaking down the long-term vision, i.e. the target of the migration, in short-term goals. This approach, as represented in Figure 13.2, is based on the lean and agile techniques, such as the Toyota Improvement Kata [31], to implement the new system step-by-step and support the continuous improvement, adaptation to changes and innovation at technical, operational and human dimensions.

The methodology adopted in FAR-EDGE to define a migration approach is described in [32]. Workshop and questionnaire results led to the identification of the important impact aspects of the FAR-EDGE reference architectures to the existing traditional production systems. Considering the identified factory dimensions of impact, an assessment tool has been realized

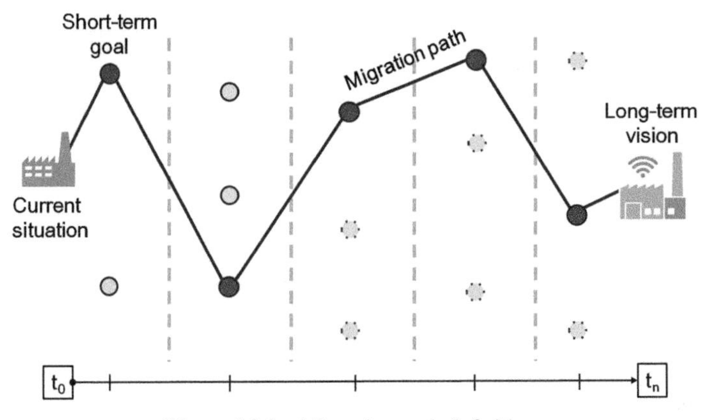

Figure 13.2 Migration path definition.

to support the analysis of the current situation and the desired ones of the manufacturing systems before defining their migration path.

Inspired by the migration process defined in [12], a methodology to define and evaluate different architectural blueprints has been defined within the FAR-EDGE project to support companies in investigating the possible technology alternatives towards the digital manufacturing automation with a positive return on investments.

First, there is a preparation phase [12] that aims at analyzing the current domain of the company, as well as the business long-term vision. Through questionnaires and workshops with people involved in the manufacturing process (i.e. production and operation management, IT infrastructure, and change management), the migration goal and starting point are defined, as well as the possible impact and the typical difficulties that the FAR-EDGE solution can have.

The scope of this phase is to have a clear picture on what should be changed in a company's business by investigating the technology and business process points of view simultaneously and deriving the implication at technical, operational and human dimensions in a holistic approach. In fact, it is important to keep in mind that the implementation of smart devices, intelligent systems, and new communication protocols has a big impact not only on the technological dimension of the factory but also on system's performance, work organization, and business strategy [32]. Therefore, a questionnaire of circa 60 questions about the technical, operational,

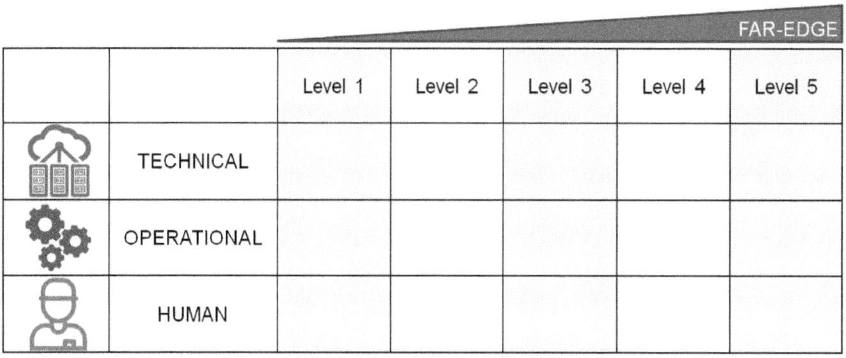

FAR-EDGE

		Level 1	Level 2	Level 3	Level 4	Level 5
	TECHNICAL					
	OPERATIONAL					
	HUMAN					

Figure 13.3 FAR-EDGE Migration Matrix.

and human factory's dimensions has been defined within FAR-EDGE to holistically analyze the current condition of the production system.

Based on the answers of this questionnaire, different migration scenarios according to the possible technology options are investigated [12] in order to identify the migration alternatives to go from the identified AS-IS situation to the TO-BE one. To this end, a tool called Migration Matrix (Figure 13.3) has been developed within the FAR-EDGE project to identify all the necessary improvements in the direction of the Industry 4.0 vision of smart factory, splitting the digital transformation in different scale levels. Thus, the matrix represents the three impact dimensions, aiming at providing a snapshot of current situation of companies and suggesting which steps should be achieved in order to reach the FAR-EDGE objective in a smooth and stepwise migration process.

The migration matrix is structured in rows and columns. The rows represent the relevant application fields selected during the preparation phase with a high potential of improvement by FAR-EDGE concepts implementation on the architecture. They refer to technology innovations, factory process maturity, and human roles. The columns describe the development steps for each application field towards a higher level of production flexibility, intelligent manufacturing, and business process in the direction of the FAR-EDGE platform implementation. As shown in Figure 13.3, the five columns represent five levels of production system's digital maturity.

These levels are based on the integrating principles of both the CMMI (Capability Maturity Model Integration) framework [24, 33, 34] and DREAMY model (Digital REadiness Assessment MaturitY) [35], which are:

- Level 1 – The production system is poorly controlled or not controlled at all, process management is reactive and does not have the proper organizational aspects and technological "tools" for building an infrastructure that will allow repeatability and usability of the utilized solutions.
- Level 2 – The production is partially planned and implemented. Process management is weak due to lacks in the organization and/or enabling technologies. The choices are driven by specific objectives of single projects and by the experience of the planner, which demonstrates only a partial maturity in managing the infrastructure development.
- Level 3 – The process is defined thanks to the planning and the implementation of good practices, management and organizational procedures, which highlight some gaps/lacks of integration and interoperability in the applications and in the information exchange because of some constraints on the organizational responsibilities and /or on the enabling technologies.
- Level 4 – The integration and the interoperability are based on common and shared standards within the company, borrowed from intra- and/or cross-industry de facto standard, with respect to the best practices in industry in both the spheres of the organization and enabling technologies.
- Level 5 – The process is digitally oriented and based on a solid technology infrastructure and a high potential growth organization, which supports business processes in the direction of Industry 4.0, including continuous improvement processes, complete integrability, organization development, speed, robustness and security in information exchange.

The main reason of this choice is that the CMMI provides a defined structure, specifying what are the capabilities, the characteristic, and the potentiality a company has at each level. Based on [35], as the five-scale CMMI provided a generic model to start from, the maturity levels have been readapted in order to be compliant and coherent with the dimensions considered by domains previously defined.

Therefore, the Migration Matrix provides a clear map of the current and desired conditions of a factory, revealing different alternatives to achieve the first short-term goal in the direction of the long-term vision. These alternatives are then evaluated according to the business strategy, considering also strengths and weaknesses points. Since FAR-EDGE aims at providing a holistic overview of the impact of edge and cloud computing solutions on the existing production environments, the developed approach supports the identification of the areas in which improvement actions are required, matching the needs of the organization and the estimation of the overall benefit of the innovative solution for the industry.

Based on the results of these phases, a migration path is defined and the solution to execute the first migration step is designed, implemented, and deployed following the migration process of [12]. In parallel, a set of guidelines and recommendations for the implementation of the FAR-EDGE solution are defined and documented.

13.4 Use Case Scenario

The industrial application example provided here describes a simple scenario in the automotive industry. The manufacturer aims to decentralize the current factory automation architecture and introduce cyber-physical system concepts in order to flexibly deploy new technologies and maximize the correlation across its technical abilities to support mass-customization. Target of the implementation of the FAR-EDGE platform is the reduction of time and effort required for deploying new applications by the automatic reconfiguration of physical equipment on different stations, according to the current operation, and its automatic synchronization among different information systems (PLM, ERP, and MES).

The factory currently presents an automation architecture compliant to ISA-95 standards with three layers: ERP, MES, and SCADA with Field devices. However, the integration of new applications at the MES level to obtain new functions at the shopfloor is very expensive because of highly dependent on the centralized control structure of the architecture. Moreover, it requires a long verification time and, consequently, a long delivery time to customers.

The factory envisioned by FAR-EDGE, according to the Industry 4.0 paradigm, is a highly networked CPS in which the modules are able to reconfigure themselves and communicate with each other via a standard I4.0 semantic protocol. As there is no central control, the system modules can identify and integrate new components automatically, negotiate their services and capabilities in some sort of social interaction. The modules have the abilities of perception, communication and self-explanation. In this way, the new modules can be integrated into the system quickly in a “Plug and Produce” fashion, and the system can reconfigure itself in the event of changes and continue the production process without additional adjustments of the overall control.

Applying this vision to the considered use case, the single physical equipment becomes a single "Plug-and-Produce" module able to configure and deploy itself without human intervention. The plugging of the module could be implemented at the edge automation component of the platform (Ref to CHAPTER 2 e chapter 4). An adapter for controlling and accessing information about the single equipment should be developed as part of the communication middleware. Data will flow to the edge automation component, which will interact with the CPS models database of the platform in order to access and update information about the location and status of the single equipment. The synchronization and reconfiguration functionalities of the platform will trigger changes to the configuration of the stations, which will be reflected in the CPS models database. The ledger automation and reconfiguration services could also be used for automating the deployment and reconfiguration of the shopfloor.

13.5 Application of the Migration Approach

13.5.1 Assessment

Table 13.1 presents the main fields of application to be considered from technical, operational and human points of view for the automation. The assessment represented in the Migration Matrix provides an overview of the current (AS-IS) situation of the factory with reference to the automation. The AS-IS situation of the considered industrial use case is depicted in red within the matrix of Table 13.1. From this Migration Matrix, it is immediately clear which are the less developed areas of a specific factory's use case, towards the implementation of digital technologies, i.e. "Plug-and-Produce" modules.

Currently, the automation control has a centralized structure that allows the vertical integration of the different architectural levels, by providing automation and analytics capabilities to entities that work in parallel. The production equipment is networked through vendor-specific API, and data can be shared from different systems. In this way, the production data can be monitored and analyzed from a MES system and the order processing, which is fully automated, and production processes are being developed to be fully integrated. However, the production system has only a very basic security and local access control. The main issue in this use case is the reconfiguration

Table 13.1 AS-IS situation of the use case for the automation functional domain

FAR-EDGE

AS-IS

Level 1	Level 2	Level 3	Level 4	Level 5
Equipment/Machinery connectivity and communication protocols				
N.A.	Basic connectivity (RS232-RS485)	Local network through LAN/WAN	Networked with vendor specific API	Networked with standard communication protocols
Security and access control mechanisms				
N.A.	Basic security or local access control	Basic security and local access control	Vendor based access control for each system	Full security and global access control
Production Data Monitoring and Processing				
N.A.	Locally, per station / equipment / machinery	Centrally available through SCADA	Available and analyzed through MES at Factory level	Available and analyzed through the Cloud
3D layouts, visualization and simulation tools				
CAD systems not related to production data	CAD systems manually feed with production data	CAD systems interfaced with other design systems	CAD systems interfaces with intelligent systems for fast development	Fully integrated CAD systems with intelligent tools for interactive design process
Reconfiguration of production equipment and processes				
Manual	Locally managed at machine level (PLC)	Centrally managed from SCADA	Centrally managed by MES	Centrally managed according to ERP
Product Optimization				
N.A.	Rare offline optimization	Offline optimization based on manual data extraction	Manual optimization based on simulation data	Automatic optimization based on simulation services
Availability of production process models				
N.A.	Models defined (Excel based) with limited use	Models defined with limited specific functions	Models defined and integrated with business functions	Models defined and integrated with several different functions
IT Operator				
N.A.	External service provider	Internal for traditional IT systems	Internal for specific digital systems	Internal for all systems from field to cloud
Impact on Operator, Product Designer and Production Engineer				
Still unclear	Identified in general terms	Analyzed	Defined	Implemented in continuous improvement

of the production equipment that is performed per equipment by configuring it at PLC level. Moreover, time-consuming reconfiguration operations can stop the production. From a human perspective, the main role to be considered in this use case is the IT Operator, who has a strong knowledge on the current IT infrastructure of the factory but not on the digital systems and Industry 4.0 concepts. Within the factory, the implications of digital technologies on IT Operator have not been addressed because they are still unclear. Furthermore, other roles are involved in the transformation: the Operator, Production Manager, Product Designer, and Production Engineer. Their tasks will change as a consequence of the automatic reconfiguration of the physical equipment, of the novel devices in the field, for the need to encompass all the necessary information within product design and production planning. However, these roles are currently performing according to the current tasks and procedures, unaware of the prospected transformation.

The manufacturer could benefit from the implementation of the FAR-EDGE architecture and components in terms of modularity and reconfigurability capabilities of the shopfloor. In fact, the implementation of Edge Nodes on the single equipment enables the identification of new entities in the shopfloor and their instantiation at Cloud level, thus being directly accessible for all IT systems that require their definition (i.e. PLM). Moreover, the decentralization of the automation architecture through the Edge and Ledger layers could increase the flexibility and reconfigurability of the architectural assets, enabling future modifications and improvements.

13.5.2 Gap Analysis

Of course, to migrate the current traditional automation system to the FAR-EDGE architecture and components, different aspects of the factory need to be evaluated to guarantee a smooth transformation of the factory at minor impact on the current production system.

13.5.2.1 Technical aspects

FAR-EDGE supports automated control and automated configuration of physical production systems using plug-and-produce factory equipment in order to enable fast adjustments of the production processes according to requirements changes. To integrate plug-and-produce capabilities within an existing shopfloor equipment, a bidirectional monitoring and control communication channel with the shopfloor equipments is required, thus not only via sensors and actuators but also with active actors (e.g. PLC) equipped with a significant processing power and with a good network connection capability, namely the Edge Nodes of the FAR-EDGE architecture that are described in earlier chapters of the book.

The connection of digital and physical worlds will also support gathering and processing Field data towards a better understanding of production processes, for example, to change an automation workflow based on changes detected in the controlled process. This requires Edge Gateways, i.e. computing devices connected to a fast LAN to provide a high-bandwidth field communication channel. Edge Gateways can execute edge processes activity, namely the local orchestration of the physical process monitored and operated by Edge Nodes. In addition, Cloud services running in a virtualized computing environment can act as entry point for external applications and provide centralized utilities to be used internally or perform activities for archiving analysis results.

The introduction of the Cloud within the production control entails full security and global access control mechanisms, which need to be increased immediately, as soon as the production information will be available at cloud level for different stakeholders in order to prevent data security and privacy issues. In addition, the automatic reconfiguration of physical equipments can be enhanced by the integration of simulation tools that provide an interactive design process leading to the optimization of the production processes. In order to improve the optimization, 3D layouts and CAD systems must be fully integrated in a common digital model by means of intelligent tools that automatically feed the simulation systems with the real production data and derive optimized solutions.

13.5.2.2 Operational aspects

Plug and Produce capability could be seen as a crucial solution to reduce the time and costs involved in not only manufacturing process (e.g. new machine/equipment/resources deployment) but also process design and process development. For this reason, it presumes the need of building an agile enterprise application platform which helps a company to be proactive in carrying out its core activities. To facilitate such tight and effective improvement in a modern enterprise, the (information and operational technology (IT/OT) integration is needed. This means, first, the integration of ERP applications, MES and shopfloor systems (i.e PLCs, SCADA, DCS) along the levels defined by ISA-95 and, second, the integration of PLM systems and MES (Level 3 and Level 4) when it comes to the transition of a ready-to-market product into production.

The latter consideration enables the integration between design and production, in terms of processes and systems, increasing product quality and process efficiency. This convergence is the source of not only product but also process definition. On one side, the Bill of Process (BoP) provides traceability to the Bill of Materials (BoM) to leverage PLM's configuration and effectiveness controls, defining the correct sequence of operations to guarantee a high level of product quality. On the other side, the Manufacturing process management carries out the documentation and the follow-up of processes in the MES, which reshapes theoretically designed processes to make them fit the reality on the shopfloor, ensuring the process efficiency. Considering this, the proper integration of systems is vital, otherwise data related to the new machine introduction or the process adjustment would "manually" be passed to MES (that coordinate and monitor the process execution).

From this consideration, the evolution to a Plug and Produce production system has to go through the information harmonization between engineering and manufacturing, coherently with a stepwise approach. To this aim, the first step is to realize an overall data backbone for all processes and products. This means to centralize the DBs and the information systems in order to integrate the information flow between manufacturing and engineering domain. Within the next step, the MES will automatically provide execution data to ensure holistic and reliable product information that, being documented and available in both systems, can be considered as a strategic asset to improve the maintenance, repair, and optimization process.

In this context, the deployment of event-driven architecture ('RT-SOA' or Real-Time Service Oriented Architecture) could facilitate the information exchange and, therefore, the seamless reconfiguration of machinery and robots as a response to operational or business events.

13.5.2.3 Human aspects

The migration towards digital manufacturing automation implies changes in the behavior of the production systems as well as in the information flows. The implications impact the work of the employees under different points of view.

The health and conditions of the operators are usually modified by the introduction of automation. In most cases, the ergonomic effort is reduced, but in some cases, additional factors, such as the introduction of robotics, have to be included in the risk management plans. The autonomy and privacy of the employees may change because of a more accurate and real-time monitoring of the operations and tracking of products and tools.

These implications need to be carefully analyzed with all the stakeholders and managed.

The role of employees can be affected by the new technological and operational landscape: on the one hand, some manual tasks or scheduling decisions are taking over by the systems; on the other hand, some new tasks are added to supervise the systems, monitor the KPIs, and address the problems. The workplace, the HMIs, the workflow, and the instructions change in several cases. It is important that the operators stay in the loop of control of the process and are aware of the states and activities of the technological systems.

The deployment of the new technologies is expected to impact not only the Production Operators, but also the Product Designers, the Production Engineers, and obviously the IT Operators. Overall, the skills requirements for each role have to be updated on the basis of the TO-BE scenario and compared with those available in the AS-IS situation, in order to identify and address the gaps, through up-skilling or recruitment initiatives. Furthermore, the job profiles and training plans need to be updated to ensure the incorporation in the standard procedures.

Although the need for these changes is perceived, they are still unclear and the size of the gap has not been evaluated yet.

13.5.3 Migration Path Alternatives

Considering the current situation of the industrial use case and the long-term vision of digital manufacturing enabled by the FAR-EDGE reference architecture, different migration path alternatives can be identified. The identified alternatives are generated on the basis of technical constraints, investment capabilities, and organizational structure. Considering different priorities and required improvements on part of the production system, these migration alternatives lead to the achievement of the first short-term goal of the migration path towards the Industry 4.0 vision.

Two main migration path (MP) alternatives have been derived according to the specific business goal of the represented factory. The first alternative (MP 1) focuses on the implementation of plug-and-produce equipment to enhance the production system reconfigurability (Table 13.2), while the second alternative (MP 2) focuses on the real-virtual automatic synchronization of the single equipment based on simulation tools to optimize the production process (Table 13.3). Both alternatives will enable the factory to improve different parts of the system towards the long-term vision of "digitalization" by implementing step-by-step some of the FAR-EDGE solution components. The manufacturer will then select the adequate solution according to the enterprise's needs, interests and constraints.

Table 13.2 MP for the implementation of reconfigurability

MP 1	Level 1	Level 2	Level 3	Level 4	Level 5 (FAR-EDGE)
	Equipment/Machinery connectivity and communication protocols				
	N.A.	Basic connectivity (RS232-RS485)	Local network through LAN/WAN	Networked with vendor specific API	Networked with standard communication protocols
	Security and access control mechanisms				
	N.A.	Basic security or local access control	Basic security and local access control	Vendor based access control for each system	Full security and global access control
	Production Data Monitoring and Processing				
	N.A.	Locally, per station/ equipment/ machinery	Centrally available through SCADA	Available and analyzed through MES at Factory level	Available and analyzed through the Cloud
	Reconfiguration of production equipment and processes				
	Manual	Locally managed at machine level (PLC)	Centrally managed from SCADA	Centrally managed by MES	Centrally managed according to ERP
	IT Operator				
	N.A.	External service provider	Internal for traditional IT systems	Internal for specific digital systems	Internal for all systems from field to cloud
	Impact of digital technologies on IT Operator				
	Still unclear	Identified in general terms	Analyzed	Defined	Implemented in continuous improvement

Color legend: red = AS-IS, yellow = intermediate step, green = TO-BE

The migration matrixes depicted for the two MPs represent two specific improvement scenarios and not the production system as a whole. In both matrixes, the maturity levels of the current situation are represented in red, while the migration steps are represented in yellow (the intermediate migration step) and in green (the final step).

MP 1: Implementation of reconfigurability. According to the business strategy, the deployment configuration should give priority to the Cloud, since the factory already planned to implement cloud technologies in the production automation control. The collection and integration of information through the Cloud will support the reconfigurability of plug and produce equipment. In fact, PLM provides the planning information about how the product will be produced and the MES serves as the execution engine to realize the plan and BoP. As a second step, the information provided by PLM

Table 13.3 MP for the implementation of simulation-based optimization

MP 2	Level 1	Level 2	Level 3	Level 4	Level 5 FAR-EDGE
	3D layouts, visualization and simulation tools				
	CAD systems not related to production data	CAD systems manually feed with production data	CAD systems interfaced with other design systems	CAD systems interfaces with intelligent systems for fast development	Fully integrated CAD systems with intelligent tools for interactive design process
	Production Optimization				
	N.A.	Rare offline optimization	Offline optimization based on manual data extraction	Manual optimization based on simulation data	Automatic optimization based on simulation services
	Availability of production process models				
	N.A.	Models defined (Excel based) with limited use	Models defined with limited specific functions	Models defined and integrated with business functions	Models defined and integrated with several different functions
	Impact of digital technologies on Product Designers and Production Engineers				
	Still unclear	Identified ingeneral terms	Analyzed	Defined	Implemented in continuous improvement

Color legend: red = AS-IS, yellow = intermediate step, green = TO-BE

needs to be reshaped. It is important to increase the amount of detail included in product information to cover machine programming, operator instructions and task sequencing. In this way, work plans, routing and BoP will serve as bridge elements between PLM and the MES [36]. In order to integrate the production systems information to the Cloud, a first improvement of the access control for each system must be immediately considered, which will be enhanced to a full security system in a second step. Moreover, because of the number of different stakeholders involved, in terms of third-part vendors and system developers, the second migration step should include also the introduction of open API to enable the standard communication among heterogeneous systems. Following this change in production systems and operations, the IT Operators must be trained in order to be able to manage the new automation control system, from the field level to the Cloud. The implications for the other roles should be analyzed in order to prepare the following steps.

MP 2: Simulation-based optimization. The virtual representation of the physical objects in cyber space can be used for optimization of the production processes. For example, the cyber modules have the ability to avoid getting stuck in local optimization extremes and are able to find the global maximum and minimum which results in high performance. Therefore, additionally to the migration steps described in MP 1, the integration of digital models must be considered. Firstly, the existing CAD systems will be interfaced to each other, and secondly, they will be fully integrated to enable the optimization of equipment reconfiguration through intelligent simulation tools. In the same way, the production will be optimized based on the integrated information derived from the CAD designs and then it will be automatically implemented through the intelligent tools. To this end, the production process models and their different layout versions will be first integrated with business functions, in order to align the process parameters with cost deployment and profitability measures. From an organizational perspective, the main implications affect the roles of product designers and production engineers: they need to increase their level of cooperation to model all the relevant aspects of the manufacturing processes into the CAD. Furthermore, the production engineers have to see that the models of the CAD are connected to the models of the actual production facilities, so that the production can be simulated, planned and monitored. Therefore, the competences of the above mentioned roles require to be enhanced with new skills concerning digitalization, modeling and simulation. Furthermore, the tasks and responsibilities of these roles have to be updated accordingly.

The migration matrixes support manufacturers by providing them with a holistic view of the required steps for migration towards the Industry 4.0 vision at different dimensions of the factory, i.e. technical, operational, and human. Based on this information and according to the business goals, the manufacturer will select the optimal scenario as first step of migration towards the long-term goal of complete digitalization of the factory. The solution identified within the selected scenario will be then designed in detail, implemented and deployed according to next process phases described in [12].

13.6 Conclusion

In conclusion, this chapter shows how the FAR-EDGE migration approach can lead a manufacturing company to achieve an improvement towards a new manufacturing paradigm following a smooth and no risk transition approach with a holistic overview.

In fact, the use case scenario points out that every part of an organization – including workforce, product development, supply chain and manufacturing – has been considered to reach more flexible and reconfigurable aspects in order to rapidly react to both endogenous and exogenous drivers that are affecting the current global market [37].

In this context, the IT/OT convergence can be seen as a first implementation of operational aspects needed to obtain a solid manufacturing layer based on the encapsulation of production resources and assets according to the existing protocols, in order to facilitate the plug-and produce readiness, and therefore, to achieve a flexible manufacturing environment.

As far as the technical dimension is concerned, the Edge Nodes and the ledger implementation can enable the realization of the overall system architecture based on information systems integration needed to obtain the seamless system reconfiguration avoiding scrap and reducing time to market and cost.

Finally, the human aspect is crucial to ensure the operation, management, and further development of the highly digitalized and automated production system. The methodology illustrated in this chapter guides manufacturing in considering the implications for skills and work organization within their migration strategy.

Only by jointly considering the technical, operational, and human aspects can a migration strategy anticipate the possible hurdles and lead to a smooth transformation towards an effective new production paradigm.

Acknowledgements

The authors would like to thank the European Commission for the support, and the partners of the EU Horizon 2020 project FAR-EDGE for the fruitful discussions. The FAR-EDGE project has received funding from the European Union's Horizon 2020 research and innovation programme under grant agreement No. 723094.

References

[1] Acatech - National academy of science and Engineering, "Recommendations for implementing the strategic initiative INDUSTRIE 4.0-Final report of the Industrie 4.0 Working Group". pp. 315–320.

[2] Acatech - National academy of science and engineering-, "Cyber-Physical Systems Driving force for innovation in mobility, health, energy and production".

[3] U. Rembold and R. Dillmann, *Computer-Aided Design and Manufacturing,* 1985.

[4] M. Foehr, J. Vollmar, A. Calà, P. Leitão, S. Karnouskos, and A. W. Colombo, "Engineering of Next Generation Cyber-Physical Automation System Architectures", in *Multi-Disciplinary Engineering for Cyber-Physical Production Systems*, Springer International Publishing, pp. 185–206, 2017.

[5] "FAR-EDGE – Factory Automation Edge Computing Operating System Reference Implementation". 2017.

[6] J. Delsing, J. Eliasson, R. Kyusakov, A. W. Colombo, F. Jammes, J. Nessaether, S. Karnouskos, and C. Diedrich, "A migration approach towards a SOA-based next generation process control and monitoring", in *IECON Proceedings (Industrial Electronics Conference),* pp. 4472–4477, 2011.

[7] C. Zillmann, A. Winter, A. Herget, W. Teppe, M. Theurer, A. Fuhr, T. Horn, V. Riediger, U. Erdmenger, U. Kaiser, D. Uhlig, and Y. Zimmermann, "The SOAMIG Process Model in Industrial Applications", in *2011 15th European Conference on Software Maintenance and Reengineering,* pp. 339–342, March 2011.

[8] S. Balasubramaniam, G. A. Lewis, E. Morris, S. Simanta, and D. Smith, "SMART: Application of a method for migration of legacy systems to SOA environments", *Lect. Notes Comput. Sci. (including Subser. Lect. Notes Artif. Intell. Lect. Notes Bioinformatics),* vol. 5364 LNCS, pp. 678–690, 2008.

[9] S. Cetin, N. I. Altintas, H. Oguztuzun, A. H. Dogru, O. Tufekci, and S. Suloglu, "A mashup-based strategy for migration to Service-Oriented Computing", in *2007 IEEE International Conference on Pervasive Services, ICPS,* pp. 169–172, 2007.

[10] P. V. Beserra, A. Camara, R. Ximenes, A. B. Albuquerque, and N. C. Mendonça, "Cloudstep: A step-by-step decision process to support legacy application migration to the cloud", in *2012 IEEE 6th International Workshop on the Maintenance and Evolution of Service-Oriented and Cloud-Based Systems, MESOCA 2012,* pp. 7–16, 2012.

[11] R. Fuentes-Fernández, J. Pavón, and F. Garijo, "A model-driven process for the modernization of component-based systems", *Sci. Comput. Program.,* vol. 77, no. 3, pp. 247–269, 2012.

[12] A. Calà, A. Luder, A. Cachada, F. Pires, J. Barbosa, P. Leitao, and M. Gepp, "Migration from traditional towards cyber-physical production systems", in *Proceedings - 2017 IEEE 15th International Conference on Industrial Informatics, INDIN 2017,* pp. 1147–1152, 2017.
[13] T. Newcomb and E. Hartley, *No TitleGroup decision and social change,* Holt. New York, 1947.
[14] S. Z. A. Kazmi and M. Naarananoja, "Collection of Change Management Models – An Opportunity to Make the Best Choice from the Various Organizational Transformational Techniques", *GSTF J. Bus. Rev.,* vol. 2, no. 4, pp. 44–57, 2013.
[15] B. H. Sarayreh, H. Khudair, and E. alabed Barakat, "Comparative study: The Kurt Lewin of change management", *Int. J. Comput. Inf. Technol.,* vol. 02, no. 04, pp. 2279–764, 2013.
[16] J. P. Kotter, "Leading change: why transformation efforts fail the promise of the governed corporation", *Harward Bus. Rev.,* no. March–April, pp. 59–67, 1995.
[17] J. Hiatt, *adkar.* Prosci Inc., 2006.
[18] A. Berger, "Continuous improvement and *kaizen*: standardization and organizational designs", *Integr. Manuf. Syst.,* vol. 8, no. 2, pp. 110–117, 1997.
[19] S. Yadav and V. Agarwal, "Benefits and Barriers of Learning Organization and its five Discipline," *IOSR J. Bus. Manag. Ver. I,* vol. 18, no. 12, pp. 2319–7668, 2016.
[20] P. M. Senge, "The fifth discipline: the art and practice of the learning organization", *5th Discipline. p. 445,* 2006.
[21] D. Romero and F. Vernadat, "Enterprise information systems state of the art: Past, present and future trends", *Comput. Ind.,* vol. 79, pp. 3–13, 2016.
[22] C. G. Worley and S. A. Mohrman, "Is change management obsolete?", *Organ. Dyn.,* vol. 43, pp. 214–224, 2014.
[23] A. M. Maier, J. Moultrie, and P. J. Clarkson, "Assessing organizational capabilities: Reviewing and guiding the development of maturity grids", in *IEEE Transactions on Engineering Management,* vol. 59, no. 1, pp. 138–159, 2012.
[24] T. Mettler and P. Rohner, "Situational maturity models as instrumental artifacts for organizational design", *Proc. 4th Int. Conf. Des. Sci. Res. Inf. Syst. Technol. - DESRIST '09. Artic. No. 22,* pp. 1–9, May 06–08, 2009.

[25] R. L. Nolan, "Managing the crises in data processing", *Harv. Bus. Rev.,* vol. 57, pp. 115–127, March 1979.

[26] P. B. Crosby, "Quality is free: The art of making quality certain", *New York: New American Library.* p. 309, 1979.

[27] P. Fraser, J. Moultrie, and M. Gregory, "The use of maturity models/grids as a tool in assessing product development capability", in *IEEE International Engineering Management Conference,* 2002.

[28] C. P. Team, "Capability Maturity Model{\textregistered} Integration (CMMI SM), Version 1.1", *C. Syst. Eng. Softw. Eng. Integr. Prod. Process Dev. Supplier Sourc. (CMMI-SE/SW/IPPD/SS, V1. 1),* 2002.

[29] R. Wendler, "The maturity of maturity model research: A systematic mapping study", *Inf. Softw. Technol.,* vol. 54, no. 12, pp. 1317–1339, 2012.

[30] M. Kerrigan, "A capability maturity model for digital investigations", *Digit. Investig.,* vol. 10, no. 1, pp. 19–33, 2013.

[31] N. Rother, "Toyota KATA - Managing people for improvement, adaptiveness, and superior results", 2010.

[32] A. Calà, A. Lüder, F. Boschi, G. Tavola, M. Taisch, P. Milano, and V. R. Lambruschini, "Migration towards Digital Manufacturing Automation - an Assessment Approach".

[33] M. Macchi and L. Fumagalli, "A maintenance maturity assessment method for the manufacturing industry", *J. Qual. Maint. Eng.,* 2013.

[34] M. J. F. Macchi M., Fumagalli L., Pizzolante S., Crespo A. and Fernandez G., "Towards Maintenance_maturity assessment of maintenance services for new ICT introduction", in *APMS-International Conference Advances in Production Management Systems,* 2010.

[35] A. De Carolis, M. Macchi, E. Negri, and S. Terzi, "A Maturity Model for Assessing the Digital Readiness of Manufacturing Companies", in *IFIP International Federation for Information Processing 2017,* pp. 13–20, 2017.

[36] Atos Scientific and C. I. Convergence, "The convergence of IT and Operational Technology", 2012.

[37] F. Boschi, C. Zanetti, G. Tavola, and M. Taisch, "From key business factors to KPIs within a reconfigurable and flexible Cyber-Physical System", in *23rd ICE/ITMC Conference- International Conference on Engineering, Technology,and Innovationth ICE/IEEE ITMC International Technology Management Conference w,* Janurary 2018.

[38] M. C. Paulk, B. Curtis, M. B. Chrissis, and C. V Weber, "Capability Maturity ModelSM for Software, Version 1.1", *Office.* 1993.

14

Tools and Techniques for Digital Automation Solutions Certification

Batzi Uribarri[1], Lara González[2], Begoña Laibarra[1] and Oscar Lazaro[2]

[1]Software Quality Systems, Avenida Zugazarte 8 1-6, 48930-Getxo, Spain
[2]Asociacion de Empresas Tecnologicas Innovalia, Rodriguez Arias, 6, 605, 48008-Bilbao, Spain
E-mail: buribarri@sqs.es; lgonzalez@innovalia.org; blaibarraz@sqs.es; olazaro@innovalia.org

The digitisation and adoption of increasingly autonomous digital capabilities in the Factory 4.0 shopfloor demands that a large number of technologies need to be integrated, while the differential value of European manufacturing, i.e. security and safety, is ensured. Industry 4.0 puts additional pressure in small and medium-sized enterprises (SMEs) in terms of navigating standards, norms, and platforms to fulfil their business ambition. The Digital Shopfloor Alliance emerges as a multisided ecosystem that provides an integrated approach and a manufacturing-centric view on the digital transformation of automation solutions. This chapter introduces a certification framework for faster system integration and validated solution deployment. The main inputs from our approach to modular Plug & Produce autonomous factory environments & Validation & Verification (V&V) framework. This chapter also discusses how a validation and verification framework in combination with certified components could become key in the development of open digital shopfloors with future digital ability extensibility, controlled return of investment on Industry 4.0 solutions. This paper discusses also how such an approach can create a virtuous cycle for digital platform ecosystems such as FIWARE for smart industry, IDSA or more commercially driven ones such as Leonardo, Mindsphere, 3DExperience, Bosch IoT Suite, Bluemix, Watson, Predix, and M3.

14.1 Introduction

In the context of Industry 4.0 and Cyber Physical Production Systems (CPPS), markets, business models, manufacturing processes, and other challenges along the value chain are all changing at an increasing speed in an increasingly interconnected world, where future workplace will present increased mobility, collaboration across humans, robots and products with inbuilt plug & produce capabilities. Current practice is such that a production system is designed and optimized to execute the exact same process over and over again.

The planning and control of production systems has become increasingly complex regarding flexibility and productivity, as well as the decreasing predictability of processes. The full potential of open and smart CPPS is yet to be fully realized in the context of cognitive autonomous production systems. In an autonomous production scenario, as the one proposed by Digital Shopfloor Alliance (DSA) [1], the manufacturing systems will have the flexibility to adjust and optimize for each run of the task. Small and medium-sized enterprises (SMEs) face additional challenges to the implementation of "cloudified" automation processes. While the building blocks for digital automation are available, it is up to the SMEs to align, connect, and integrate them together to meet the needs of their individual advanced manufacturing processes. Moreover, SMEs face difficulties to make decisions on the strategic automation investments that will boost their business strategy.

Within the AUTOWARE project [3], new digital technologies including reliable wireless communications, fog computing, reconfigurable and collaborative robotics, modular production lines, augmented virtuality, machine learning, cognitive autonomous systems, etc. are being made ready as manufacturing technical enablers for their application in smart factories. Special attention is paid to the interoperability of these new technologies between each other and with legacy devices and information systems on the factory floor, as well as to providing reliable, fast integration, and cost-effective customized digital automation solutions. To achieve these goals, the focus has been set on open platforms, protocols, and interfaces, providing a Reference Architecture for the factory automation, and on a specific certification framework, for the validation not only of individual components but of deployed solutions for specific purposes, to help SMEs and other manufacturing companies to access and integrate new digital technologies in their production processes.

This chapter aims to review the certification framework, tools and techniques proposed within the global vision of DSA ecosystem, with a clear focus on enabling the digital transformation process on manufacturing SMEs through the adoption of digital automation solutions in their shopfloors.

Section 14.2 presents Safety as a main asset of European manufacturing industry that is challenged by autonomous operations and which represents a big challenge for SMEs in terms of regulation and level of integration across technologies and platforms. Section 14.3 presents a global vision of the DSA initiative and ecosystem, while Section 14.4 presents the alignment of DSA ecosystem with AUTOWARE Reference Architecture (RA), Technical and Usability Enablers to leverage digital abilities in the Shopfloor. This chapter also introduces the main strategic services to be provided. Next, Section 14.5 presents the V&V framework and component/system certification that constitutes the basis for the Digital Automation Technologies Validation framework. Section 14.6 elaborates in-depth DSA ecosystem players, approach, benefits, and services towards a win-win model for the multi-sided ecosystem.

14.2 Digital Automation Safety Challenges

SMEs are a focal point in shaping enterprise policy in the European Union (EU). In order to preserve and increase competitiveness in the global market, the SMEs need to digitalize their processes through the adoption of CPPS technologies in Digital Automation Solutions. After the analysis of new trends and challenges in SME manufacturing towards digital production paradigm by accessing new CPPS technologies and tools, we focused on new emerging technologies and paradigms such as Internet of Things, Industry 4.0, machine learning and artificial intelligence, robotics, Virtual/Augmented Reality, cloud computing, Cyber Physical Production Systems, and particularly on their impact on the SME production.

All these technologies that can be deployed in SME manufacturing and low-volume production are beginning to emerge and were proved to be beneficial to gain a competitive edge. However, the adoption of these technologies in actual SME production is still limited and needs to be sped up. Two main barriers preventing wider usage of these digital solutions were identified. On the end-users' side, the lack of knowledge and the time and cost constraints are dominant. On the supply side, there is a need to move from application orientation towards integrated solutions that will better support small enterprises, both in terms of customized and flexible applications. An effective measure to overcome problems related with the application of new smart technologies in the SMEs is to provide easy access to them through an ecosystem with integrated tools and techniques for Digital Automation

Solutions certification. This section forms a report on identified demands and challenges faced by manufacturing SMEs with regard to safety and certification areas.

The fourth Industrial Revolution for EU Manufacturing Industry **(Industry 4.0)** is generally associated with the full adoption of digital technologies in production and for having an exclusive focus on smart factory automation. This was at the basis of the Industrie 4.0 initiative in Germany when it started back in 2011. However, the most recent evolutions of the Industry 4.0 paradigm have considerably extended the scope and characteristics of Industry 4.0 projects, embracing and addressing new ways of conceiving products, production, and running manufacturing-oriented business models. During the **World Manufacturing Forum** 2016 in Barcelona, Roland Berger [1] presented the main transformations of new Industry 4.0, such as from *mass production* to *mass customization*, from *volume scale effect* to localized and *flexible production units*, from *make to stock* static and hierarchical supply chains to *make to order* dynamic reverse supply networks, from *product oriented* economy to *service and experience economy*, from hard *Taylorism-driven workplaces* to attractive and *adaptive workspaces*. The result of materializing the newly identified **"Industry 4.0"** is the identification of characteristics, where extensions of the traditional Smart Production model (well represented by the RAMI 4.0 Reference Architecture) are required.

The Digital Shopfloor Alliance (DSA) is a manufacturing-driven approach to digital transformation and it is the response to such new production paradigms. Production lines are in the process of migrating from production lines into autonomous work cell environments where increased autonomy and flexibility in operations are the key features. However, such flexible environments yet need to retain the same safety features as the traditional production chains. Hence, there is a demand for the development of digital platforms that will support the engineering, commissioning, and safe and secure operation of such advanced autonomous production strategies.

The smart factories of the future are built on a modular basis. With standardized interfaces and cutting-edge information technology, they enable the establishment of flexible automated manufacturing reflecting the "plug and produce" and autonomous production principles. Initiatives such as Industry 4.0 and the European digital shopfloor alliance (DSA) are developing the concept of **modular certification scheme and control in real time** to include a specific approach which takes into account all specific requirements for adaptive, configurable systems. All plant manufacturers need to develop a safety and security concept for their equipment and confirm that their equipment complies with legal requirements. Modular certification and self-assessment schemes are critical elements in the operation of autonomous equipment variants. This ensures that the equipment is automatically certified when a module is replaced, or a new line configuration is set by the integration of autonomous equipment in modular manufacturing settings and thus continues to be in conformity with the legal requirements and/or the standard.

Currently, industrial automation is a consolidated reality, with approximately 90 per cent of machines in factories being unconnected. These isolated and static systems mean that product safety (functional safety and security) can be comfortably assessed. However, the connected world of Industry 4.0's smart factories adds a new dimension of complexity in terms of machinery and production line safety challenges. IoT connects people and machines, enabling bidirectional flow of information and real-time decisions. Its diffusion is now accelerating with the reduction in size and price of the sensors, and with the need for the exchange of large amount of data. In today's static machinery environment, the configuration of machines and machine modules in the production line is completely known at the starting point of the system design. However, if substantial changes are made, a new conformity assessment may be required. It is an employer's responsibility to ensure that all machinery meet the requirements of the Machinery Directive and Provision and Use of Work Equipment Regulations (PUWER), of which risk assessments are an essential ingredient. Therefore, if a machine has a substantial change made, a full CE marking and assessment must be completed before it can be returned to service. Any configuration change in the production line requires re-certification of the whole facility.

However, the dynamic approach of Industry 4.0's autonomous robotic systems means that with a simple press of a button, easily configurable machinery and production lines can be instantly changed. As it is the original configuration that is risk assessed, such instant updates to machinery mean that the time-hungry, the traditional approach of "risk assessment as you make changes" will become obsolete. The risk assessment process therefore needs to be modified to meet the demands of the more dynamic Industry 4.0 approach. This would mean that all possible configurations of machines and machine modules would be dynamically validated during the change of the production line. Each new configuration would be assessed in real time, based on digital models of the real behavior of each configuration, which would be based upon the machinery manufacturer's correct (and trusted) data. The result would be a rapidly issued digital compliance certificate.

This Section discuss the challenges that such approach would entail from the context of safe operation of modular manufacturing, reconfigurable cells, and collaborative robotic scenarios.

14.2.1 Workplace Safety and Certification According to the DGUV

The Deutsche Gesetzliche Unfallversicherung (DGUV) is the German statutory accident insurance. The DGUV has published a requirements document that addresses workplace safety and certification aspects concerning collaborative robots. On conventional industrial robot systems, safeguards, such as protective fences and light curtains, prevent the access of people to hazardous areas. Collaborative robot systems, however, represent a link between fully automated systems and manual workplaces. The fact that Smart Manufacturing tends towards smaller batch sizes is one reason why collaborative robots are taking on greater significance. In an almost fenceless operation, which is dependent on the type of collaboration, the robot can thus support workers on manual tasks. This relieves the worker which is of benefit to the company managers in the medium to long term, since it results in less downtimes and an enhanced health situation of employees.

The DGUV provides the information "Collaborative robot systems – Design of systems with Power and Force Limiting" function for free download [4]. This information is intended to give an initial overview on the procedures when planning collaborative robot systems. The implementation of the AUTOWARE Use Case 3 – Industrial Cooperative Assembly of Pneumatic Cylinders necessitates the compliance with the workplace safety standards. In doing so, the DGUV requirements are considered in the design of the collaborative workspace. The fulfilment of the requirements provided by the DGUV is an essential prerequisite for the legal operation of collaborative robot systems in factories and their certification.

14.2.2 Industrial Robots Safety According to ISO 10218-1:2011 & ISO 10218-2:2011

The main current, i.e. published, standards regarding security as relevant to industrial robots (in contrast to personal care robots), are ISO 10218-1:2011 and ISO 10218-2:2011. The ISO 10218-1:2011, "Robots and robotic devices – Safety requirements for industrial robots – Part 1: Robots"

[5] specifies requirements and guidelines for the inherent safe design, protective measures, and information for use of industrial robots. It describes the basic hazards associated with robots and provides requirements to eliminate, or adequately reduce, the risks associated with these hazards.

The ISO 10281-2:2011, "Robots and robotic devices – Safety requirements for industrial robots – Part 2: Robot systems and integration," specifies safety requirements for the integration of industrial robots and industrial robot systems as defined in ISO 10218-1 with industrial robot cell(s) [6]. The integration includes the following:

- the design, manufacturing, installation, operation, maintenance, and decommissioning of the industrial robot system or cell;
- necessary information for the design, manufacturing, installation, operation, maintenance, and decommissioning of the industrial robot system or cell; and
- component devices of the industrial robot system or cell.

ISO 10218-2:2011 describes the basic hazards and hazardous situations identified with these systems, and it also provides requirements to eliminate or adequately reduce the risks associated with these hazards. It also specifies requirements for the industrial robot system as part of an integrated manufacturing system. The design of experiments in AUTOWARE JSI reconfigurable robotics cell will take into account these two standards.

14.2.3 Collaborative Robots Safety According to ISO/TS 15066:2016

As flexible, fast reconfigurable robot tasks nowadays often include collaborative activity between human operators and robots (and such application will only increase), a very important standard we take into account is ISO/TS 15066:2016, "Robots and robotic devices – Collaborative robots." ISO/TS 15066:2016 specifies the safety requirements for the collaborative industrial robot systems and the work environment. It supplements the requirements and guidance on collaborative industrial robot operation given in ISO 10218-1 and 10218-2. Two main newly introduced points are that 1) in essence, we have to obtain a safe collaborative application – the robot per se is not enough to guarantee a safe robot application and 2) the safety is specified by limited physical values that can be exerted in relation to humans (e.g. limited contact forces) rather solely by adoption of some technical type of safety solution. JSI reconfigurable robotics cell makes use of robots certified for collaboration with humans.

14.3 DSA Ecosystem Vision

Until recently, digital products for SME businesses were nothing more than products for large enterprises, with reduced functionalities. This has resulted in a first opportunistic rather than the strategic adoption of CPPS by SMEs, which handicaps the sustainable growth of such industries. To accelerate the adoption of CPPS by SMEs as producers of CPPS or as users of CPPS, the barriers to translate the benefits of CPPS into core business values, need to be reduced.

There are several European initiatives under the framework of FoF-11 H2020 DEI call that are working on providing platforms and solutions for the acceleration of digital automation engineering processes and the development of the necessary building blocks to realize full support to fog/cloud-based manufacturing solution in the context of Industry 4.0.

Based on the common approach of H2020 AUTOWARE [3], DAEDALUS [7], and FAR-EDGE [8] projects for the European digitisation of SMEs, the DSA has been defined with the common objective of providing reliable, cost-effective integrated solutions to support small enterprises, both in terms of customized and flexible applications.

The DSA is an open ecosystem of certified applications that will allow the ecosystem partners to access different components to develop smart digital automation solutions (the so-called shopfloor digital abilities) for their manufacturing processes. This ecosystem is aimed at reducing the cost, time, and effort required for the deployment of digital automation system on the basis of validated & verified components for specific configurations and operation profiles.

The three projects provide a complete CPPS solution allowing SMEs to access all the different components in order to develop digital automation cognitive solutions for their manufacturing processes. AUTOWARE provides a complete CPPS ecosystem, including a reference architecture that perfectly fits with FAR-EDGE architecture based on splitting the computing in the field (considering the decentralized automated shopfloor defined inside DAEDALUS), the edge, and the cloud. DAEDALUS also defines an intermediate layer (Ledger) to synchronize and orchestrate the local processes. Finally, AUTOWARE also enriches the different technical enablers to make easier the adoption of CPPS by SMEs as well as reliable communications and data distribution processes.

This combined solution reduces the complexity of the access to the different isolated tools significantly and speed up the process by which multi-sided partners can meet and work together. Moreover, the creation of added value products and services by device producers, machine builders, systems integrators, and application developers will go beyond the current limits of manufacturing control systems, allowing the development of innovative solutions for the design, engineering, production, and maintenance of plants' automation.

AUTOWARE has defined a complete open framework including a novel modular, scalable, and responsive Reference Architecture (RA) for the factory automation, defining methods and models for the synchronization of the digital and real world based on standards and certified components. AUTOWARE RA aligns several cognitive manufacturing technical enablers, which are complemented by usability enablers, thereby making it easy to access and operate by the manufacturing SMEs. The third key element in the ecosystem is the certification framework for the fast integration and customization of digital automation solutions.

The DSA proposes to go beyond a mere marketplace, (see Figure 14.1) and provide an integrated approach that on the development side ensures the provisioning of qualified CPPS components, certified systems and solutions thereby reducing the integration and customization costs. Moreover, the operational conditions and performance expected from Systems of Systems (SoS) operations can be managed in a controlled manner that ensures that machine and co-botic EU safety requirements can be addressed in the context of increased flexibility and system reconfiguration. On the demand side, the acquisition and operation costs are reduced based on shorter deployment cycles and customization on the basis of certified components already qualified with concrete working and development conditions validated for a specific purpose.

DSA ecosystem aim to ease the digital transformation process for manufacturing SMEs and it is based on an integrated approach, aligned with AUTOWARE goals of leveraging autonomous solutions through digital abilities, which includes a set of tools, techniques and services:

- **DSA experts network** helping manufacturing SMEs to define and evaluate their digital transformation strategy, and providing support for its implementation;
- **DSA RA**, aligned with widely established open HW and open platforms technologies, based on AUTOWARE RA;
- the provisioning of **DSA compliant:**
  - **Technological components** (from well-known technology providers and aligned to open HW, SW and platforms);

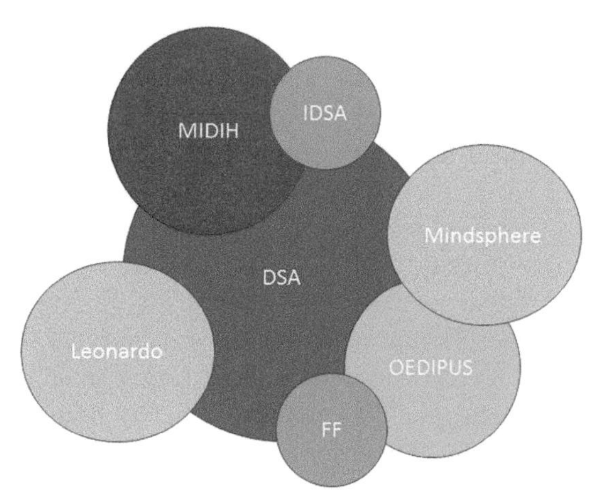

Figure 14.1 DSA manufacturing multisided ecosystem.

  - **Core Products** (architectural, functional, non-functional, normative and S&S compliant, validated for a purpose VPP);
  - **Certified solutions** (safety compliant: certified Components and Core Product validated for a specific application/service); and
  - **Validated deployments,** developed by trained professional integrators, for SME's customized automation solutions;
- access to trial-oriented **testbeds and neutral facilities** to offer a quick access and hands-on demonstrations of already validated solutions;
- the **Digital Automation Technology Validation (DATV) framework** for technologies, tools and services validation for a specific use under certain conditions, normatives, and standards based on AUTOWARE V&V enablers; and
- the access to an **homologated professional network** of integrators trained by Core Products owners and expertise in DSA technologies.

This approach enables DSA to offer **both top-down and bottom-up vision to implement safe digital transformation strategies and secure I4.0 digital automation systems in manufacturing SMEs.** In contrast to the top-down vision of known large technology providers focused on its core products or focused on its core architecture; DSA lowers the barrier that hinders the adoption of the latest technologies for the implementation of digital shopfloors in manufacturing SMEs.

DSA approach will focus on easing and enabling the digital transformation strategy for key application areas & services of competitive interest for manufacturing SMEs:

Figure 14.2 DSA ecosystem objectives.

- Energy efficient manufacturing services;
- condition-based monitoring & predictive maintenance services;
- zero defect manufacturing services;
- factory logistics management services;
- workplace augmentation, training, and human decision support services;
- digital twin modeling and simulation services; and
- Big Data Analytics for production planning and optimisation services.

The DSA bottom-up vision, based on the access to DATV-validated Core Products and Solutions, DSA services, and professionals, helps fulfil the DSA main goals of reducing the cost, time, and effort required to implement safe digital processes and products and secure Industry 4.0 digital automation solutions, in line to its objectives, see Figure 14.2 above:

- **Maximize Industry 4.0 RoI**, DSA services and DATV solutions will help optimize the SMEs investment for digital shopfloor implementation.
- **Keep integration time under control**, DSA-established methods and framework ease the adoption of digital solutions through validated deployments and access to certified testbeds.
- **Ensure future digital shopfloor extendibility**, relying on DATV-validated and standard-compliant components and DSA RA to safely plan the digital transformation strategy towards a future digital shopfloor.

On the business dimension, the DSA ecosystem is offering **a set of services to support SMEs in defining and executing their own digital transformation strategy** (see Figure 14.3), including:

Figure 14.3 DSA ecosystem strategic services.

- **DSA profiling**; DSA experts offer SMEs support on digital shopfloor profile selection, and ROI assessment of their digital shopfloor strategy.
- **DSA certification**; DATV framework application ensures safe operation of customised DSA deployments in modular/reconfigurable manufacturing cell or collaborative robotic workplace.
- **DSA integration**; DSA network of expert integrators offers suitable support for the safe and secure deployment of the digital shopfloor services.
- **DSA-ready products**; DATV HW components and SW solutions and infrastructures validated for purpose (VPP) helps reduce the ramp-up time of digital shopfloor services.

This set of services oriented to manage and support the digital transformation strategy for manufacturing SMEs' shopfloors is based on the AUTOWARE technical usability and V&V enablers and exploitable results. The DSA digitisation strategy's first steps will comprise a **digital transformation status assessment** that will enable the **digital transformation strategy and an action plan definition** through an **investment proposal aligned with the manufacturing SME global strategy and situation**, ensuring future extendibility of the deployments in the shopfloor and maximizing the Industry 4.0 ROI. The next steps will be supported by both catalogue of the DATV Core Products and validated deployments for specific purposes and

Figure 14.4 DSA-aligned open HW & SW platforms.

the Integrators network services, eased by the access to trial-ready testbeds in neutral facilities offered by the Autoware partners and manufacturing DIHs.

On the technological dimension, the DSA is centred in the AUTOWARE-based RA and aligned with main open HW and SW Platforms groups and initiatives in the digital automation area for Industry 4.0, as can be seen in Figure 14.4.

DSA catalogue of Core Products (DSA-CP) will offer, thanks to DATV certification framework, a complete description including the classification of the different DSA technology levels (visualization, security, connectivity, and open standards) achieved by the DSA-CP, its set of components, main features, DSA-RA mapping, component providers, qualified integrators availability (training level backed by CP owner, own homologation and expertise), estimated investment cost & deployment time table depending on complexity level of deployment.

14.4 DSA Reference Architecture

The RA aligned AUTOWARE manufacturing technical enablers, i.e. robotic systems, smart machines, cloudified control, secure cloud-based planning systems, and application platforms to provide cognitive automation systems as solutions while exploiting cloud technologies and smart machines as a common system. The AUTOWARE RA goal is to have a broad industrial applicability, map applicable technologies to different areas and guide technology and standard development.

The AUTOWARE RA has four levels, which target all relevant layers for the modeling of CPPS automation solutions (as depicted in Figure 14.5):

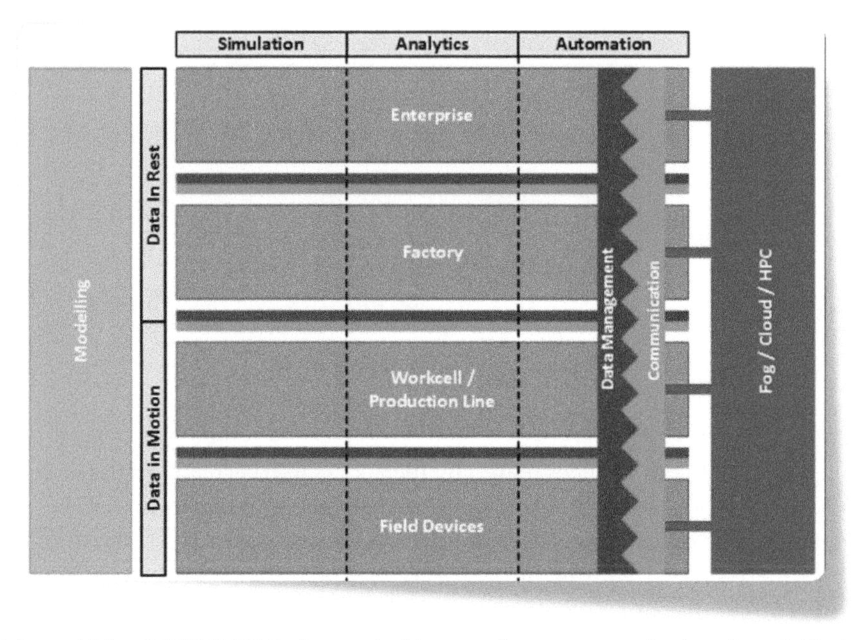

Figure 14.5 AUTOWARE Reference Architecture (layers, communication, & modelling).

Enterprise, Factory, Workcell, and Field devices. To uphold the concept of Industry 4.0 and move from the old-fashioned automation pyramid, the communication pillar enables direct communication between the different layers by using Fog/Cloud concepts. Finally, the last part of the RA focuses on the actual modelling of the different technical components inside the different layers. Additionally, to maintain compliancy with the overall AUTOWARE Framework, the reference architecture of the Software Defined Autonomous Service Platform (SDA-SP) broadens the overall AUTOWARE RA (see Figure 14.6) with the mapping of main technologies and CPPS services identified:

- A reconfigurable workcell that demonstrates solutions typical for robot automation tasks, e.g. robotic assembly using multiple robots;
- A mixed or dual reality supported automation to illustrate an automation solution that builds upon and benefits from intensive use of technologies like Virtual Reality (VR), Augmented Reality (AR), and Augmented Virtuality (AV). This system will be used to demonstrate the application of these technologies for automatic assembly of custom-ordered pneumatic cylinders.

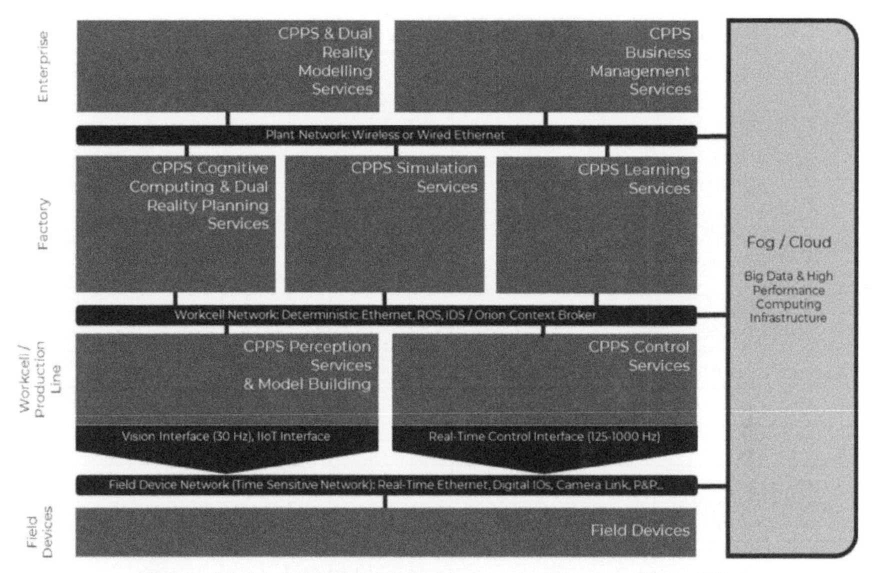

Figure 14.6 AUTOWARE Reference Architecture (SDA-SP).

- A multi-stage production line, where configuration, production, and traceability is built upon use of digital product memory technologies and functionalities.

Table 14.1 presents a list of AUTOWARE enablers mapped with the AUTOWARE-based RA different layers, network levels, and dimensions identified in Figure 14.6.

14.5 AUTOWARE Certification Usability Enabler

AUTOWARE will improve the European manufacturing industry situation by opening the door to new digital and digitally modified business opportunities with immediate global reach. Moreover, it will provide the enablers for putting innovation faster in the market with better streamlined customer processes and customer insights. The adoption of the CPPS technologies by SMEs is a well-known issue in which AUTOWARE has a major role in the automatization process by SMEs, facilitating that the SMEs build more sustainable and innovative business models. In addition, it also allows SMEs to focus both on the development or exploitation of the personalized applications and on the services to operate their strategic business assets (brand, culture, distribution, sales, production, and innovation).

Table 14.1 AUTOWARE enablers aligned to DSA-RA

Reference Architecture / **AUTOWARE Enablers**	**Enterprise Services**	**Factory Network**	**Factory Services**	**Workcell Network**	**Workcell Services**	**Field Device Network**	**Field Devices**	**OpenFog & Cloud dimension, BigData & HPC**	**Ecosystem (Innovation, validation & standardization)**
Validation & Certification usability enabler									x
FIWARE for Industry usability enabler		x		x				x	
Smart data distribution mechanisms		x		x		x			
Development of a gripping concept					x		x		
Development of a safety concept for the collaborative workplace					x				
Active Digital Object Memories (ADOMe)					x				
Neutral experience facility – Automation processes	x	x	x	x	x	x	x	x	x
Deep learning & high performance computing for reconfigurable robotic workcells			x		x			x	
Efficient robot tasks deployment using novel programming by demonstration framework			x		x				
OpenFog Architecture		x		x		x		x	
Human-Robot Collaborative Workplace	x		x		x		x		x
Faster (re-)training of vision systems based on deep neural networks					x		x		
Hierarchical Communication and Data Management Architecture for Industry 4.0		x		x		x			
Scalable & Self-Organizing Industrial Wireless Networks		x		x		x			
Deterministic industrial 5G communications		x		x		x			
Reliable industrial wireless communication						x			
Mobile robotic pilot for intra-logistic operations						x			

The impact on traditional SMEs, as shown in Figure 14.7, is immediate since technological complexity is decoupled from business value and a simple path towards maximizing the business value of advanced CPPS is facilitated. AUTOWARE hides the complexity of automatization to allow Future Internet SMEs and entrepreneurs to devote their resources and energies to effective and efficient business operation and value generation.

The number of technologies and platforms that need to be integrated to realize a cognitive automation service for Industry 4.0 is significantly high and complex. To this end, the AUTOWARE-proposed RA is rooted on solid foundations and intensive large-scale piloting of technologies for the development of cognitive digital manufacturing in autonomous and collaborative

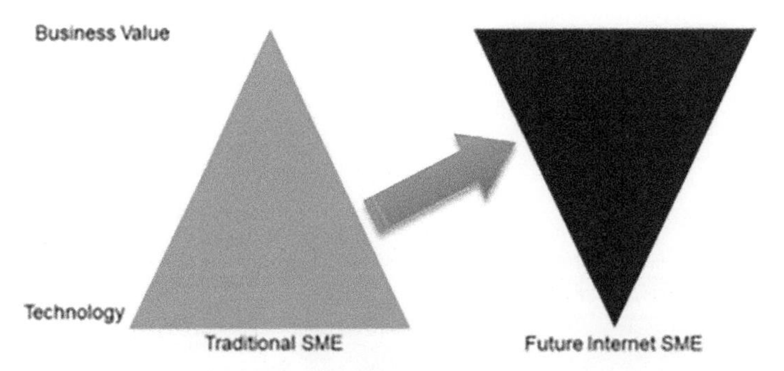

Figure 14.7 AUTOWARE business impact on SMEs.

robotics as an extension of ROS and ReconCell frameworks and for modular manufacturing solutions based on the RAMI 4.0 Industry 4.0 architecture.

The digital convergence of the traditional industries is increasingly leading towards the disappearance of the boundaries between the industrial and service sectors. In March 2015, Acatech [9], through the Industry-Science Research Alliance's strategic initiative "Web-based Services for Businesses," proposed a layered architecture to facilitate a shift from product-centric to user-centric business models. At a technical level, these new forms of cooperation and collaboration will be enabled by new digital infrastructures.

Smart spaces are the smart environments, where smart, internet-enabled objects, devices and machines (smart products) connect to each other. The term "smart products" not only refers to actual production machines but also encompasses their virtual representations (CPS digital twins). These products are described as "smart" because they know their own manufacturing and usage history and are able to act autonomously. Data generated on the networked physical platforms is consolidated and processed on software-defined platforms. Providers connect to each other via these service platforms to form digital ecosystems. AUTOWARE extends those elements, which are critical for the implementation of the autonomy and cognitive features. AUTOWARE also extends those reference models adopting the layered structure suggested by the Industry 4.0 Smart Service Welt initiative [10] (shown in Figure 14.8) for digital business ecosystem development based on industrial platforms (smart product, smart data and smart service).

AUTOWARE at the smart product level leverages enablers for deterministic wireless CPPS connectivity (OPC-UA and Fog-enabled analytics). At the smart data level, the AUTOWARE technical approach is to develop cognitive planning and control capabilities supported by cloud tools

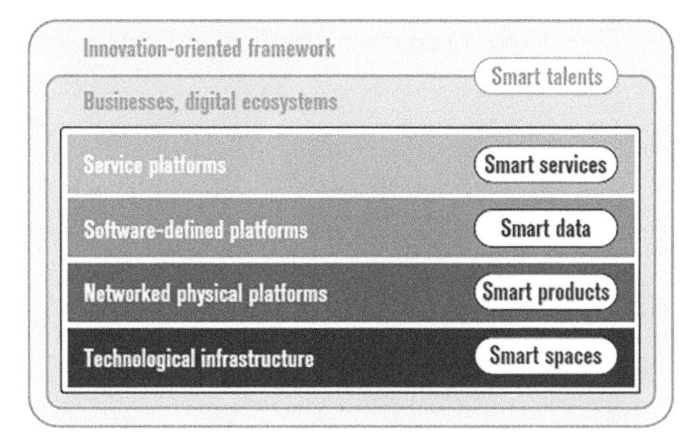

Figure 14.8 Smart service welt data-centric reference architecture.

and services and dedicated data management systems that will contribute to meet the real-time visibility and timing constrains of the cloudified planning and control algorithms for autonomous production services. Moreover, at the smart service level, AUTOWARE provides secure CPS capability exposure and trusted CPPS system modeling, design, and (self) configuration. In this latter aspect, the incorporation of the TAPPS CPS application framework, coupled with the provision of a smart automation service store, will pave the way towards an open service market for digital automation solutions which will be "cognitive by-design." The AUTOWARE cognitive operating system makes use of a combination of reliable M2M communications, human-robotics-interaction, modelling and simulation, and cloud and fog-based data analytics schemes. In addition, taking into account the mission-critical requirements, this combination is deployed in a secure and safe environment, which includes validation and certification processes in order to guarantee its correct operation. All of this should enable a reconfigurable manufacturing system that enhances business productivity.

14.5.1 AUTOWARE Certification Techniques

As previously stated, including validation and certification processes in AUTOWARE, Open CPPS ecosystem offers an easy adoption, secure environment, and greater credibility to SMEs. The planning and control of production systems has become increasingly complex regarding flexibility and productivity as well as regarding the decreasing predictability of processes. It is well accepted that every production system should pursue the following three main objectives:

- Providing capability for rapid responsiveness,
- Enhancement of product quality, and
- Production at low cost.

These requirements can be satisfied through highly stable and repeatable processes. However, they can also be achieved by creating short response times to deviations in the production system, the production process, or the configuration of the product in coherence to overall performance targets. In order to obtain short response times, a high process transparency and the reliable provisioning of the required information to the point of need at the correct time and without human intervention is essential. As a result, variable and adaptable systems are needed resulting in complex, long, and expensive engineering processes.

Although CPPS are defined to correctly work under several environment conditions, in practice, it is enough if it properly works under specific conditions. In this context, certification processes help guarantee the correct operation under certain conditions making the engineering process easier, cheaper, and shorter for SMEs that want to include CPPS in their businesses.

In addition, certification can increase the credibility and visibility of CPPS, as it guarantees its correct operation even following specific standards. If a CPPS is certified to follow some international or European standards or regulation, it is not necessary to be certified in each country, so the integration complexity, cost, and duration highly reduce. Nowadays, security and privacy are one of the major concerns for every business. SMEs with no specific knowledge need to be able to quickly assess, if an item provides confidence that required security and privacy is provided. For example, a minimal required barrier may need to be set to deter, detect, and respond to distribution and use of insecure interconnected items throughout Europe and beyond.

Security certification as a means of security assurance demonstrates conformance to a security claim for an item and eases the adoption of CPPS. Many certification schemes exist, each having a different focus (product, systems, solutions, services, and organizations) and many assessment methodologies also exist (checklists and asset-based vulnerability assessment). Some of the most important standards related to security are as follows:

- **ISO 10218-1:2011:** It is the standard that specifies the requirements and guidelines for the inherent safe design, protective measures, and information for use of industrial robots. It describes the basic hazards associated with robots and provides requirements to eliminate, or adequately reduce, the risks associated with these hazards. It does not address the robot as a complete machine. Noise emission is generally not considered a significant hazard of the robot alone, and consequently noise is excluded from the scope of ISO 10218-1:2011.

- **ISO 10218-2:2011:** It is the standard that specifies safety requirements for the integration of industrial robots and industrial robot systems as defined in ISO 10218-1, and industrial robot cell(s). The integration includes the following:
  - the design, manufacturing, installation, operation, maintenance and decommissioning of the industrial robot system or cell;
  - necessary information for the design, manufacturing, installation, operation, maintenance and decommissioning of the industrial robot system or cell; and
  - component devices of the industrial robot system or cell.

ISO 10218-2:2011 describes the basic hazards and hazardous situations identified with these systems and provides requirements to eliminate or adequately reduce the risks associated with these hazards. ISO 10218-2:2011 also specifies requirements for the industrial robot system as part of an integrated manufacturing system.

- **ISO/TS 15066:2016:** It is the standard that specifies the safety requirements for collaborative industrial robot systems and the work environment and supplements the requirements and guidance on a collaborative industrial robot operation given in ISO 10218-1 and ISO 10218-2.

Various methods can be used to systematically test and improve the security of CPPS systems. Apart from testing individual software components for security-related errors, all components of the CPPS infrastructure can also be tested, and the associated processes can be systematically examined and improved.

Depending on the initial situation, technical security tests may start at various testing stages, from all phases of the engineering or development cycle to integration testing and acceptance of the production infrastructure. It is possible to identify and eliminate security faults and the resulting risks at an early stage for relatively little cost, saving money, improving the accuracy of the planning and staying one step ahead of potential hackers.

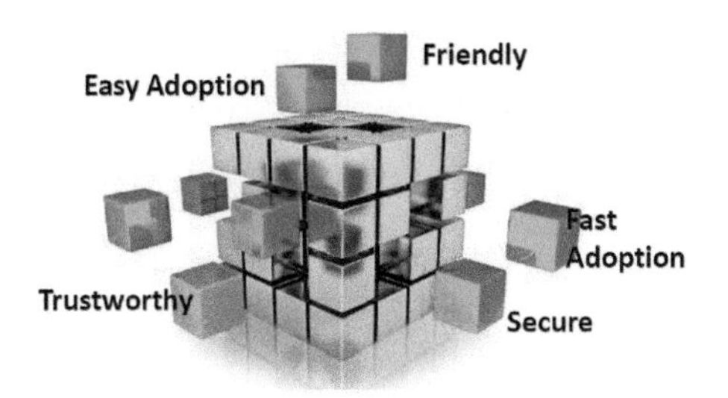

Figure 14.9 Main characteristics of CPPS solutions that are desired by SMEs.

Summarizing, it is well known that SMEs choose CPPS solutions that assure a correct operation, are easy and cheap to adapt, as well as safe & secure (see Figure 14.9). In addition, the CPPS solutions have greater credibility if they are made with certified tools, guaranteeing their correct operation under specific conditions defined according to the specific application requirements. Moreover, certification increases the solution visibility and makes the maintenance operation easier.

14.5.2 N-axis Certification Schema

Once the AUTOWARE solution is finished, a certification process is needed in order to guarantee the solution's correct operation and assure its compliance with the regulation. As a result, the engineering, integration, and launching processes are easier, cheaper, and shorter for SMEs. The AUTOWARE-proposed certification methodology consists of the following different stages.

14.5.2.1 Data collection

In this step, all the data useful for the certification process is collected. For example, documentation, which are the components, which technologies are used, what risks exist, etc. In the case of the components, it is also necessary to determine if they are critical, security, technological or commercial components, and if they are already certified or not. Table 14.2 shows an example of a possible template for obtaining data related to the solution/production. This information can be directly provided by the client or obtained by the certifying team during an ocular review.

Different options can be considered for the data collection, such as customer surveys, product/solution inspection, interviews, videos, etc. All of them are compatible and complementary to each other and their results can be combined.

Table 14.2 Data collection template for the certification process

	Component 1	Component 2	Component 3				Component n
Critical							
Certificate							
Security Level							
Technology							
Commercial							

14.5.2.2 Strategy

An appropriate strategy must be determined depending on the specific product/solution and the data obtained during the data collection process. For this purpose, the following questions must to be answered.

- Which tools are the most appropriate?
- How far the certification process has to go?
- What type of tests should be defined?

Depending on the data obtained in the data collection process, an appropriate series of tests must be defined encompassing as much as possible all the different possibilities: functional tests per component, integrity tests, unit tests per component, complete functional tests, etc.

14.5.2.3 Test execution

During this phase, the different tests defined during the strategy process are executed using the selected tools.

14.5.2.4 Analysis & reports

The results obtained from the test execution process are analysed in order to detect possible errors and indicate the level of criticality. The results obtained from the data analysis are gathered in a relevant report for the customer.

This four-phase process applied to the different system components and considering the different kind of components has to be combined with different fields of action (medicine, aviation, etc.) and with different standards (ISO-15066, ISO10218-1, ISO-10218-2). For this reason, the AUTOWARE certification scheme must be a multi-axis certification scheme such as that shown in Figure 14.10.

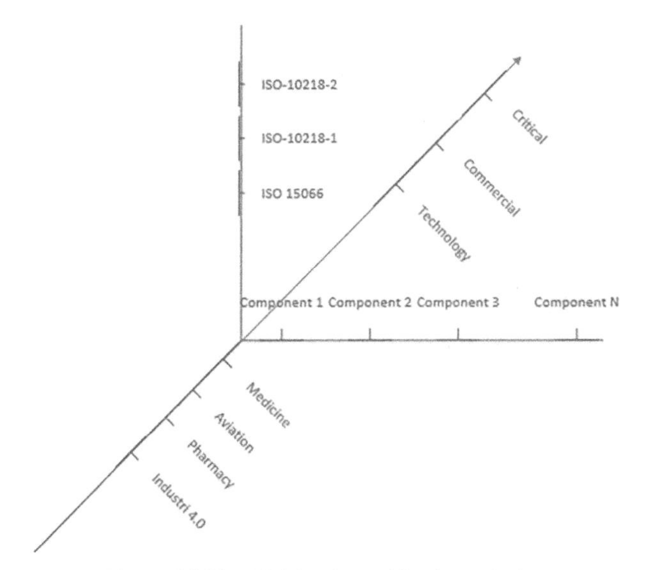

Figure 14.10 Multi-axis certification solution.

14.6 DSA Certification Framework

DSA RA compliant components provided by large well-known providers of technologies are the base for the development of Core Products designed and validated for a purpose (previously defined as predictive maintenance, zero-defect manufacturing, energyefficient manufacturing, etc.). Each **DSA Core Product** (as shown in Figure 14.11) should be composed by a set of DSA RA-compliant components with their matching datasheets (features and performance specifications), configuration & programming profiles and validation for purpose profiles (VPP), **as guidance to ensure DATV** when integrated in future solutions. Thanks to the support and training offered by Core Product owners, **Medium & Small Integrators** within the DSA network of experts will offer their services for the implementation of a **validated deployment** with customized Core Products for the specific application demanded by the manufacturing SMEs.

AUTOWARE promotes the use of open protocols & standards such as HW platform (openFog), connectivity (MQTT, TSN, iROS), control (IEC61499) data protocol (OPC-UA), data sharing (IDS, FIWARE/ETSI Context Information Framework), and data security. Individual components should support relevant open standards, APIs and specifications to become part of the AUTOWARE framework. However, AUTOWARE does not promote the simple certification of individual components but moreover the

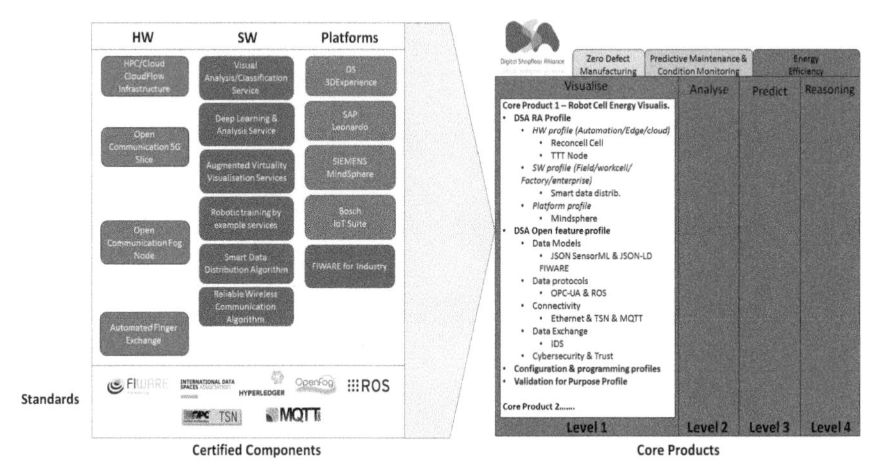

Figure 14.11 DSA-integrated approach for Digital Automation Solutions Certification.

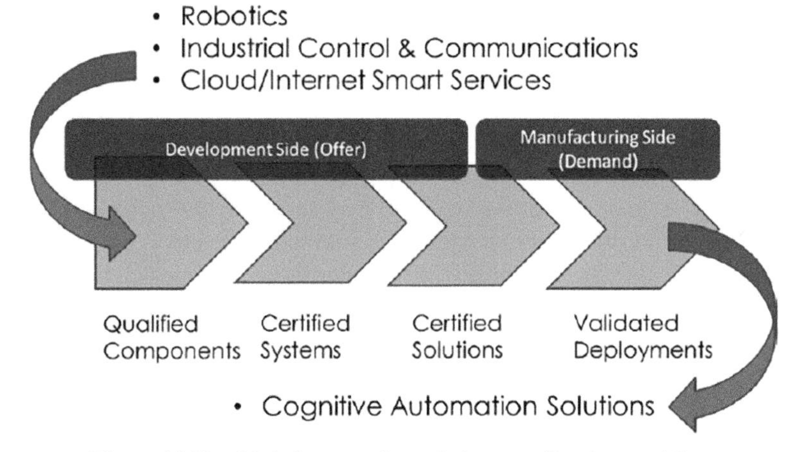

Figure 14.12 Digital automation solutions certification workflow.

availability of **core products** (HW infrastructure and software services and digital platforms) that are **constructed following the DSA RA architecture**; built **for a purpose** (visualisation, analysis, prediction, reasoning) **in the context of specific digital services** (energy efficiency, zero defect manufacturing, and predictive maintenance) **for manufacturing lines** (collaborative workspaces, robots, reconfigurable cells, modular manufacturing), as can be seen in Figure 14.12.

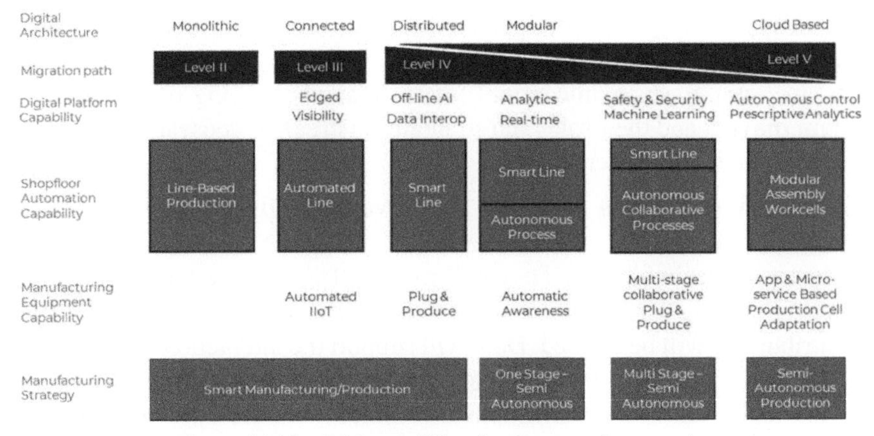

Figure 14.13 DSA capability development framework.

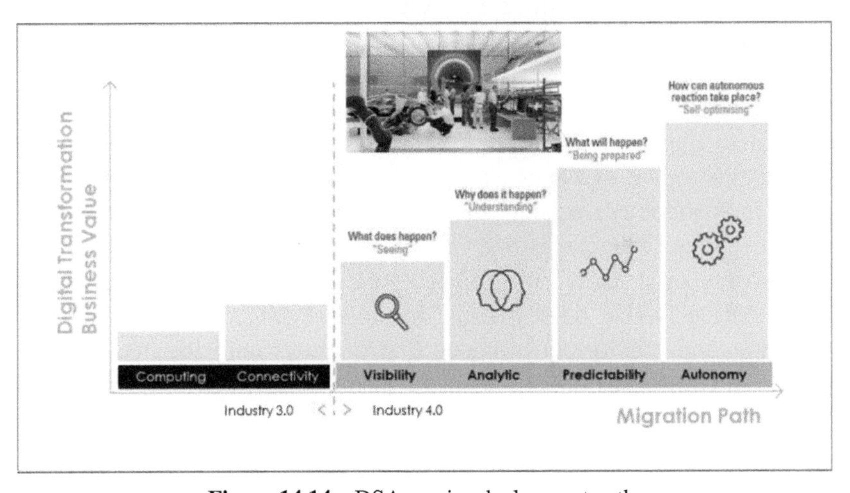

Figure 14.14 DSA service deployment path.

The DSA approach, based on the access to DATV Core Products and Solutions and DSA expert professionals and services, will reduce considerably the integration and customization costs of validated deployments. Through the proposed certification framework and DATV tools, the DSA aims to maximize the Industry 4.0 ROI and ensure the future scalability/extendibility of the digital shopfloors, by the implementation of a capability development framework (shown in Figure 14.13) and a service deployment path (shown in Figure 14.14) that guide SMEs in their Digital Transformation strategy in order to leverage their automation solutions visibility, analytic, predictability and autonomy.

After a preliminary vision in joint exploitation description of some of the main DSA Ecosystem Players (i.e. providers of the DSA products and services, medium & small integrators…), a detailed mapping of the **DSA ecosystem players** and their roles and strategies in the DSA ecosystem, based in a win-win model, is presented here:

- **Open HW and Open Platform initiatives and groups** will provide the required support for the established DSA RA and open source Industry 4.0 technologies. They will join the DSA Ecosystem as members by signing a Memorandum of Understanding, MoU, where their contributions will be defined. DSA will support the interaction of the rest of the ecosystem players like integrators to ease the adoption of these technologies and the certifications associated with the open technologies. This category could integrate interested universities and research centres working on these open technologies.
- **Manufacturing Champions**, the key Large Manufacturing Companies will have an essential role in the DSA as main tractors of the manufacturing sector, since they define the regulations and standardizations required to their providers' network. DSA ecosystem will study and analyse the sector demands and needs to ensure the Manufacturing Champions endorsement align their activities in the right direction. As DSA members, Manufacturing Champions pay a fee that will allow them access to a DSA-validated network of homologated providers implementing DATV deployments that ensure safe operation, energy efficiency, and quality performance first contacts with the main manufacturing companies will be done for instance through the Boost 4.0 Lighthouse European project.
- **Technology Providers**, the main technology providers (i.e. Siemens, Bosch, ATOS) will join the DSA Ecosystem as Core Product owners offering these DATV solutions ready to be customized and integrated in the shopfloor, associated maintenance, and support & training services. Prior to join DSA, technology providers' components and Core Product should be DSA open SW, open HW and open Platform compliant, providing associated datasheet, configuration & programming profiles and validation for purpose profile. As DSA members, the Technology Providers will pay a fee that provide them an alternative access to SME manufacturing companies market, not profitable for the direct sale of their SW packages, platforms and services. The Technology Providers will be able to offer adjusted prices to developers and thus access this SME market.

- **Development Partners** will form a network of small and medium integrators, qualified for the implementation of validated deployments for customized solutions for manufacturing SMEs. DSA will search a first contact with the Digital SME Alliance to have access to potential development partners. Prior to joining the DSA network of experts, these small and medium integrators should comply with specific training on the Core Products and DSA technologies, architecture and strategy, and provide a signed SLA. The DSA network provide an homologation methodology based on the training levels on the different DSA technologies and their expertise that, together with a cost/time estimation table for different deployment complexities, will give them visibility and a way to improve their competitive position.
- **Manufacturing SMEs**, DSA offers them not only the services and technologies for digital transformation and implementation of Digital Shopfloor technologies, but the visibility as DSA homologated providers to large manufacturing companies. Manufacturing SMEs main access to DSA will be not only the DSA platform/ecosystem web, but through the activities and services offered in clusters and DIH focused on manufacturing sector, and other agents like the Trilateral Alliance cooperation between German Plattform Industrie 4.0, French Alliance Industrie du Futur and Italian initiative Piano Industria 4.0, or Spanish Industria Conectada 4.0.

DSA will also work for the **integration of standardization methodologies** in DSA solutions and deployments, considering not only the technological aspects but other aspects like data protection and GDPR.

As a summary Table 14.3 is presented with the initial players identified in DSA ecosystem within AUTOWARE project, and the potential players that will conform the DSA ecosystem in a future.

14.7 DSA Certification Methodology

The DSA intends to promote the appropriate ecosystem to develop and commercialize Innovative Solutions that respond and can be adapted to end-user needs. When defining the mission pursued by the DSA, a reflection has been made on the key aspects when starting an initiative of this kind:

- WHY: A different way to commercialize I4.0 solutions; Implement DT to Industry; Foster the creation of innovative products

Table 14.3 Identification of DSA players

DSA Players	Preliminary Stage (AUTOWARE)	Future Stages (DSA Potential Players)
Open HW and Open Platform initiatives and groups	UMH, CNR, Fraunhofer, imec, INNO (open initiatives alignment role)	FIWARE, IDS, openFog, iROS, OPC-UA... OpenForum Europe
Manufacturing Champions	Fraunhofer, INNO, Blue Ocean, SQS (manufacturing sector alignment role)	Key manufacturing large companies from different sectors (i.e. Boost4.0 champions) National Manufacturing Enterprise Associations (CONFINDUSTRIA, It's OWL, FrenchTech...)
Technology Providers	TTTech, JSI, Robovision	EIT Digital, AIOTI Siemens, Bosch, SAP, Huawei, Telefonica, Azzure, CloudFlow, Dassault, ESI Group... Digital SME Alliance
Development Partners	SmartFactoryKL, JSI, Tekniker (Neutral Experimental Facilities as integrators)	Digital SME Alliance for Small & Medium IT Integrators...
Manufacturing SMEs	SMC, Stora Enso (industrial Use Cases)	Sectorial clusters & DIH, German Plattform Industrie 4.0, French Alliance Industrie du Futur, Italian initiative Piano Industria 4.0, Spanish AIC...

- HOW: Offering solutions vs technologies; creating an ecosystem of beneficiaries from stakeholders from research to end-users;
- WHAT: Consultancy; Certification; Solutions; Integration

This analysis has led us to the definition of four key sets of services to be offered within DSA ecosystem: Consultancy, Certification, Integration, and Solutions as shown in Figure 14.15, with the support of the certification framework that ensures easy configuration and operation of reliable scalable open based Digital Automation Solutions with low cost and fast RoI deployment. As shown in Figure 14.16, DSA certification methodology covers the Core Product key aspects to successfully support a manufacturing SME in its digital transformation strategy:

- FUNCTION: Identified key functionality aspects, defining processes and customisation, global, interoperability, and standard features of the core product

Figure 14.15 DSA key services.

Figure 14.16 DSA key services.

- TECHNOLOGY: Identified open SW/HW components, RA alignment, validation, and configuration tools
- DELIVERY: Identified network of local experts for integration & training, pricing model, and operations & support services

Moreover, the integration of diverse stakeholders in the DSA ecosystem fosters the adoption of I4.0 & leverages the creation of innovative products:

- TECHNOLOGY PROVIDERS: Provide core components and technologies, Open HW and Open Platform Initiatives and Groups, and Research & Private
- SOLUTION PROVIDERS: Core products/solution developers
- INTEGRATORS: Adapt core products to end-user needs
- VALIDATORS & CERTIFIERS: Validate core products/solutions and their adaptations Service Providers V&V
- STANDARDIZATION BODIES: Technology and process related
- MANUFACTURING: Large (Prescription & Customers), SMEs (end-users/Customers), Clusters (Prescription), and Industrial Associations (Prescription)
- CONSULTANCY: Digital transformation consultancy experts.

14.8 Conclusion

This section has presented the foundation of the DSA and the associated validation and verification framework as the basis to develop a manufacturing driven multi-sided ecosystem. The DSA is originated as a means for SMEs to navigate and exploit the large set of tools and platforms available for the development of digital solutions for the digital shopfloor. This paper has discussed how the DSA approach can nurture synergies across multiple stakeholders for the benefit of SME digitization and the gradual integration of the digital abilities in the digital shopfloor with a business impact. This paper has presented main standardization and compliance drivers, for instance, digital shopfloor safety in advanced robotic systems as one of the multipliers for adoption and the need for a DSA ecosystem that facilitates navigation across standards, platforms, and services with a focus on business competitiveness. This paper has also presented the fundamental services envisioned for such DSA and the dimensions that need to be validated to ensure that digital abilities such as automatic awareness can be fully realized in the context of cognitive manufacturing digital transformation.

Acknowledgments

This work has been funded by the European Commission through the FoF-RIA Project *AUTOWARE: Wireless Autonomous, Reliable and Resilient Production Operation Architecture for Cognitive Manufacturing* (No. 723909).

References

[1] Max Blanchet, The Industrie 4.0 transition. How it reshuffles the economic, social and industrial model http://www.ims.org/wpcontent/uploads/2017/01/2.02_Max-Blanchet_WMF2016.pdf
[2] Digital Shopfloor Alliance website. Available at: https://digitalshopflooralliance.eu/, last accessed September 2018.
[3] H2020 AUTOWARE project website. Available at: http://www.autoware-eu.org/, last accessed September 2018.
[4] Kollaborierende Robotersysteme. Planning von Anlagen mit der Funktion "Leistungs- und Kraftbegrenzing", [Online], Available at: https://www.google.de/url?sa=t&rct=j&q=&esrc=s&source=web&cd=2&ved=0ahUKEwjdlry0wNrZAhUuyqYKHXb0BWcQFghAMAE&url=http%3A%2F%2Fwww.dguv.de%2Fmedien%2Ffb-holzundmetall%2Fpublikationendokumente%2Finfoblaetter%2Finfobl_deutsch%2F080_roboter.pdf&usg=AOvVaw1UxaEcsQ9K4lWZXq3W3UDy, last accessed: March 2018.
[5] ISO 10218-1:2011, "Robots and robotic devices – Safety requirements for industrial robots – Part 1: Robots" Available at: https://www.iso.org/standard/51330.html
[6] ISO 10218-1 with industrial robot cell(s) Available at: https://www.iso.org/standard/41571.html
[7] H2020 DAEDALUS project website. Available at: http://daedalus.iec61499.eu, last accessed September 2018.
[8] H2020 FAR-EDGE project website. Available at: http://www.faredge.eu/#/, last accessed September 2018.
[9] ACATECH NATIONAL ACADEMY OF SCIENCE AND ENGINEERING Available at: https://en.acatech.de/, last accessed September 2018.
[10] Industry 4.0 Smart Service Welt initiative Available at: https://www.digitale-technologien. de/DT/Navigation/EN/Foerderprogramme/Smart_Service_Welt/smart_service_welt.html, last accessed September 2018.

15

Ecosystems for Digital Automation Solutions an Overview and the Edge4Industry Approach

John Soldatos[1], John Kaldis[2], Tiago Teixeira[3], Volkan Gezer[4] and Pedro Malo[3]

[1]Kifisias 44 Ave., Marousi, GR15125, Greece
[2]Athens Information Technology, Greece
[3]Unparallel Innovation Lda, Portugal
[4]German Research Center for Artificial Intelligence (DFKI), Germany
E-mail: jsol@ait.gr; jkaldis@ait.gr; tiago.teixeira@unparallel.pt; Volkan.Gezer@dfki.de; pedro.malo@unparallel.pt

Stakeholders' collaboration is a key to successful Industry 4.0 deployments. The scope of collaboration spans the areas of solution development and deployment, experimentation, training, standardization, and many other activities. To this end, Industry 4.0 vendors and solution providers are creating ecosystems around their project's developments, which allow different stakeholders to collaborate. This chapter reviews some of the most prominent ecosystems for Industrial Internet of Things and Industry 4.0 solutions, including their services and business models. It also introduces the Edge4Industry ecosystem portal, which is a part of the ecosystem building efforts of the H2020 FAR-EDGE project.

15.1 Introduction

The advent of the fourth industrial revolution (Industrie 4.0) is enabling a radical shift in manufacturing operations, including both factory automation operations and supply chain management operations. CPS (Cyber Physical Systems)-based manufacturing facilitates the collection and processing of large volumes of digital data about manufacturing processes, assets, and operations, towards improving decision-making, driving the efficiency of processes such as production scheduling, quality control, asset management, maintenance, and more. In addition to access to CPS and Industrial Internet of Things (IIoT) platforms that realize these improvements, both manufacturers and providers of industrial automation solutions need a lot of support in testing, validating, and integrating novel applications in the factories. In support of these needs, a wide range of online platform and services have emerged, including:

- Online platforms for IIoT services, notably public cloud IoT services. These enable solution integrators to develop, deploy, and validate innovative services for manufacturers. Moreover, these platforms come with a wide range of support services, which are offered to the communities of developers, solution providers, and manufacturers working around them.
- Testbed platforms for manufacturers and automation solution providers, which enable them to test and validate solutions prior to actual deployment, while supporting them in research and knowledge acquisition.
- Software/middleware library providers. Instead of providing a complete online platform with a pool of related services, these providers focus on the provision of middleware services that could help other organizations to establish the CPS/IIoT infrastructure.

These online platforms and services enable the formation of entire ecosystems around them. A business ecosystem is generally defined as an economic community that is supported by a range of interacting organizations and individuals. This community produces goods, services, and knowledge that provide value to the customers of the ecosystem, who are also considered members of the ecosystem along with suppliers, producers, competitors, and other stakeholders [1].

The development of such ecosystems is a key success factor for the successful adoption of platforms such as the ones listed above. In this context, IIoT and Industry 4.0 projects and initiatives (such as our H2020 FAR-EDGE project that is described in previous chapters), should also undertake similar ecosystem building initiatives. In particular, one of the main objectives of FAR-EDGE is to create an ecosystem of manufacturers, factory operators, IT solutions integrators, and industrial automation solution providers around the project's results, which will facilitate access and sustainable use of the project's assets. The FAR-EDGE ecosystem services will be provided as part of an on-line platform, which will operate like a multi-sided market platform

(MSP), which will bring together supply and demand about digital factory automation services based on the edge-computing paradigm. A wide range of solutions and services will be provided by FAR-EDGE to its ecosystem community, including industrial software and middleware-related services (e.g., automation and analytics solutions), as well as business and technical support services (e.g., support on solutions migration).

This chapter aims at providing insights on the IIoT ecosystems in general and the FAR-EDGE ecosystem in particular. The presentation of the existing ecosystems provides a comprehensive overview of the different types of services that they provide, as well as of their business models. Likewise, the presentation of FAR-EDGE ecosystem portal (www.edge4industry.eu) provides an overview of the solutions, services, and the knowledge base that are provided as part of the project and are made available to the community. The chapter is structured as follows:

- Section 15.2 following the chapter's introduction presents a review of some of the most representative Industry 4.0 and IIoT ecosystems and their services;
- Section 15.3 provides a comparative analysis of the presented ecosystems, including a description of their business models;
- Section 15.4 introduces the Edge4Industry ecosystem portal and describes its structure and services; and
- Section 15.5 is the final and concluding section of the chapter.

15.2 Ecosystem Platforms and Services for Industry 4.0 and the Industrial Internet-of-Things

In the following paragraphs, we describe a representative sample of the IIoT platforms and their ecosystems, as well as a range of other Industry 4.0 platforms and testbeds, including their validation and experimentation services. Each ecosystem platform is presented both in terms of its technical/technological characteristics as well as in terms of its business model.

15.2.1 ThingWorx Foundation (Platform and Ecosystem)

The ThingWorx Foundation (www.thingworx.com) that is now part of PTC (https://www.ptc.com) provides a platform for the development and deployment of enterprise-ready, cloud-based IoT solutions. It is an end-to-end solution, which provides access to all elements that comprise an IoT application. Its main value proposition lies in the provision of a simple and seamless way for developing IoT applications, which reduces the development and deployment efforts.

ThingWorx's services are accessible to the developers via a developers' portal and can be classified as follows:

- Connectivity Services (Make): Based on the ThingWorx platform, one can connect devices, sensors, and systems, among themselves but also with other systems. Connectivity and information exchange is facilitated in order to reduce the time and effort needed for rapid development of integrated solutions.
- Data Analytics (Analyze): The ThingWorx platform provides the means for analyzing the data derived from connected IoT devices.
- Development/Coding Services (Code): The platform offers development tools and APIs, which provide development flexibility and increase the overall productivity of solution integrators.

While ThingWorx is a general-purpose platform for IoT solutions, smart manufacturing is explicitly listed as one of the primary markets of application. In this direction, ThingWorx provides a wide range of functionalities for interconnected assets within factories, plants and supply chains with business information systems. Moreover, some of the components of the platform, such as its AR (Augmented Reality) and IoT-based immersion module, are demonstrated as a part of the manufacturing scenarios such as industrial maintenance.

Around the Thingworx platform, the foundation has been building an ecosystem, which is providing a complete set of integrated IoT-specific development tools and capabilities in order to ease the delivery of IoT solutions. The ThingWorx ecosystem comprises the following participants, concepts, structures and associated stakeholders' roles:

- Partners: Enterprises are offered the opportunity to join the ThingWorx ecosystem as partners on the basis of a variety of different (partnership) programmes, which cover various needs. In particular, the partner programmes are available for: (i) Enterprises building IoT solutions based on the ThingWorx platform; (ii) Companies that build products that are certified by ThingWorx and made available through the ThingWorx marketplace; (iii) Professional service providers who opt to offer consulting, solution design and technical delivery services based on the ThingWorx IoT platform. These partners are called "services partners" and are provided with cumulative educational attainment; and (iv) Reseller of ThingWorx's based technologies, which participate in the "ThingWorx Channel Advantage" program and can benefit from earning margins for reselling ThingWorx solutions.

- Marketplace: The ThingWorx Marketplace provides access to everything needed in order to build and run ThingWorx-based IoT applications, including extensions, apps, and partners that can facilitate the development of IoT solutions based on the platform. The marketplace component of the ecosystem is therefore a means for the extensibility of the ecosystem.
- Academic Programme: An academic programme is also offered to students, researchers, makers, universities, and trainers. It is an IoT education programme, which is built over the platform, leveraging its practical features and content.

15.2.2 Commercial Cloud-Based IIOT Platforms

All major IT and industrial automation vendors are offering cloud-based IIoT services. Likewise, they are also building ecosystems around these platforms or in most cases expanding their existing ecosystems in the IIoT space. A detailed analysis of each of the public IIoT services providers is beyond the scope of this chapter. Nevertheless, we can make a broad ballpark classification of the available services to the following:

- **General purpose public IoT cloud services**, which are typically offered by IT vendors. These include, for example, Microsoft's Azure IoT Suite, IBM's Watson IoT platform, SAP's HAN Cloud platform with IoT support and extensions, Amazon AWS IoT, LogmeIN's Xively platforms, and more. These platforms are not tailored to a specific vertical industry. Rather, they provide scalable and cost-effective cloud infrastructures for IoT, which can be used to develop, deploy, and operate solutions in the different industries.
- **IIoT services for industrial automation**, which are typically offered by industry leaders in industrial automation solutions including SIEMENS, Bosch, and ABB. In several cases, there are partnerships between IIoT vendors and providers of IT (IoT/cloud) infrastructure services as evident in the case of ABB and Microsoft, but also in the fact that Bosch's IoT services run over various digital plumbing platforms such as Amazon's. These partnerships are overall indicative of the distinction of business roles.

The scope of these services includes connectivity services along with the offering of tools for rapid and cost-effective application developments.

Each of the above-listed platforms is associated with an ecosystem of developers, solution providers, and business partners. In most cases, the above-listed vendors act as ecosystem expanders in the IoT/IIoT space, given that they primarily expand the ecosystem of their existing accounts, customers, and business partners in the area of IoT. Access to the IIoT services, including consulting, technical support, training, and hosting, but mainly turn-key solution deployments is provided on a commercial basis with appropriate SLA (Services Level Agreements). Both public cloud services and private cloud services are offered. Public cloud services are charged in pay-per-use modality (e.g., pay-per-use and pay-as-you-go services are offered by Microsoft Azure, Amazon AWS IoT, and Xively).

15.2.3 Testbeds of the Industrial Internet Consortium

The Industrial Internet Consortium (IIC) is an open-membership, international not-for-profit consortium that is leading the establishment of architectural frameworks and overall directions for the Industrial Internet. Its members represent large and small industries, entrepreneurs, academics, and government organizations. The Industrial Internet Consortium is a global, member-supported organization that promotes the accelerated growth of the IIoT by coordinating the ecosystem initiatives to securely connect, control, and integrate assets and the systems of assets with people, processes, and data using common architectures, interoperability, and open standards, in order to deliver transformational business and societal outcomes across industries and public infrastructure.

The IIC scope includes the identification and location of sensor devices, the data exchange between them, control and integration of collections of heterogeneous devices, data extraction, and storage plus data and predictive analytics. The challenge for the IIC is to ensure that these efforts come together into a cohesive whole. The IIC Working Groups coordinate and establish the priorities and enabling technologies of the Industrial Internet in order to accelerate market adoption and drive down the barriers to entry. There are currently 19 Working Groups and teams, broken into seven broad areas, including Business Strategy and Solution Lifecycle, Legal, Liaison, Marketing, Membership, Security, Technology, and Testbeds.

One of the areas of focus of the IIC is the development of Testbeds. A testbed is a controlled experimentation platform that:

- Implements specific use cases and scenarios,
- Produces testable outcomes to confirm that an implementation conforms to expected results,

- Explores untested or existing technologies working together (interoperability testing),
- Generates new (and potentially disruptive) products and services, and
- Generates requirements and priorities for standards organizations supporting the Industrial Internet.

Testbeds are a major focus and activity of the IIC and its members. The Testbed Working Group accelerates the creation of testbeds for the Industrial Internet and serves as the advisory body for testbed proposal activities for members. It is the centralized group which collects testbed ideas from member companies and provides systematic yet flexible guidance for new testbed proposals. Testbeds are where the innovation and opportunities of the Industrial Internet – new technologies, new applications, new products, new services, and new processes – can be initiated, thought through, and rigorously tested to ascertain their usefulness and viability before coming to market.

15.2.4 Factory Automation Testbed and Technical Aspects

One type of Testbed known as Platform as a Service (PaaS) for Factory Automation (FA), is expected to facilitate the integration of the IoT systems to connect the manufacturing sites and head offices for strengthened operations, such as the globalization of supply chains and improved production quality, delivery time, and productivity when responding to sudden changes in markets. The FA testbed provides connectivity between the Factory and Cloud, a data analytics platform, and security resources, in order to ease the FA application development for Application Providers and FA Equipment Vendors. Based on the facilities of the testbed, the Application Providers and FA Equipment Vendors have the opportunity to develop and provide solutions to the manufacturers and factory operators at minimum effort and cost, by engaging in the development of the core logic of each application only, rather in the development of industrial middleware as well. Overall, the Testbed provides the following features to reduce application development process:

- Connectivity between Factory and Cloud where architectures differ;
- APIs specialized in FA, which are re-usable for FA applications: Edge Applications, Cloud Applications, and Domain Applications;

- Security to protect the Factory brown field from the outside network; and
- Integration of data from the Business Systems.

IIC testbeds are privately funded by member companies or publicly funded by government agencies, while Hybrid models involving both public and private funding are also possible.

15.2.5 Industry 4.0 Testbeds

As part of the platform "Industrie 4.0" in Germany, several testbeds have been established at specialist centres within universities and research institutions in Germany. These testbeds enable the testing and validation of complex production and logistics systems under realistic conditions. They are intended to be used by mechanical and plant engineering companies, notably Small and Medium Enterprises (SMEs). The latter are provided with facilities for testing their I4.0 developments in real-life nearly operational conditions, prior to their deployment in actual production environments. The testbeds are also addressed to factory operators wishing to take advantage of CPS manufacturing in a way that reduces barriers and risks.

As already outlined, the Industry 4.0 testbeds is a public sector-supported/funded initiative for evaluating innovative approaches to CPS manufacturing. This initiative is addressed to equipment manufacturers and operators. Along with access to the testbeds infrastructure, members of the Industry 4.0 platform are offered access to a range of advisory and coordination services. A central coordination office at the Federal Ministry of Education and Research (BMBF) provides funding support for testing innovative Industry 4.0 components by SMEs at the various testbeds. As part of the offered advisory services, BMBF provides SMEs with advice about the most appropriate testbeds to be used, while at the same time undertakes the focused dissemination of the results towards specialist communities. In this way, BMBF's initiatives complement the activities undertaken by the Centres of Excellence (CoE) funded by the Federal Ministry for Economic Affairs and Energy (BMWi). The latter CoEs are primarily destined to support operators of machine and plant equipment.

15.2.5.1 SmartFactory pilot production lines – testbeds

The FAR-EDGE project partner SmartFactory participates in the provision of various testbed services for Industry 4.0, which will herewith be presented as indicative examples. In particular, the SmartFactory provides several production lines to integrate, customize, test, and demonstrate CPS solutions in a realistic industrial production setup. All of its experimental production lines are designed to be strictly modular and are comprised of devices coming from several different vendors, being identical to those found in most modern industrial plants. The open and modular design facilitates the usage as test-bed for various experiments. Several demonstrators have been built along four main production lines:

15.2.5.2 Industry 4.0 production line

The first test-bed is a multi-vendor, highly modular factory system with "plug n' play" module extension. The independent modules are thereby fulfilling vendor-independent standards defined by SmartFactory, which are based on the widely accepted communication protocols. This test-bed representing the key concept of "Industry 4.0" has the following features: 1) Service-oriented production line with modular CPS-based field devices, 2) Multi-vendor, highly modular factory system with "plug n' play" module extension, and 3) Demonstration platform for distributed processes based on communicating component. As shown in Figure 15.1, items 1 to 10 are production modules, while 11 to 15 are infrastructure boxes connecting with 16 to 22 into an integrated IT system.

15.2.5.3 SkaLa (scalable automation with Industry 4.0 technologies)

In today's market, the customers do not only need products that they can configure individually, but they also desire products that are cost-effective and readily available. Meeting these requirements calls for a flexible and efficient approach to manufacturing. One way to meet these challenges is provided by "SkalA", a demonstration unit that offers a scalable automation process.

The mobile demo unit can, if necessary and depending on the situation, be scaled to the automation process. The unit's scalability is based on

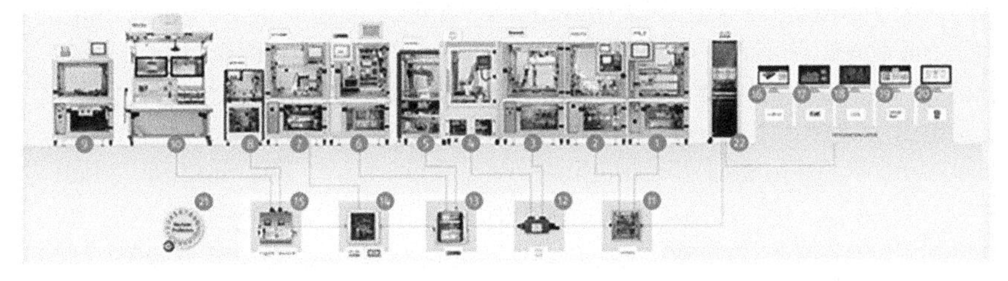

Figure 15.1 SmartFactory's Industrie 4.0 production line.

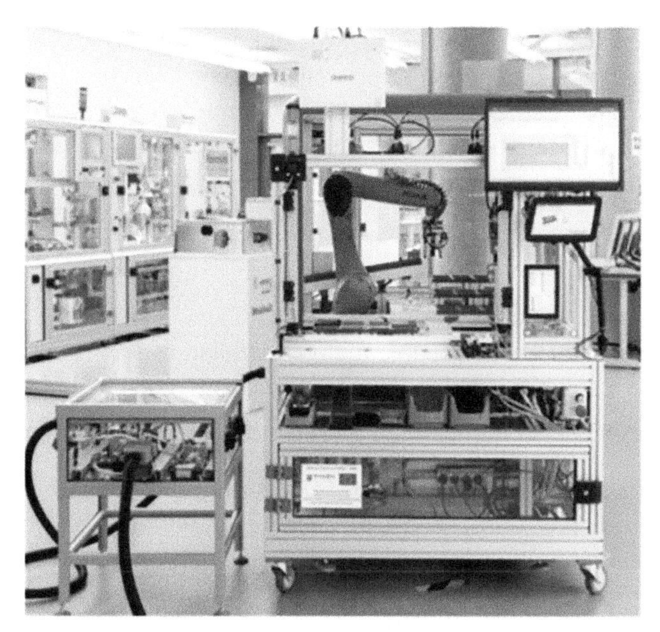

Figure 15.2 SkaLa production line.

a fully decentralized, controlled manufacturing process, made possible by cyber-physical systems (CPS). For each work step, independently operating modules are used, which communicate with each other and control the process. The system can be expanded with a robot module via standardized interfaces to add an automated production component. In the manual mode, workers are provided with support in the form of projected recommendations for work steps. For improved flexibility, both order management and service activities are supported via mobile devices.

15.2.5.4 Key finder (The keyfinder production line from SmartFactoryKL)

SmartFactoryKL has presented a unique demonstration plant as the central exhibit of the Forum Industrial IT together with the German Research Center for Artificial Intelligence (DFKI) at the Hannover Messe industrial trade fair in Hanover. On the basis of a complete production line, the relevant aspects of the fourth industrial revolution were exemplified for the first time using innovative information and communication technologies. The modular plant shows the flexible, customized manufacturing of an exemplary product, the components of which (housing cover, housing base and circuit board) are handled, mechanically processed and mounted.

15.2.5.5 SME 4.0 competence center kaiserslautern

The SME 4.0 Competence Center Kaiserslautern is one of the several regional centers of excellence launched by the Federal Ministry for Economics and Technology (BMWi). The aim of this nationwide funding initiative is to highlight the importance of Industry 4.0 for the future of SMEs, to inform SMEs about the great opportunities in this area, and to actively support them with the implementation of projects.

As part of its mission, the SME 4.0 Competence Center Kaiserslautern assists companies from Rhineland-Palatinate and Saarland. The aim is to assist, offer an extensive, up-to-date knowledge base and valuable practical experience in the area of Industry 4.0. Focus, in particular, is on sharing know-how from many years of research and implementation with small and medium enterprises.

The SME 4.0 Competence Center Kaiserslautern consists of four partners, namely Technology Initiative SmartFactoryKL e.V., the German Research Center for Artificial Intelligence GmbH, the Kaiserslautern University of Technology and the Institute for Technology and Work e.V.

15.2.6 EFFRA Innovation Portal

To foster information exchange and collaboration between innovation projects and the EC, the European Factory of the Future Research Association (EFFRA) has created an Innovation Portal, which serves as a single entry to point to information about FoF projects and their results. The EFFRA Innovation Portal stimulates clustering, maps projects on the 'Factories of the Future 2020' roadmap, and allows for project monitoring and impact measurement. Within the portal, each project profile provides a summary of the project work and information on its consortium.

The portal is currently accessible to EFFRA project members. However, it also contains publicly accessible pages. It is maintained by EFFRA with support by EU projects involved in the association.

15.2.7 FIWARE Project and Foundation

The FIWARE's Community led by the EU industry and supported by the academic community, has built an open sustainable ecosystem and several implementation-driven software platform standards that could ease the development of new Smart Applications in multiple sectors. Its main goal is to enable an open community of developers including entrepreneurs, application sponsors and platform providers. FIWARE provides one of the most prominent operational Future Internet platforms in Europe. Its platform provides a rather simple yet powerful set of open public APIs that ease the development of applications in multiple vertical sectors. The implementation of a FIWARE Generic Enabler (GE) becomes a building block of a FIWARE instance. Any implementation of a GE is made up of a set of functions and provides a concrete set of APIs and interoperable interfaces that are in compliance with open specifications published for that GE. The FIWARE project delivers reference implementations for each defined GE, where an abstract specifications layer allows the substitution of any Generic Enabler with alternative or custom made equivalents.

FIWARE's main contribution is the gathering of the best available design patterns, emerging standards and open source components, putting them all to work together through well-defined open interfaces. There is a lot of knowledge embedded, lowering the learning curve and mitigating the risks of bad architecture designs. The scope of the platform is also very wide, covering the whole pipeline of any advanced cloud solution: connectivity to the IoT, processing and analyzing Big data, real-time media, cloud hosting, data management, applications, services, security, etc. But FIWARE does not only accelerate the development of robust and scalable cloud based solutions, it also establishes the basis for an open ecosystem of smart applications. In the FIWARE sense, be SMART means to be Context Aware and to be able to interoperate with other applications and services; and this is where FIWARE excels.

FIWARE has over the years developed an ecosystem of developers, integrators and users of FIWARE technologies, which includes several SMEs. An instrumental role for the establishment and development of the FIWARE ecosystem has been played by the FIWARE Acceleration Programme, which promoted the take up of FIWARE technologies among solution integrators and application developers, with special focus on SMEs and start-ups. Around this programme, the EU has also launched an ambitious campaign where SMEs, start-ups and web entrepreneurs can get a funding support for the development of innovative services and applications using FIWARE technology. This support intends to be continuous and sustainable in the future, engaging accelerators, venture capitalists and businesses who believe in FIWARE.

The FIWARE ecosystem is supported and sustained by the FIWARE Foundation, which is the legal independent body providing shared resources to help achieve the FIWARE mission. The foundation focuses on promoting, augmenting, protecting, and validating the FIWARE technologies, while at the same time organizing activities and events for the FIWARE community. The latter empower its members (end users, developers, integrators and other stakeholders in the entire ecosystem.

Note that the FIWARE Foundation is open, as anybody can join and contribute to a transparent governance of FIWARE activities. The foundation operates on the basis of the principles of openness, transparency and meritocracy.

15.2.8 ARROWHEAD ARTEMIS JU Project and ARROWHEAD Community

The Arrowhead project implemented a framework for developing service-oriented industrial automation solutions in five business domains, namely: production (process and manufacturing), smart buildings and infrastructures, electro mobility, energy production and virtual markets of energy. The project's framework ensures the interoperability between different systems and approaches for implementing Service-Oriented Architecture (SOA)-based solutions in the target industries. To this end, Arrowhead provides and enables the following:

- A system to make its services known to service consumers;
- A system for service consumers to discover the services that they want/need to consume;
- Authorized use of services provided by some service provider to a service consumer; and
- Orchestration of systems, including control of the provided service instances that a system shall consume.

The Arrowhead Framework contains common solutions for the core functionality in the area of Information Infrastructure, Systems Management, and Information Assurance as well as the specification for the application services carrying information vital for the process being automated.

Arrowhead is a recently concluded project, which offers a range of industrial middleware solutions to developers and deployers of industrial automation systems. It also provides resources that facilitate developers to develop, deploy, maintain, and manage Arrowhead compliant systems, including technical resources that boost a common understanding of how the Services, Systems, and System-of-Systems are defined and described. The latter resources include design patterns, documentation templates, and guidelines that aim at helping systems, newly developed or legacy, to conform to the Arrowhead Framework specifications.

Arrowhead has managed to establish around its framework an ecosystem of solution developers, along with end-users for the target industry areas, as well as associated use cases where the framework has been deployed and used.

15.3 Consolidated Analysis of Ecosystems – Multi-sided Platforms Specifications

15.3.1 Consolidated Analysis

In the following paragraphs, we perform a consolidation of the services and business models which have been outlined in the previous section. The following table provides a high-level taxonomy of the services that are presented in the following paragraphs, including the different ecosystems that offer them.

The business and sustainability models of the various ecosystems are essential for their longer-term viability. The main monetization strategies are as follows:

- **Revenues from sales or use of services on a commercial basis (licensed or pay-as-you-go models):** The ecosystems of the large vendors provide commercial services for end-users and providers of the IIoT

Ecosystem /Services	Hosting & Support of IIoT solutions	Solution Design and Integration Services	Training & Education	Advisory & Consulting Services	Solution Experimentation and Validation Services	Standardization Services	Access to software/middle ware libraries	Information and News Updates
ThingWorx and IIoT/cloud platforms	X	X	X	X			X	
IIC Testbeds			X		X			
I4.0 Testbeds			X		X			
EFFRA Innovation Portal				X				X
FIWARE							X	
Standards Bodies						X		

Figure 15.3 Overview of Services offered by various IIoT/Industry 4.0 ecosystems and communities.

solutions. The services are provided based on either licensed models or pay-as-you-go models. The latter is the primary monetization modality for public cloud services, yet they are considered as part of the private cloud services that vendors build for manufacturers.

- **Sales of complementary services:** Complementary services (notably training, education, advisory, and consulting services) are also provided on a commercial basis as part of the presented ecosystems). These services are offered separately or bundled with IIoT solution development, hosting, and deployment services.
- **Public funding support services:** Several of the services (such as some of the testbed services) are financed by public funding (including projects) or even by the combination of private and public funding sources.
- **Membership fees:** In foundations (such as FIWARE) and associations (such as EFFRA) there income is also generated from membership fees.

There are different types of legal entities that support the above-listed monetization models. These include commercial entities, associations and non-profit foundations.

Based on the analysis of the above ecosystem platforms and services, it is important to highlight some important considerations for anyone attempting a similar ecosystem building initiative:

- **Critical Mass:** The formation of a critical mass of stakeholders is a prerequisite for establishing an ecosystem.
- **Viability of Service Offerings:** In addition to creating a range of services, ecosystems should ensure that the offered services are viable.
- **Business Models and Sustainability of Service Offerings:** A viable business model should also support the sustainability of the ecosystem services.

15.3.2 Multi-sided Platforms

It should be also outlined that the reviewed platforms provide services for both demand-side stakeholders (i.e. users of IIoT/Industry 4.0 services) and the supply-side ones, i.e. vendors and solution providers. As such, these platforms offer a range of base features such as a catalogue of services, services for registering and managing participants, authentication and authorization (as a prerequisite for accessing these services) and more. A basic set of such functionalities has been listed in the following figure (Figure 15.4) and illustrated in the literature (e.g., [2–4]).

MSP Platform Functionality	Short Description
Registering Participants & Business Entities	Registration of participants to the ecosystem (i.e. manufacturers, factory operators and factory automation solution providers)
Publishing service offerings	Publication and presentation of the ecosystem services (notably the services listed in the following subsection)
Search and discovery of service offerings	Search engine for discovering available services based on appropriate metadata for the services descriptions
Review and rating of service offerings	Tools for rating service offerings from the end-users / participants viewpoints
Provision of recommendations	Context aware proposition of relative service offerings
Pricing and Payments Support	Services for pricing services and supporting payment modalities
Manage and tracking registered services	Access to the status of subscriptions and services
Authentication and Authorization	Ensuring authenticated and authorized access to the various services
Localization	Support for an international environment through appropriate localization of the services including currency and language support

Figure 15.4 Baseline functionalities of a Multi-sided market platform.

15.4 The Edge4Industry Ecosystem Portal

The FAR-EDGE Ecosystem portal (publicly accessible at www.edge4industry.eu) is a vertical IIoT ecosystem on factory automation, focusing on FoF/I4.0 applications for manufacturers, with the objective to ensure EU's leadership in the manufacturing sector. It presents all the research work and innovation developed in the FAR-EDGE project and aims to advance the competitiveness of the participants, manufacturers, and providers of the industrial automation solutions. Figure 15.5 presents the home page of the ecosystem portal.

As the goal for the Edge4Industry Ecosystem portal is to remain active, functional, and independent beyond the FAR-EDGE project, having broader adoption aspirations, a new unique brand and domain name has been specified to support the ecosystem evolution and branding beyond the duration of the FAR-EDGE project. Figure 15.6 provides a mind-map with the structure of the portal that includes the FAR-EDGE services and solutions, a knowledge-base, a blog, and a registration/sign-in section. These pages can be accessed through the main menu and contain the following information:

- **Services:** Provides all relevant information about each FAR-EDGE service.
- **Solutions:** Provides information and access to the FAR-EDGE solutions.

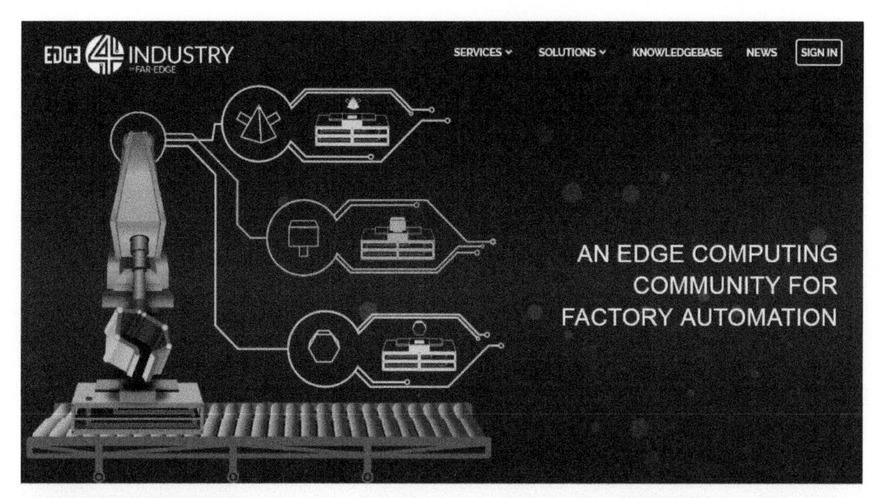

MISSION
STATEMENT
FAR-EDGE provides a blueprint solution for industrial automation based on edge computing, which emphasizes the use of blockchain technologies to synchronize digital information models with the actual status of the factory.
Industry 4.0 is the integration of Information Technology and Operational Technology into what is sometimes called the "smart and connected factory". The goal is having more efficient and flexible production systems on the one side, enabling new business models on the other. This fourth industrial revolution was started by the convergence, over time, of several innovative concepts like IoT, Cloud Computing, Cognitive Computing and Digital Simulation.
Today, FAR-EDGE is raising the stakes: reshaping the highly hierarchical structure of factory systems (the "automation pyramid") into a peer-to-peer collaboration of local actors on the shop floor. The catalysts of such transformation are, once again, new paradigms: Edge Computing and Blockchain. FAR-EDGE follows a unique approach, using Edge Gateway devices to inject Edge Computing powers into the legacy factory with a minimum of disruption. A virtual, decentralized coordination layer is provided by means of Blockchain and Smart Contracts.
The FAR-EDGE Platform is the tangible result of this effort. It is a reference implementation of the FAR-EDGE Reference Architecture with baseline functionalities tailored to automation, simulation and analytics solutions. Around the FAR-EDGE Platform, a multi-sided Ecosystem is growing that brings together developers, integrators and final users, all sharing the common goal of advancing the smart and connected factory to the next level. This site supports this Ecosystem by providing resources and services to the community.

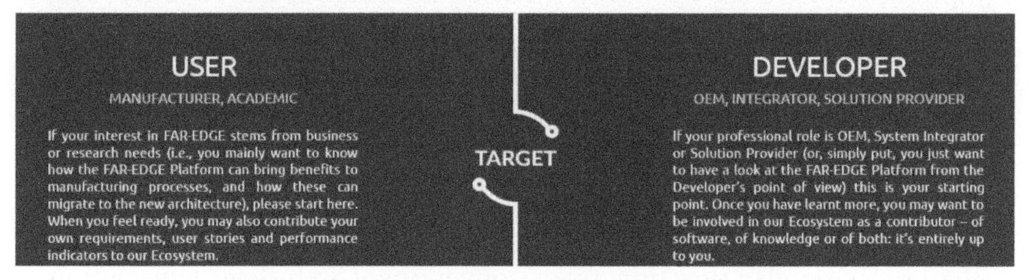

USER
MANUFACTURER, ACADEMIC
If your interest in FAR-EDGE stems from business or research needs (i.e., you mainly want to know how the FAR-EDGE Platform can bring benefits to manufacturing processes, and how these can migrate to the new architecture), please start here. When you feel ready, you may also contribute your own requirements, user stories and performance indicators to our Ecosystem.
TARGET
DEVELOPER
OEM, INTEGRATOR, SOLUTION PROVIDER
If your professional role is OEM, System Integrator or Solution Provider (or, simply put, you just want to have a look at the FAR-EDGE Platform from the Developer's point of view) this is your starting point. Once you have learnt more, you may want to be involved in our Ecosystem as a contributor – of software, of knowledge or of both: it's entirely up to you.

Figure 15.5 Home page of Edge4Industy portal.

- **Knowledgebase:** This is a dedicated page with articles, training and presentations regarding the project.
- **Blog:** This section provides articles, news, and latest publications about the Edge4Industry community.
- **Sign in:** This is a sign in area that enables users' registration/login.

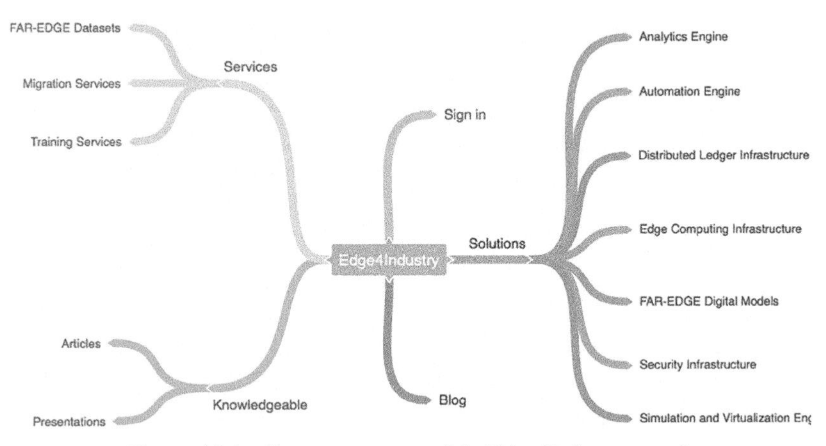

Figure 15.6 Content structure of the Edge4Industry portal.

15.4.1 Services

The Services section can be easily accessed through the main menu by clicking in the Services button and intends to present to the users community all the available FAR-EDGE services. At this stage, the following services are available:

- **FAR-EDGE Datasets:** Provides access to open datasets that can be used for experimentation and research. The first datasets provided include data related to individual production modules such as their power consumption, their status, operating mode (maintenance, active, etc.). The datasets include all module production-related information, including Module ID, module description, production status, conveyor status, operating status, error status, uptime information, power consumption, order number, process time etc.
- **Migration Services:** The FAR-EDGE Migration Services supports manufacturers, plant operators and solutions integrators in planning and realizing a smooth migration from conventional industrial automation systems (like ISA-95 systems) into the emerging Industry 4.0 ones (like edge computing systems). The service provides a Migration Matrix Tool, which includes all the essential improvement steps and plans needed to enable a smooth migration from traditional control production systems towards the decentralised control automation architecture based on edge computing, CPS, and IoT technologies.
- **Training Services:** This service delivers technical, architectural, and business training to Industry 4.0-related communities, as a means of raising awareness about digital automation in general and FAR-EDGE

solutions in particular. It includes specific courses and training presentations. The latter are appropriate for stakeholders that wish to understand opportunities stemming from the deployment of edge computing and distributed ledger infrastructure for industrial automation use cases.

15.4.2 Solutions

Similar to the Services section, the Solutions section intends to present all the available FAR-EDGE solutions and can be accessed too through the main menu by clicking in the Solutions button. At this stage, the FAR-EDGE solutions that are available are as follows:

- **Analytics Engine:** The Analytics Engine solution is a middleware component for configurable distributed data analytics in industrial automation scenarios. Its functionalities are accessible through an Open API, which enables the configuration and deployment of various industrial-scale data analytics scenarios. It supports processing of large volumes of streaming data, at both the edge and the cloud/enterprise layers of digital automation deployments. It also supports data analytics at both the edge and the cloud layers of a digital automation system. It is extremely flexible and configurable based on the notion of Analytics Manifests (AMs), which obviate the need for tedious data analytics programming. AMs support various analytics functionalities and are amenable by visual tools. Note that the Analytics Engine is provided with an open source license.
- **Automation Engine:** This solution provides the means for executing automation workflows based on an appropriate Open API. It enables lightweight high-performance interactions with the field for the purpose of configuring and executing automation functionalities. It provides field abstraction functionalities and therefore supports multiple ways and protocols for connecting to the field. It also facilitates the execution of complex automation workflows based on a system-of-systems approach. It offers reliable and resilient functionalities at the edge of the plant network, based on Arrowhead's powerful local cloud mechanism. Finally, it leverages a novel, collaborative blockchain-based approach to synchronizing and orchestrating automation workflows across multiple local clouds.
- **Distributed Ledger Infrastructure:** This solution results in a runtime environment for user code that implements decentralized network services as smart contracts, which are used for plant-wide synchronization of industrial processes. It enables the synchronization of several

edge analytics processes, as well as various edge automation processes. The solution is a first of a kind implementation of permissioned ledger infrastructure for the reliable synchronization of distributed industrial processes.

- **Edge Computing Infrastructure:** The Edge Computing Infrastructure solution is a pool of components, which provide the means for high-performance connectivity and data acquisition at the edge of the industrial automation network. The solution leverages the capabilities of popular connectivity protocols (like MQTT) and high-performance data streaming frameworks (like Apache Kafka). It also enables dynamic connectivity and data acquisition for the field, in order to facilitate edge computing configurations. Its implementation is containerized (i.e. Docker based), which facilitates usage and deployment.
- **FAR-EDGE Digital Models:** This solution offers the means for representing, exchanging and sharing information in the scope of an edge computing system for industrial automation. Also support is provided for the development of digital twins for field configurations and digital simulations. These Digital Models are based on ideas from several standards for plant modeling, while being tailored to the needs of edge computing for factory automation. They are among the few publicly available digital models for edge computing implementations of industrial automation systems.
- **Security Infrastructure:** This solution is a system designed following the principles of the Industrial Internet Security Framework (IISF) of the Industrial Internet Consortium (IIC) that provide superior integrity of distributed security functions within an Edge Computing based system. It can operate in conjunction with the Distributed Ledger in order to host the security policy and to provide consistent security across various edge analytics and edge automation processes. It is a first of a kind distributed ledger implementation of an IISF compliant security system.
- **Simulation and Virtualization Engine:** This solution provides the means for configuring and executing digital simulations. It includes a real-to-digital synchronization tool, which allows simulation services providers and integrators to improve the reliability of simulations predictions and develops synchronization functionalities between physical world elements and their digital twin. This tool regards any related data source based on appropriate digital models while offering all steps necessary to translate the messages from the physical world element format to the data model format used by the simulation.

15.4.3 Knowledge Base

The Knowledgebase section is a dedicated area of the portal that provides direct access to articles and presentations concerning the latest research and innovation work provided by the FAR-EDGE project and the Edge4Industry community.

The goal of this section is to enable the Edge4Industry members to acquire an in-depth knowledge regarding all the FAR-EDGE project issues, the information, and the resources available; while the access to them is user-friendly and dynamic.

15.4.4 Blog

The blog section presents to the ecosystem community publications about topics that are related to the industry, including those that have been published by members of the Edge4Industry community as well as other sources such as other blogs and electronic magazines. Similar to the Knowledgebase section, access to the Edge4Industry Blog section publications is user-friendly and dynamic.

15.4.5 Sign-in and Registration

The Edge4Industry portal includes a user management system that enables access to different user's types and determines which portal resources are applicable and authorized for each user. At this stage are two user types:

- **Guest**, which is assigned to unauthenticated users and grants lowest-level permission within the portal.
- **Registered member**, which is assigned to members that can access all the relevant resources that are provided in the knowledgebase.

The Edge4Industry Register members can authenticate in the portal by the Sign in section. Members can use a set of different authentication tools to access the Edge4Industry portal.

15.5 Conclusions

In the era of digitization, the development of proper ecosystems is as important as the development of digital platforms. In many cases, most of the value of a digital platform lies in its ecosystem and the opportunities that it provides to stakeholders' in order to collaborate and advance the digital transformation of modern organizations. Digital automation platforms are no exception, which is the reason why all major vendors of IIoT and Industry 4.0 platforms have established ecosystems around their products and services. Likewise, several public and private funded initiatives have established testbeds, where industrial organizations can experiment with digital technologies without disrupting their production operations.

As part of this chapter, we have reviewed several IIoT/Industry 4.0 ecosystem building efforts, including ecosystems established around commercial platforms, experimental testbeds and community portals. Moreover, we have provided the key building blocks and success factors of multi-sided platforms. Furthermore, we have presented the Egde4Industry portal, which is providing a single point of access to the full range of digital automation results of the FAR-EDGE project, including results presented in previous chapters such as the project's analytics engine, digital models and approach to supporting smooth migration from ISA-95 to decentralized automation.

The Edge4Industry community is gradually growing in size and expanding in terms of stakeholders' engagement. In support of this growth, we plan to provide a range of collaboration and engagement features, which will also be supporting its growth based on an ambitious dissemination and communication plan during the next couple of years.

Acknowledgments

This work has been carried out in the scope of the FAR-EDGE project (H2020-703094). The authors acknowledge help and contributions from all partners of the project.

References

[1] J. Moore 'The Death of Competition: Leadership & Strategy in the Age of Business Ecosystems' New York: HarperBusiness. ISBN 0-88730-850-3, 1996.

[2] T. Eisenmann, G. Parker and M. W. Van Alstyne 'Strategies for Two-Sided Markets,' Harvard Business Review, November 2006.

[3] A Hagiu. 'Two-Sided Platforms: Pricing, Product Variety and Social Efficiency' mimeo, Harvard Business School, 2006.

[4] Leslie Brokaw, (2014) "How to Win With a Multisided Platform Business Model", MIT Sloan Business School (blog), May 20, 2014.

16

Epilogue

At the dawn of the fourth industrial revolution, the benefits of the digital transformation of plants are gradually becoming evident. Manufacturers and plant operators are already able to use advanced CPS systems in order to increase the automation, accuracy, and intelligence of their industrial processes. They are also offered opportunities for simulating processes based on digital data as a means of evaluating different scenarios (i.e. "what-if" analysis) and taking optimal automation decisions. These capabilities are empowered by the accelerated evolution of digital technologies, which is reflected in rapid advances in areas such as cloud computing, edge computing, Big Data, AI, connectivity technologies, block chains and more. The latter digital technologies form the building blocks of the state-of-the-art digital manufacturing platforms.

In this book, we have presented a range of innovative digital platforms, which have been developed in the scope of three EU projects, namely the AUTOWARE, DAEDALUS, and FAR-EDGE projects, which are co-funded by the European Commission in the scope of its H2020 framework programme for research and innovation. The presented platforms emphasized the employment edge computing, cloud computing, and software technologies as a means of decentralizing the conventional ISA-95 automation pyramid and enabling flexible production plants that can support mass customization production models. In particular, the value of edge computing for performing high-performing operations close to the field was presented, along with the merits of deploying enterprise systems in the cloud towards high performance, interoperability, and improved integration of data and services. Likewise, special emphasis was paid in illustrating the capabilities of the IEC 61499 standard and the related software technologies, which can essentially allow the implementation of automation functionalities at the IT rather than the OT part of the production systems.

Special emphasis has been put in the presentation of some innovative and disruptive automation concepts, such as the use of cognitive technologies for increased automation intelligence and the use of the trending block chain technologies for the resilient and secure synchronization of industrial processes within a plant and across the supply chain. The use of these technologies in automation provide some characteristic examples about how the evolution of digital technologies will empower innovative automation concepts in the future.

In terms of specific Industry 4.0 functionalities and use cases, our focus has been put on systems that boost the development of flexible and high-performance production lines, which boost the mass customization and reshoring strategies of modern manufacturers. A distinct part of the book was devoted to digital simulation system and their role in digital automation. It is our belief that digital twins will play a major role in enhancing the flexibility of production lines, as well as in optimizing the decision-making process for both production managers and business managers.

Nevertheless, the successful adoption of digital automation concepts in the Industry 4.0 era is not only a matter of deploying the right technology. Rather, it requires investments in a wide range of complementary assets, such as digital transformation strategies, new production processes that exploit the capabilities of digital platforms (e.g., simulation), training of workers in new processes, and many more. Therefore, we have a dedicated a number of chapters to the presentation of such complementary assets such as migration strategies, ecosystem building efforts, training services, development support services, and more. All of the presented projects and platforms pay emphasis to the development of an arsenal of such assets as a means of boosting the adoption, sustainability and wider use of these solutions.

Even though this book develops the vision of a fully digital shopfloor, it should be outlined that we are only in the beginning and far from the ultimate realization of this concept. In particular, we have only marginally discussed integration and interoperability issues, which are at the heart of a fully digital shopfloor. Moreover, we have not presented how different components and modular solutions can be used to address the different needs of manufacturers and plant operators. Our Digital Shopfloor Alliance (DSA) initiative (https://digitalshopflooralliance.eu/) aims at bringing these issues into the foreground, but also in creating critical mass for successfully confronting them.

Industry 4.0 will be developed in a horizon that spans across the next three to four decades, where digital platforms will be advanced in terms of intelligence and functionalities, while becoming more connected. In particular, the following developments are likely to take place over state-of-the-art digital platforms presented in this book:

- **The establishment of industrial data spaces**, which will provide the means for interoperable data exchanges between different platforms and stakeholders. As a characteristic example, industrial data spaces that allow supply chain stakeholders to exchange production orders and materials information without only minimal effort for integrating their enterprise systems with the industrial data space infrastructure.
- **The enhancement of machines and equipment with intelligence features**, based on the integration of advanced digital technologies such AI. As a prominent example, future machines will be able to identify and in several cases repair defect causes on-line i.e. without a need for stopping production.
- **The development and establishment of open APIs** for accessing capabilities and datasets within these platforms. Such APIs will greatly facilitate their integration and access in the scope of end-to-end applications. For example, they will provide the means for processes that span multiple stations and platforms within a factory.
- **The provision of support for smart objects such as smart machines and industrial robots**. Smart objects feature (semi)autonomous behaviour and are able to operate as stand-alone systems in the shopfloor. Occasionally, they will be able to synchronize their state with the state of digital automation platforms that control the shopfloors. Hence, they will be able to co-exist with digital platforms in order to perform collaborative tasks in the plant.
- **The implementation of strong security features**, which will ensure secure operations for both IT and OT systems of the plant. Strong security and data protection will be required as a result of the expanding scope of the digital automation platforms, but also as a result of their interconnection with other CPS, IT and OT systems.

Overall, Industry 4.0 will be certainly an exciting journey for plant operators, providers of industrial automation solutions, IIoT solution providers and many other stakeholders. In this book we have provided knowledge and insights about where we stand in this journey, while trying to develop a vision for the future. We really hope you will enjoy the journey and will appreciate our efforts to help you get started with the right foot.

Index

About the Editors

Dr. John Soldatos (http://gr.linkedin.com/in/johnsoldatos) holds a Phd in Electrical & Computer Engineering from the National Technical University of Athens (2000) and is currently Associate Professor at the Athens Information Technology (2006 to present) and Honorary Research Fellow at the University of Glasgow, UK (2014 to present). He was also Adjunct Professor at Carnegie Mellon University, Pittsburgh, PA (2007–2010). Dr. Soldatos is a world-class expert in Internet-of-Things (IoT) technologies and applications, including IoT's applications in smart cities and the fourth industrial revolution (Industry 4.0 & Industrial Internet of Things).

Dr. Soldatos has played a leading role in the successful delivery of more than 50 (commercial-industrial, research and consulting) projects for both private and public sector organizations, including some complex integrated projects. He is co-founder of the open-source platforms OpenIoT (https://github.com/OpenIotOrg/openiot) and AspireRFID (http://wiki.aspire.ow2.org). He has published more than 150 articles in international journals, books and conference proceedings. He has also significant academic teaching experience, along with experience in corporate training. Dr. Soldatos is regular contributor to various international magazines and blogs, as well as in social media, where he is among the influencers of the IoT community. Moreover, Dr. Soldatos has received national and international recognition through appointments in standardization working groups, expert groups and various boards.

Dr. Oscar Lazaro is the Managing Director of Innovalia Association, the Associated Research Lab founded by the Innovalia Alliance, one of the three strategic technology groups in the Basque Country. Oscar has more than 20 years of experience in the ICT and manufacturing field. He is also Visiting Professor at the Electrical and Electronic Engineering Department of the University of Strathclyde in the area of wireless & mobile communications. Also, he is permanent representative of Innovalia in EFFRA and he has also served to the Future Internet Advisory Board and the Sherpa Group on 5G Action Plan.

Dr. Oscar Lazaro has been one of the three experts appointed in the high-level group supporting the EC in the analysis of the 15 national initiatives in Digitising European Industry. He has been supporting the activities of the I4MS Programme since its very beginning and leads the I4MS Competence Centre for Advanced Quality Control Services in the Zero Defect Manufacturing DIH at the Automotive Intelligence Centre in the Basque Country. He is also part of the Smart Industry 4.0 Technical Committee of the FIWARE Foundation and regular contributor to the activities of the Industria Conectada 4.0 DIH and Platform working groups. Since January 2018, he has been coordinating the European lighthouse initiative BOOST 4.0 on Big Data Platforms for Industry 4.0.

Dr. Franco Cavadini has a PhD in aerospace engineering on the subject of artificial intelligence applied to robotic manipulation tasks at Politecnico di Milano. He is Chief Technical Officer at Synesis, a small Italian company whose mission is the technology transfer of advanced automation solutions from the research to the market. With a specific focus on the research and development of technologies for the optimization of production systems and control under the constraints of high-energy efficiency and low environmental impact, he has guided Synesis technical department throughout several Horizon 2020 and industrial projects, providing both technical and project management contributions. Dr. Cavadini is an expert in the field of distributed real-time automation and on the design of complex control architectures. He is currently the project coordinator for the European Daedalus initiative for Industry 4.0 (www.daedalus.eu) and technical coordinator of the DEMETO project (www.demeto.eu).